Recent Advances in **Nuclear Explosion Monitoring**

Edited by
Andreas Becker
Bernd Schurr
Martin B. Kalinowski
Karl Koch
David Brown

Previously published in *Pure and Applied Geophysics* (PAGEOPH), Volume 167, Nos. 4–5, 2010

Birkhäuser

Editors

Andreas Becker
Vienna International Centre
P.O. Box 1200
1400 Vienna
Austria
andreas.becker@ctbto.org

Bernd Schurr
Helmholtz Centre Potsdam
GFZ German Research Centre
for Geosciences
Telegrafenberg, C 223
14473 Potsdam
Germany
bernd.schurr@gfz-potsdam.de

Martin B. Kalinowski
University of Hamburg
Carl Friedrich von Weizsäcker Centre
for Science and Peace Research (ZNF)
Beim Schlump 83
20144 Hamburg
Germany
martin.kalinowski@uni-hamburg.de

Karl Koch
Federal Institute for Geosciences & Natural
Resources
Stilleweg 2
30655 Hannover
Germany
karl.koch@bgr.de

David Brown
CTBTO, IDC
P.O. Box 1200
1400 Vienna
Austria
david.brown@ctbto.org

ISBN 978-3-0346-0370-6   e-ISBN 978-3-0346-0372-0
DOI 10.1007/978-3-0346-0372-0

Library of Congress Control Number: 2010926921

Cover illustration: Composed of elements of a graphic from the article „Backtracking of noble gas measurements taken in the aftermath of the announced October 2006 event in North Korea by means of PTS methods in nuclear source estimation and reconstruction" by A. Becker, G. Wotawa, A. Ringbom, and P.R.J. Saey and of a seismogram of the closest record of the October 2006 nuclear test explosion in North Korea taken at station KS31 in South Korea.

Printed on acid-free paper

Springer Basel AG is part of Springer Science+Business Media

www.birkhauser.ch

Contents

Pure Appl. Geophys. 167 (2010), 373–379
© 2010 Birkhäuser Verlag, Basel/Switzerland
DOI 10.1007/s00024-009-0039-7

Introduction to the Topical Volume: Recent Advances in Nuclear Explosion Monitoring

ANDREAS BECKER,[1] BERND SCHURR,[2] MARTIN B. KALINOWSKI,[3] KARL KOCH,[1,4] and DAVID BROWN[1]

Following a scientific symposium on "CTBT: Synergies with Science 1996–2006 and beyond" hosted by the Comprehensive Nuclear-Test-Ban Treaty Organization (CTBTO) Preparatory Commission (PrepCom) in 2006 at the Hofburg Palace in Vienna (CTBTO, 2006) staff members of CTBTO PrepCom were encouraged to instigate a dialogue with the scientific community on research relevant to nuclear explosion monitoring. For this purpose they proposed to the European Geosciences Union (EGU) to hold a special session on "Research and Development in Nuclear Explosion Monitoring" within the next EGU General Assembly which was then convened in April 2007 in Vienna. With an initial contribution in 2007 of 42 research papers (EGU, 2007) covering all relevant fields in treaty monitoring including seismology, infrasound, hydroacoustics, nuclear physics dealing with particulate and noble gas radionuclides, and atmospheric backtracking, followed by 48 papers (EGU, 2008) submitted in 2008, the guest editors were encouraged to collect this research work within a proceedings volume in a scientific publication.

The present volume can be considered a follow-up to research published in a series of topical volumes in Pure and Applied Geophysics in the years 2001–2002 (PAGEOPH 2001a, b, c, d, e; PAGEOPH, 2002a, b) following the opening for signature of the CTBT in 1996. These publications covered all technologies relevant for CTBT monitoring of seismo-acoustic events, and particularly comprehensively seismology, but had no emphasis on those technologies introduced from nuclear physics and atmospheric sciences. This imbalance was of course due to the fact that the former technologies were omnipresent during treaty negotiations, while at that time radionuclide monitoring was in a build-up stage, or in its infancy as noble gas monitoring is concerned. These circumstances found their expression in the purely seismic network that was already operated for many years at the prototype International Data Centre (pIDC) in Arlington, Virginia, USA. Consequently there was little radionuclide monitoring data yet accumulated with the potential to reveal new insights, e.g. on the global occurrences and background of CTBT relevant radionuclides, and on related scientific aspects of global radionuclide monitoring and event categorization.

As such, the papers presented in the current volume may be a reflection of two main developments in nuclear explosion monitoring since 2002. The first is the formation of the CTBTO PrepCom since 1997 with its Provisional Technical Secretariat (PTS), including an International Data Centre (IDC), which meanwhile has assumed the responsibility for establishing the International Monitoring System (IMS) of 321 sensing stations. In this context, the IDC has taken over the operation, data collection, data processing and event analysis of the primary and auxiliary seismic network from the pIDC. Secondly, the build-up of the IMS has, by now, resulted in a network of more than 50 radionuclide and 20 radioxenon stations, fully

---

[1] CTBTO Preparatory Commission, Provisional Technical Secretariat, Vienna International Centre, P.O. Box 1200, 1400 Vienna, Austria. E-mail: andreas.becker@ctbto.org; david.brown@ctbto.org

[2] Section 3.1, Lithosphere Dynamics, GFZ German Research Centre for Geosciences, Telegrafenberg, 14473 Potsdam, Germany. E-mail: schurr@gfz-potsdam.de

[3] Carl Friedrich von Weizsäcker Centre for Science and Peace Research, University of Hamburg, Beim Schlump 83, 20144 Hamburg, Germany. E-mail: martin.kalinowski@uni-hamburg.de

[4] *Present Address:*
Federal Institute for Geosciences and Natural Resources, Stilleweg 2, 30655 Hannover, Germany. E-mail: karl.koch@bgr.de

operating and sending monitoring data for further IDC processing and event analysis.

In all sensor technologies and relevant scientific disciplines, significant advances in nuclear explosion monitoring were recently achieved. This progress in the development and testing of new methods improves the capabilities in detection, location and characterization of CTBT relevant events. In particular the latter poses a challenge for smaller events, where natural or man-made but CTBT irrelevant sources can generate false positive events. The efficient discrimination of these events pursued at a minimum risk of missing a relevant event is the overall challenge.

The 15 papers of this volume can be structured into four seismo-acoustic studies and eight in the field of radionuclide monitoring and related atmospheric backtracking. Finally three papers are dealing with studies on well known past cases relevant in the context of CTBT monitoring, namely the Chernobyl disaster and the announced nuclear test explosion of October 2006 in North Korea. These latter studies demonstrate that the progress made in the relevant research fields has contributed to a good understanding of the characteristics of these events of very different nature and also serves as a reminder that supplementary methods, e.g. satellite observations, are offering CTBT member states additional and enhanced national technical means in their efforts for compliance monitoring.

1. Seismo-acoustic Studies

To improve seismo-acoustic signal detection and source location algorithms, the use of a well-calibrated data set of signals or events, so-called ground-truth data, is an established approach. The utility of such data sets is demonstrated and highlighted in two papers in this volume.

In the first, *Gibbons* et al., motivated by the requirement of detecting and classifying low-magnitude seismic events at regional distances by relatively small aperture seismic arrays, consider the recording of mining and military explosions close to the ARCES seismic array in Northern Norway. The authors perform f–k analyses on the collected waveforms, concluding that the frequency band chosen for processing that optimizes the SNR, which is the band-pass often chosen by analysts, can generate significant mislocations in the hypocentral solution. It is shown that by using a frequency-optimized template for a station, mislocations can be significantly reduced, further suggesting that event association algorithms may be tuned by biasing the hypocentral location toward the contribution from one or two arrays that have been selectively optimized for a given target region, at the expense of more distant stations.

The second paper in this volume that uses well-constrained ground-truth data to potentially improve seismo-acoustic source location algorithms is *Koch's* analysis of the acoustic observations of 5 years of fixed-location firings of the Ariane-5 rocket engines, recorded at the German IMS infrasound array IS26 in the Bavarian forest. In this data-set the receiving station is close to the first stratospheric bounce, at a distance of around 300 km, making the data-set valuable for infrasonic detection and source location studies. Clearly identified in this analysis is the influence on infrasound by the seasonal nature of the stratospheric winds.

These seasonal wind effects have a dominant influence on the propagation of atmospheric infrasound, controlling the extent to which acoustic signals are propagated to a receiver over distances of hundreds or thousands of kilometres. As well as modifying the amplitude of the received signal, the travel time and, most importantly, observed backazimuth at the receiving station are affected. These influences are a primary consideration in the operation of automatic infrasound signal detection and source location algorithms, such as that being operated by the PTS of the CTBTO PrepCom in Vienna.

For recurring sources of the same type, such as mining or quarry explosions, resulting in waveforms with nearly wiggle-to-wiggle correspondence cross correlation is nowadays a standard technique in waveform analysis. Newly occurring events of these sources are readily recovered with high confidence. *Gibbons and Ringdal* present a study where the detected seismo-acoustic events are most likely from a closely-spaced source area, but where the correlation of the (seismic) waveforms is not very high.

They propose a new processing technique for identifying such recurring events using seismic array data. They also discuss the seismic signatures, including those from different recurring sources in the Scandinavian arctic region, and the acoustic observations which are compatible in their source location estimates. More than 50 explosion times are given and may represent a useful ground-truth database for future infrasound studies.

The paper by *Drob* et al. outlines a method that allows the temporal nature of infrasound propagation to be accommodated in the automatic infrasound processing system in operation at the IDC of the CTBTO PrepCom. These authors take a unique atmospheric data set formed by the seamless integration of a near-real time weather prediction model, for altitudes below 55 km, with a seasonal climatological model for higher altitudes, together with the τ–p model for acoustic travel-time prediction, adapted to include winds by GARCÉS *et al.* (1998), to perform more accurate backtracking to potential acoustic source locations in the IDC's automatic global association (GA) algorithm. Several examples are provided that demonstrate a useful tool that, as well as improving hypocentral location estimates in automatic processing, could facilitate the interactive analysis of infrasound events.

2. Radionuclide Monitoring Studies

The present topical volume on advances in CTBT monitoring uniquely features papers dealing with phenomenology, detection, analysis and categorization of measurements of radionuclides (particulates and noble gases) as well as atmospheric backtracking methods applied in the interpretation of radioisotope observations. While the CTBT bans nuclear testing in all environments, compliance monitoring for underground nuclear testing poses the biggest challenge for the radionuclide component of the verification regime. For this case four factors are most crucial for the monitoring capability:

i. the degree of containment of the nuclides yielded in the cavity, which depends on the particular release scenario,

ii. the dilution of the nuclear debris during its atmospheric transport from the source location to the detecting IMS stations,

iii. the detection capability of the particulate or noble gas radionuclide system, and

iv. the capability to distinguish a nuclear explosion signature from legitimate civilian sources.

It is well known that for a well contained underground nuclear explosion hardly any radionuclide particulates can escape the cavity. However, for noble gases even with sophisticated containment technology a significant chance remains for accidental rapid venting or operational releases that can under favourable meteorological conditions be detected by the IMS. This can be concluded from the Nevada Test Site release data (KALINOWSKI, 2010). For 433 out of 824 underground tests radioactive releases were reported.

The study by *Dubasov* extends the important knowledge about containment to two other test sites. It covers almost 500 underground nuclear tests conducted by the Soviet Union at the two major test sites at Semipalatinsk and Novaya Zemlya in the 30 years from 1961 to 1990. The data are presented for the first time in an English language peer reviewed journal here, and demonstrates the large variability of noble gas containment encountered and the high percentage of apparently fully contained underground tests. High containment poses a challenge as the related nuclear test might stay undetected by the network of 80 IMS radionuclide stations, despite its unprecedented detection capability. Even if we assume that the xenon measurements *Dubasov* reports were taken in the close vicinity of the test locations with devices being orders of magnitudes less sensitive than today's state-of-the-art xenon sampling systems (thoroughly discussed in this volume by *Auer* et al.), the frequent occurrence of fully contained underground tests stresses the need for on-site inspections (OSIs), to provide the evidence to characterize an event as nuclear. This remains true even if the IMS sensors already provide indications to this effect.

Fortunately, for initially fully contained tests, OSI measurements will have a chance of detecting noble gases at the surface above the test location. Since the so-called non-proliferation experiment at the Nevada test site in 1993, it is known that natural subsoil gas transport processes can move highly diluted and rapidly decaying xenon radioisotopes to the surface within weeks (CARRIGAN *et al.*, 1996). Therefore, the detection of radioxenon traces in sub-soil samples is likely to become an accepted method to be applied during an OSI campaign, but sub-soil xenon measurements cannot be interpreted with regard to their treaty relevance without a thorough understanding of the natural occurrences of radioxenon in the lithosphere.

The paper by *Hebel* investigates the quantities and isotopic ratios that can be expected as a result of natural uranium or thorium content in the sub-surface strata. These are promising for OSI measurements of radioxenon in soil gas. The lithospheric radioxenon production is dominated by spontaneous fission of ^{238}U. In the presence of minerals with a high uranium concentration, the radioxenon concentrations in the soil gas could be above the detection threshold, if the permeability is high enough. *Hebel* demonstrates that the isotopic ratio method described by *Kalinowski* et al. can resolve any doubt about the origin of radioxenon in soil gas.

Adding xenon monitoring to every second station of the 80 station particulate network will result in a permanently growing database of highly accurate measurements of the four short-lived isotopes ^{133}Xe, ^{133m}Xe, ^{131m}Xe, and ^{135}Xe. In this context the paper of *Auer* et al. serves as an excellent introduction to the advances that were made in the past 10 years in the field of atmospheric measurements of noble gases.

In 1999 the testing of a new generation of radioxenon monitoring equipment designed for automated and continuous operation started in a laboratory setting. Since 2004, an increasing number of stations is operated in unattended mode in the field, within an international cooperation, the so-called 'International Noble Gas Experiment' (INGE). *Auer* et al. report on the operational experience and performance parameters fully meeting and partly exceeding the stringent requirements for CTBT monitoring. In particular, the sensitivity is very high with the minimum detectable concentrations for ^{133}Xe ranging from 0.1 to 1 mBq/m^3.

The significant improvement in measurement capabilities allows for a better understanding of the radioxenon distribution resulting from nuclear facility emissions. Two papers present first collections of atmospheric radioxenon observations in different regions of the world. *Dubasov and Okunev* report on the ^{85}Kr and radioxenon levels from 2 years of measurements in the vicinity of St. Petersburg and Cherepovets (Russia). Both locations are known to be affected by krypton and xenon releases from regional nuclear facilities as well as from sources at larger distances. The ^{85}Kr data fill an important gap in the published data for its atmospheric levels with relevance for nuclear arms control agreements other than the CTBT because it can serve as indicator for unreported plutonium production. Considerably more interesting are the results for radioxenon. Due to very large sampled air volumes (several thousand of m^3), *Dubasov and Okunev* achieve the highest sensitivity ever reached for ^{133}Xe with a minimum detectable concentration of 0.008 mBq/m^3. The minimum ^{133}Xe activity concentration observed during the 2 years period was 0.2 mBq/m^3 at St. Petersburg and 0.09 mBq/m^3, at the Cherepovets city. Therefore, for the first time, monitoring of ^{133}Xe was achieved over an extended period of time without being limited by the sensor sensitivity. *Dubasov and Okunev* discuss possible sources for elevated concentrations both by analyzing air trajectories and the isotopic activity ratio of ^{135}Xe/^{133}Xe.

In view of these developments in noble gas monitoring, the CTBTO Prepcom is building up its radioxenon monitoring capability in the framework of INGE. To understand region specific backgrounds, the xenon stations have been organized in regional mininetworks, with the European cluster being one with the largest number of stations (6). It is also the region with the highest density of civilian sources that interfere with CTBT verification. To provide a comprehensive catalogue on the environmental radioxenon levels in Europe, *Saey* et al. have reviewed 5 years of measurements between 2003 and 2008 at the European network. An elevated ^{133}Xe background is typical for these stations, e.g. Schauinsland, Germany, also examined by *Auer* et al. As these levels could be normally categorized as anomalous, atmospheric backtracking of irrelevant events is crucial to avoid false positive events. *Saey* et al. demonstrate state-of-the-art atmospheric backtracking for two xenon detections that can be attributed to releases from radiopharmaceutical facilities (RPF). One of the major surprises in radioxenon monitoring is that not the hundreds of nuclear power plants (NPPs) are the main challenge in event characterization but the few RPFs that each can easily have batch emissions of up to three orders of magnitude larger than the mean daily emission of all NPP sites.

This situation demands for reliable methods to distinguish CTBT relevant sources of noble gas emissions from civilian sources. Three different approaches for categorizing a radioxenon observation regarding its relevance for the CTBT have been suggested. (a) Anomalous observations with respect to the history of concentrations found at the same site, (b) the characteristics of isotopic ratios and (c) attribution to known sources by atmospheric transport simulations. Important contributions for the first and the latter approach are made by the studies of *Saey* et al., (a), *Dubasov and Okunev* (a and c), as well as by two papers discussed below, *Wotawa* et al. (a and c) and by *Plastino* et al. (c). The second approach is the only method that cannot only provide an indication but a solid proof for a nuclear test as the origin of a radioxenon observation. This one is introduced by the paper of *Kalinowski* et al. that can be considered as the methodological foundation for source discrimination by isotopic activity ratios. It presents a fundamental methodology that is backed by simulated and empirical data. Based on a large data set of observations at INGE stations in Europe and North America, *Kalinowski* et al. provide a proof-of-principle for source discrimination based on isotopic activity ratios. The strength of the proposed method depends on the availability of certain xenon isotopes in the measurement. In the relatively rare case of all four xenon isotopes being detected, the isotope ratios allow for a clear-cut separation of the nuclear test domain from the nuclear reactor domain. For the case of less than four isotopes being quantified, *Kalinowski* et al. examine all possible options with various combinations of isotopes and present their usability and their limitations.

A study to enhance the understanding of the global xenon background concentrations is presented by *Wotawa* et al. who apply a folding technique, i.e. a linear combination of a global emission inventory of all known sources (NPPs and RPFs) with a database of so-called source-receptor sensitivity (SRS) fields. This yields a model capable of reproducing the xenon backgrounds encountered at all IMS stations and which is shown to work very well for monthly averages. The authors demonstrate through a multiple-linear regression approach the prevailing role of two RPFs for the backgrounds at the xenon stations in operation so far, and the minor role NPP emissions play. Finally they diagnose and quantify the adverse impact of the two largest ^{133}Xe sources on the detection capability of the xenon network. It is geographically constrained to the well monitored mid-latitudes, and therefore tolerable. However, if new RPFs were erected in low latitude regions to diversify the global ^{99}Mo production, the situation would be very different. In general the study demonstrates the important role of reliable and accurate atmospheric backtracking calculations based on a good and well updated understanding of the global meteorology for assessment of the radionuclide networks coverage and monitoring threshold. It is also crucial to flag xenon samples of being potentially categorized false positive detections due to interference of CTBT irrelevant xenon sources.

The last radionuclide monitoring study of this volume by *Plastino* et al. also examines the role of the meteorology to understand and correctly categorize ^{133}Xe detections. However, the study is constrained to the consideration of the local meteorology at a station. For the two IMS stations in St. Johns, Canada (CAX17) and Charlottesville, USA (USX75) the radioxenon time series are detrended with the corresponding local meteorological patterns and the residuals are assessed to what extent they can provide sufficient information to identify false positive samples. The proposed statistical method works well for CAX17 whereas the method yields ambiguous results for USX75 in terms of the required pure white noise time series residual. The authors argue that other than the local meteorological patterns would have to be taken into account for USX75, hence pointing to the fact that the aforementioned SRS information containing the regional atmospheric transport information should also be taken into account. On the other hand, the method proposed is fast and can be done on basis of the data collected at the station alone. Further studies could identify those IMS xenon stations, besides CAX17, where the approach may work well in the xenon sample categorization task.

3. Case Studies

Any advance in nuclear explosion monitoring techniques is best proven through real world cases, where all processes governing the detection, location and identification of relevant events are acting comprehensively together. In this volume two well known such cases are examined: one constituting a strong release of radionuclides into the atmosphere not only at surface level but in a vertical column, i.e. the Chernobyl disaster of 1986, and one with a fairly weak signature.

The Chernobyl case is still interesting and elucidating in many aspects crucial in the radionuclide verification context of the CTBT. In the study of *Pakhomov and Dubasov* the question is addressed to what extent radioxenon observations and their backtracking can be utilized to reveal information on the release scenario of an event whose source location and time is well known. The authors make the ambitious attempt to estimate the explosion yield of the Chernobyl accident from isomeric ^{133}Xe ratios of krypton–xenon mixture samples taken in Cherepovets in the period from 22 April to 6 May 1986 and to shed new light on the physical processes that actually took place during this catastrophic event. Aside from these new insights the case study serves as an example for the challenges that radionuclide verification approaches face in terms of event characterization, regarded as an important element of CTBT verification tasks.

The other case, considered in the final two papers of this volume, represents *the* first real test for the CTBT verification system being established by the PTS: the underground nuclear test of 9 October 2006 in North Korea announced by the Democratic People's Republic of Korea (DPRK) on 3 October 2006. With regard to this event a particular surprise was that the few already operational radioxenon stations were capable of picking traces of ^{133}Xe at a remote IMS station in Yellowknife, Canada, at a distance of several thousand kilometres. This conclusion was

supported by atmospheric transport modelling (ATM) calculations that, however, could not be more accurate than stating that the DPRK test site is located within a vast possible source region. The paper by *Becker* et al. constitutes an enhanced backtracking analysis of this case, as it considers additional measurements of a mobile xenon sampling system that was located close to the demarcation line between DPRK and the Republic of Korea (South Korea), thus putting stronger constraints on the nuclear source estimation and reconstruction results currently retrieved from the ATM methods implemented at the PTS. While demonstrating the crucial role of the measurements from the mobile systems the paper clarifies their minor role for the watchdog role of the PTS for unannounced tests. The authors conclude that only a fully fledged network of sufficient station density can assure that the requirement specifications for CTBT compliance monitoring are met. For the 2006 DPRK nuclear test explosion the currently scheduled 40 stations network design would have been sufficient, as the closest station could have picked traces of the event on the next day. It should be noted that the final decision on the scope and size of the xenon network has yet to be taken by the policy-making organs of the CTBTO after entry into force of the treaty.

What is enough then? Being to a large extent a political question to be evaluated in the general CTBT framework, the study by *Schlittenhardt* et al. finally is most interesting in this context. It shows an example for supplementing the standard PTS means with available satellite observations, which could be part of so-called national technical means that every member state can use during the clarification process on any suspicious event of CTBT non-compliance. An interesting aspect of this paper is the method of pixel- and object-based change detection that is already in the process of being established in the context of IAEA nuclear safeguards (NIEMEYER and NUSSBAUM, 2006). It is demonstrated here for CTBT verification purposes and makes use of commercially available satellite imagery data that is public domain. Besides showing the usefulness of the approach with historical nuclear explosions on the Indian and Nevada test sites, the authors examine the DInSAR method to precisely monitor co-seismic or post-seismic changes in elevation. The authors discuss the capabilities and also the constraints of the method; the latter imposed *inter alia* by slopes in mountainous terrain that are too steep, so direct evidence of the event was ambiguous and could have been the result from preparation activities. It can be generalized as a conclusion that the civil society on its own can use these emerging data and software capabilities in the public domain, to contribute to monitoring, thus increasing transparency of cases of interest.

4. Summary

The papers compiled for this topical volume on "Recent Advances in Nuclear Explosion Monitoring" summarize a significant number of research papers (EGU, 2007, 2008) that were presented during the 2007 and 2008 EGU General Assembly special sessions on this topic. The editors hope that the breadth of the information given therein will prove a useful reference for future research in nuclear explosion monitoring.

Acknowledgments

We are grateful for the contributions of all authors and the helpful and constructive advices received during the peer review process from the following 33 scientists: Juan-Antonio Añel, Matthias Auer, Anders Axelsson, Dmitry Bobrov, Ted Bowyer, Lars Ceranna, Sabine Eckhardt, Lars-Erik De Geer, Laslo Evers, Ross Heyburn, Christian Igel, Gerald Kirchner, Frank Krüger, Giovanni Laneve, Raquel Nieto, Alexis Le Pichon, Harry Miley, Franca Padoani, Roland Purtschert, Anders Ringbom, Sigrid Rössner, Paul R.J. Saey, Jörg Schlittenhardt, Clemens Schlosser, Martin Schultz, Jeff Schneider, Dmitry Storchak, Joachim Schulze, Brian Stump, Curt Szurbela, Rick Tinker, Laura de la Torre, and Kurt Ungar. We are also grateful for the smooth collaboration with the editor-in-chief, Renata Dmowska. We would also like to express our appreciation to the publisher Birkhäuser for hosting this proceedings volume, and to the European Geosciences Union for the opportunity to hold special sessions on "Research and Development in Nuclear Explosion Monitoring" within the EGU General Assembly since 2007. The work of one of the authors (MBK) is supported by the German Foundation for Peace Research.

REFERENCES

CTBTO (2006), *CTBT: Synergies with science 1996–2006 and beyond.* Symposium website on http://www.ctbto.org/the-organization/ctbt-synergies-with-science1996-2006-and-beyond/.

EGU (2007), Abstracts of all accepted contributions to the nuclear explosion monitoring session on http://www.cosis.net/members/meetings/sessions/accepted_contributions.php?p_id=235&s_id=4563.

EGU (2008), Abstracts of all accepted contributions to the nuclear explosion monitoring session on http://www.cosis.net/members/meetings/sessions/accepted_contributions.php?p_id=296&s_id=5593.

CARRIGAN, C.R., HEINLE, R.A., HUDSON, G.B., NITAO, J.J., and ZUCCA, J.J. (1996), *Trace Gas Emission on Geological Faults as Indicators of Underground Nuclear Testing*, Nature 202, 528–531.

GARCÉS, M.A, HANSEN, R.A., and LINDQUIST, K.G. (1998), *Travel Times for Infrasonic Waves Propagating in a Stratified Atmosphere*, Geophys. J. Int. *135*, 255–263.

KALINOWSKI, M.B. (2010), *Characterisation of Prompt and Delayed Atmospheric Radioactivity Releases from Underground Nuclear Tests at Nevada as a Function of Release Time*, J. Environ. Radioactivity (accepted).

NIEMEYER, I. and NUSSBAUM, S. (2006): *Change Detection: The Potential for Nuclear Safeguards.* In *Verifying Treaty Compliance,* (Avenhaus, R., Kyriakopoulus, K., Richard M. and Stein, G., eds.). Springer Berlin, 2006.

PAGEOPH (2001a), *Monitoring the Comprehensive Nuclear-Test-Ban Treaty: Source Location* (Ringdal, F. and Kennett, B.L.N., eds.), Pure Appl. Geophys. *158*, No.1/2.

PAGEOPH (2001b), *Monitoring the Comprehensive Nuclear-Test-Ban Treaty: Hydroacoustics* (deGroot-Hedlin, C. and Orcutt, J., eds.), Pure Appl. Geophys. *158*, No. 3.

PAGEOPH (2001c), *Monitoring the Comprehensive Nuclear-Test-Ban Treaty: Regional Wave Propagation and Crustal Structure* (Patton, H.J. and Mitchell, B.J., eds.), Pure Appl. Geophys. *158*, No. 7.

PAGEOPH (2001d), *Monitoring the Comprehensive Nuclear-Test-Ban Treaty: Surface Waves* (Levshin, A.L. and M.H. Ritzwoller, eds.), Pure Appl. Geophys. *158*, No. 8.

PAGEOPH (2001e), *Monitoring the Comprehensive Nuclear-Test-Ban Treaty: Source Processes and Explosion Yield Estimation* (Ekström, G., Denny, M. and Murphy, J.R., eds.), Pure Appl. Geophys. *158*, No. 11.

PAGEOPH (2002a), *Monitoring the Comprehensive Nuclear-Test-Ban Treaty: Seismic Event Discrimination and Identification* (Walter, W.R. and Hartse, H.E., eds.), Pure Appl. Geophys. *159*, No. 4.

PAGEOPH (2002b), *Monitoring the Comprehensive Nuclear-Test-Ban Treaty: Data Processing and Infrasound*, (Der, Z.A., Shumway, R.H. and Herrin, E.T., eds.), Pure Appl. Geophys. *159*, No. 5.

Pure Appl. Geophys. 167 (2010), 381–399
© 2009 Birkhäuser Verlag, Basel/Switzerland
DOI 10.1007/s00024-009-0024-1

▌Pure and Applied Geophysics

Considerations in Phase Estimation and Event Location Using Small-aperture Regional Seismic Arrays

STEVEN J. GIBBONS,[1] TORMOD KVÆRNA,[1] and FRODE RINGDAL[1]

Abstract—The global monitoring of earthquakes and explosions at decreasing magnitudes necessitates the fully automatic
detection, location and classification of an ever increasing number
of seismic events. Many seismic stations of the International
Monitoring System are small-aperture arrays designed to optimize
the detection and measurement of regional phases. Collaboration
with operators of mines within regional distances of the ARCES
array, together with waveform correlation techniques, has provided
an unparalleled opportunity to assess the ability of a small-aperture
array to provide robust and accurate direction and slowness estimates for phase arrivals resulting from well-constrained events at
sites of repeating seismicity. A significant reason for the inaccuracy
of current fully-automatic event location estimates is the use of $f-k$
slowness estimates measured in variable frequency bands. The
variability of slowness and azimuth measurements for a given
phase from a given source region is reduced by the application of
almost any constant frequency band. However, the frequency band
resulting in the most stable estimates varies greatly from site to site.
Situations are observed in which regional P- arrivals from two sites,
far closer than the theoretical resolution of the array, result in
highly distinct populations in slowness space. This means that the
$f-k$ estimates, even at relatively low frequencies, can be sensitive
to source and path-specific characteristics of the wavefield and
should be treated with caution when inferring a geographical
backazimuth under the assumption of a planar wavefront arriving
along the great-circle path. Moreover, different frequency bands
are associated with different biases meaning that slowness and
azimuth station corrections (commonly denoted SASCs) cannot be
calibrated, and should not be used, without reference to the
frequency band employed. We demonstrate an example where
fully-automatic locations based on a source-region specific fixed-
parameter template are more stable than the corresponding analyst
reviewed estimates. The reason is that the analyst selects a frequency band and analysis window which appears optimal for each
event. In this case, the frequency band which produces the most
consistent direction estimates has neither the best SNR or the
greatest beam-gain, and is therefore unlikely to be chosen by an
analyst without calibration data.

Key words: Automatic event locations, CTBT monitoring,
parameter estimation, regional arrays.

1. Introduction

The design, implementation and operation of
seismic arrays has been motivated greatly by a need
to detect, locate and identify low-magnitude seismic
events which could potentially be clandestine
underground nuclear explosions (DOUGLAS, 2002;
RINGDAL and HUSEBYE, 1982). The initial emphasis
was on medium and large-aperture arrays designed
for the detection of weak P arrivals at teleseismic
distances. In the 1980s, the requirements of nuclear
explosion monitoring were extended to encompass
the detection and classification of low magnitude
seismic events which were only observable at regional distances. This motivated the development of a
new generation of smaller aperture seismic arrays
which facilitated the detection and measurement of
the high-frequency regional Pn, Pg, Sn, and Lg phases (e.g., MYKKELTVEIT et al., 1983; MYKKELTVEIT and
BUNGUM, 1984; MYKKELTVEIT et al., 1990). The 1990s
saw a large increase in the number of array stations
and the development of a global seismic array
network whereby regional phases detected and identified at several different array stations could be
associated and interpreted to produce fully automatic
seismic event bulletins (e.g., RINGDAL and KVÆRNA,
1989).

In the context of the comprehensive nuclear-test-
ban treaty (CTBT), which was adopted by the United
Nations in 1996, a worldwide international monitoring
system (IMS) of seismic, infrasound, hydroacoustic

---
[1] NORSAR, P.O. Box 53, 2027 Kjeller, Norway. E-mail:
steven@norsar.no

and radionuclide sensors is being deployed in order to detect the occurrence of any possible violation of such a treaty. The seismic component of the IMS comprises a relatively sparse network of arrays and three-component stations. A decrease in the required event magnitude detection threshold is associated with a large increase in the number of seismic events which require processing. The exponential increase in the number of earthquakes per unit time with decreasing magnitude (GUTENBERG and RICHTER, 1954) is compounded by an enormous increase in seismicity from industrial and other artificial sources. Many IMS seismic stations are within local or regional distances of mines or quarries and all events from which signals are detected need to be located automatically with as great an accuracy as possible.

The generalized beamforming (GBF) algorithm (RINGDAL and KVÆRNA, 1989) considers all the phase detections made by a network of regional seismic arrays in a given time interval and assesses how well the pattern of phase determinations is matched by each of many event hypotheses on a grid of trial epicenters. Fully-automatic solutions from the GBF form the basis for the reviewed regional event bulletins at NORSAR. Each of the phase determinations has an associated slowness estimate (primarily used to identify the phase) and a backazimuth estimate (indicating the apparent direction of the wavefront), both measured using frequency-wavenumber $(f-k)$ analysis (CAPON, 1969). The stability of azimuth estimates from small-aperture arrays was investigated by KVÆRNA and DOORNBOS (1986) and KVÆRNA and RINGDAL (1986) who concluded that the estimates obtained using the narrow-band approach of Capon (1969) were less stable than so-called broadband estimates whereby the Fourier coefficients are averaged over a wide range of frequencies. KVÆRNA and DOORNBOS (1991) examined $f-k$ measurements of Pn phase arrivals at the NORES array generated by repeating events from a single source location. They found that not only did the stability of the estimates vary with the frequency band applied, but that a significant bias of up to several degrees was observed to vary continuously with frequency. This indicates that for stable direction estimates for phase arrivals from a given single source, it would be better to use a fixed frequency band than variable frequency bands.

Based upon network event location estimates, the azimuthal bias and variance for regional phases was investigated as a function of frequency band and sensor configuration for the NORES (BAME et al., 1990) and ARCES (CARR, 1993) arrays. SCHWEITZER (2001b) exploited a far larger set of events, better constrained in source location and origin time, to calculate tables of slowness corrections for all the Fennoscandian seismic arrays which have led to a reduction in network event location errors for this region. The need to calibrate arrays for seismic event location in the CTBT context resulted in the calculation of so-called slowness-azimuth station corrections (SASCs) for IMS array stations for both teleseismic (BONDÁR et al., 1999) and regional (BONDÁR and NORTH, 1999; BEN HORIN et al., 2004) phases.

Our ability to assess the accuracy of automatic location estimates obtained using GBF (and other related phase association algorithms) has improved greatly recently due to an immense improvement in "Ground Truth" (GT) information for seismic events (whereby the origin time and location are essentially known exactly). Figure 1 shows the locations of a number of sources of repeating man-made seismicity in northern Fennoscandia and NW Russia, together with fully automatic and reviewed event location estimates. The automatic location estimates (Fig. 1b) are sufficiently accurate that they can form the basis for subsequent analyst review, although constraints on human resources mean that only a very small number of events can be treated in this way. The estimates are not accurate enough to be able to attribute the signals to any of the displayed repeating sources without further analysis. There are several well-understood reasons for the large spread in location estimates in Fig. 1b:

1. Many of the industrial events from sites displayed in Fig. 1 have very complicated source-time functions with multiple ripple-fired firing sequences. This can lead to the concealing of some arrivals within the coda of previous events and a spurious association of P and S phases from different events.

2. The GBF associates phases whose parameters fall within given bounds. For the defining rules, a

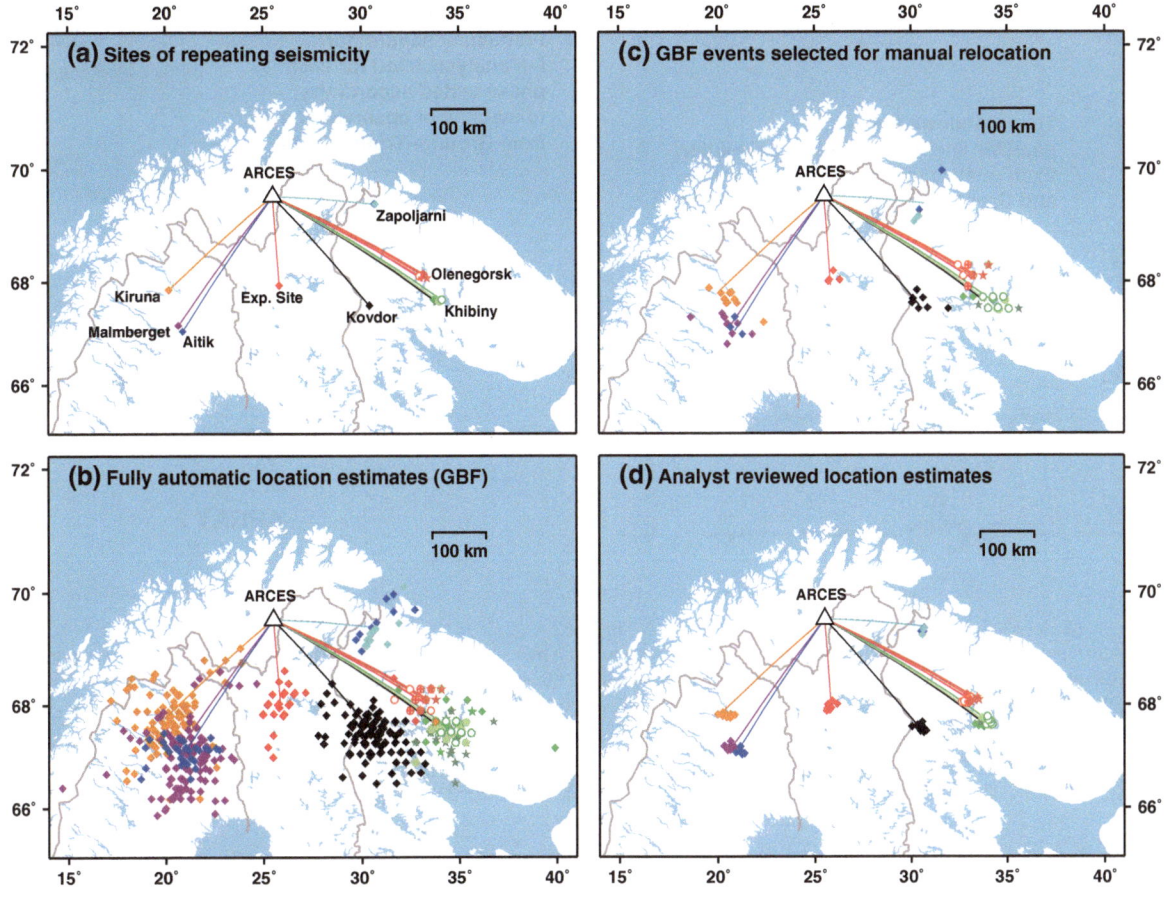

Figure 1

a Locations of sites which are the source of repeating seismic events within 500 km of the IMS primary seismic array ARCES. Kiruna, Malmberget and Aitik are large-scale mining operations in northern Sweden with all events taking place within approximately 2 km of the symbols shown. The explosion site in northern Finland is used for the routine destruction of outdated ammunition. Zapoljarni, Olenegorsk and Khibiny are regions on the Russian Kola peninsula each containing several mines. **b** Fully automatic GBF location estimates for events known to have taken place at each of the sites in **a**. **c** Automatic locations of (*larger*) events from **b** selected for analyst relocation. The reviewed location estimates of these events are shown in **d**. The coordinates for the sites are provided in Table 1. The two mines in the Zapoljarni cluster are represented by two different shades of *blue* and the five reporting mines from each of the Khibiny and Olenegorsk regions are represented by *different symbols* of *green* and *red shades*, respectively

single event hypothesis at a spurious location may match the observed set of automatically detected phases better than two subsequent event hypotheses located elsewhere which more closely represent the true event history.

3. All azimuth and slowness estimates are made using $f-k$ analysis in variable frequency bands. (Many of the more marginal signals are only observed above the background noise in relatively narrow frequency bands).

The mitigation of each of these causes motivates the implementation of "template-based post-processing"

for the identification and location of events from a specific source region as illustrated in Fig. 2. KVÆRNA and RINGDAL (1994) demonstrated that vast improvements could be made to uncalibrated event location estimates by applying carefully calibrated diagnostic tests to the waveforms from different stations at times fixed relative to an event hypothesis. The template prevents the use of a spuriously associated phase in locating the event (the whole identification procedure is largely insensitive to signals which do not appear within the short time-windows specified by the site-template) and ensures

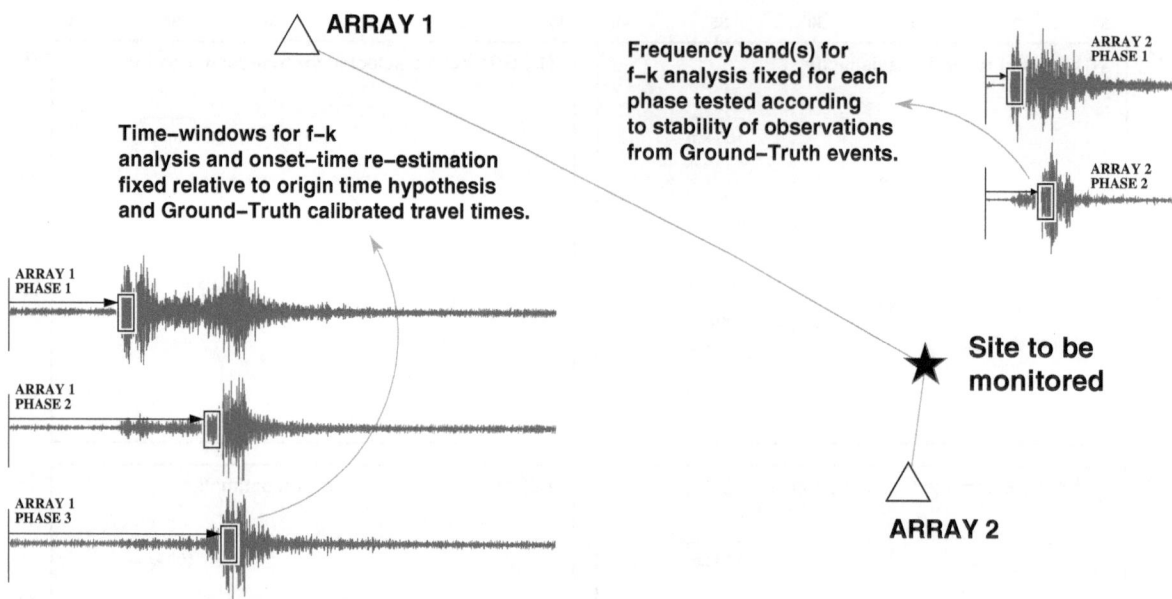

Figure 2
Schematic representation of template-based reprocessing. We wish to test a hypothesis that a seismic event occurred at a given time at or close to our site of interest. A "site-template" defines which phases should be observed at which times at which stations. For all phases for which identical diagnostic tests for successive events produce stable results, a "phase-template" (indicated by the *framed boxes*) defines processing parameters and bounds of acceptance. If a sufficient number of phase templates return a positive match (i.e., an indication that the current measurements are consistent with observations of that phase from previous events from the same site), the site-template will indicate that there is a high likelihood that the event hypothesis is correct

that the parameters used for the direction estimates are optimal for the given monitoring situation. GIBBONS *et al.* (2005) describe in detail a template procedure for a single-array monitoring scenario, a case of increasing importance given the need to monitor events of ever smaller magnitude with a relatively sparse station network. Whereas the tuned relocation algorithm of KVÆRNA and RINGDAL (1994) required that we already had a location estimate sufficiently close to the site of interest, the algorithm in GIBBONS *et al.* (2005) is triggered by only a single phase detection. This means that the system can identify and locate an event successfully even in cases where other factors (such as an interfering signal or a spurious phase association) preclude a preliminary location estimate which is sufficiently close to the source location (see Fig. 1b). It should be noted that template-based event identification is not limited to array processing and that alternatives (based primarily on waveform similarity) exist for networks and single stations (see, for example,

SCHULTE-THEIS and JOSWIG, 1993; MASSA *et al.*, 2006; MACCARTHY *et al.*, 2008).

The purpose of the current paper is not to discuss details of template-based algorithms since the principles involved in the procedure of, for example, GIBBONS *et al.* (2005) are readily extended to different source regions and station configurations by modifying selection rules and measurement techniques. The success rate (how often actual events from the site being monitored are correctly identified) and false alarm rate (how often an event hypothesis is accepted in the absence of such an event) will vary from situation to situation. However, in the calibration of analogous location procedures for other sites, we have encountered caveats which need to be taken into account when optimizing both source-specific and general event detection and location procedures. Since our focus is on array-processing with small-aperture arrays, the kind most widely employed within the IMS, this paper will focus on the measurement of slowness and backazimuth using frequency-wavenumber analysis. GT data obtained in recent years has provided an unparalleled database of well-constrained observations whereby array measurements can be compared directly with predicted values. GIBBONS *et al.* (2005) considered routine mining events which were known to have occurred at the Kovdor mine, allowing a systematic study of the accuracy and stability of slowness and azimuth estimates in different frequency bands for the various phases observed. In the current paper, we will consider such observations from many different sites in order to investigate how source-specific monitoring and screening procedures need to be. The present GT database exceeds greatly that which was available for the earlier studies of CARR (1993) and SCHWEITZER (2001b).

In Sect. 2 we describe briefly how the lists of events from each of the sites in Fig. 1 were obtained. Section 3 describes how slowness and azimuth estimates for defining seismic phase arrivals from events at the sites in Fig. 1 vary with the frequency band applied. Section 4 covers one specific site in detail and describes possible consequences of the observations for both automatic and analyst reviewed event-location procedures.

2. Collection of Ground-Truth Information on Mining and Other Repeating Seismic Sources

A project funded by the United States Department of Energy began in October 2001 with the goal of collecting "Ground Truth" data for routine mining explosions on the Kola Peninsula of NW Russia for seismic calibration purposes (HARRIS *et al.*, 2003). The term Ground Truth needs to be applied with caution. Mining explosions are designed to displace large volumes of material in a carefully controlled manner and often consist of hundreds of small detonations in long and irregular ripple-fire sequences over large areas of a mine or quarry. It is neither practical to expect the mining authorities to provide accurate coordinates for each charge, together with the exact times of detonation, or meaningful to represent such complicated events as point sources in time and space. The mining companies typically

recorded that an event took place at a given mine during a given interval of several minutes. A more meaningful origin time was then seismically inferred by colleagues at the Kola Regional Seismological Center (KRSC) in Apatity, most often using an on-site instrument. In this paper, "Mining Ground Truth" refers to the confirmation that a given event took place at a specified mine with a time corresponding approximately to the start of the ripple fire sequence.

Table 1 provides a list which includes all the mines for which Ground Truth information was collected. Many of these mines are very closely spaced within the clusters labelled in Fig. 1 but are administered as separate units. In addition to the Russian mines, three sites in Sweden and one in Finland are listed in Table 1. The Swedish ore mines at Kiruna and Malmberget are sites of continual industrial production and are of interest due to the large numbers of signals generated (Malmberget means "ore-mountain" in Swedish). Aitik is a quarry approximately 30 km to the Southwest of Malmberget and is the site of large ripple-fired events approximately once per week. Ground Truth information of the form provided for the Russian events was not available for these sites, but, using a handful of events confirmed by the mine operators, a bootstrap waveform correlation procedure was able to identify large numbers of subsequent events which due to waveform similarity are assumed to have occurred in close proximity of each other (see, for example, GELLER and MUELLER, 1980; HARRIS, 1991).

The final site is a military installation in the north of Finland where expired ammunition is destroyed in controlled surface explosions. Unlike the mining explosions, where the waveforms can vary significantly from event to event, the signals from the Finnish explosions all display great similarity. The high correlation coefficients indicate very closely-spaced source locations and the military authorities responsible for these events have confirmed that they are all separated by less than a few hundred meters. These explosions are easily identified from the automatic detection lists since they are recorded with a high signal-to-noise-ratio (SNR) at ARCES and occur at very characteristic times of the day (i.e., always to within a few seconds of a full-hour or

Table 1

List of sites of repeating seismicity considered in this paper

Identifier	Site name	Latitude (°N)	Longitude (°E)	BAZ (ARCES)	Distance (km)	Description
ZP1	Zapadny (Zapoljarni)	69.404	30.682	91.7	203	Quarry
ZP2	Central (Zapoljarni)	69.397	30.742	91.8	205	Quarry
OL1	Olenegorsk (Olenegorsk)	68.154	33.192	112.8	346	Quarry
OL2	Kirovogorsk (Olenegorsk)	68.106	32.996	114.3	341	Quarry
OL3	Bauman (Olenegorsk)	68.057	33.145	114.5	349	Quarry
OL4	Oktjabrsk (Olenegorsk)	68.078	33.106	114.3	347	Quarry
OL5	Komsomolsk (Olenegorsk)	68.075	33.385	113.4	356	Quarry
KH1	Kirovsk (Khibiny)	67.670	33.729	118.0	393	Mine
KH3	Rasvumchorr (Khibiny)	67.631	33.835	118.0	400	Mine
KH4	Central (Khibiny)	67.624	33.896	118.0	403	Quarry
KH5	Koashva (Khibiny)	67.632	34.011	117.5	406	Quarry
KH6	Norpakh (Khibiny)	67.665	34.146	116.6	409	Quarry
KV1	Kovdor	67.557	30.425	135.4	298	Quarry
FES	Finnish explosion site	67.934	25.832	175.6	179	Military site
AIT	Aitik	67.060	20.900	216.7	335	Quarry
MAL	Malmberget	67.179	20.675	219.4	329	Mine
KIR	Kiruna	67.849	20.196	231.4	286	Mine

quarter-). The high level of waveform similarity also means that they are readily picked out by a correlation detector of the type described by GIBBONS and RINGDAL (2006). These events were investigated by GIBBONS et al. (2007) because they more often than not generate infrasound signals recorded on the seismic sensors of the ARCES array.

Most of the events from the sites listed in Fig. 1 are observed with a high SNR at ARCES and most are between magnitudes 1.0 and 2.0. Some of the mining-induced seismic events are somewhat larger, sometimes close to magnitude 3.0, although events of this size are not common. The times of mining events and the Finnish explosions used to calibrate the templates are provided in the appendix of KVÆRNA et al. (2007).

3. Accuracy and Stability of Slowness and Azimuth Estimates for Phase Arrivals from Well-Constrained Industrial Seismic Events

One of the primary reasons identified for the large scatter of the automatic event locations in Fig. 1 (b,c) is the variability of backazimuth estimates. Figure 3 displays slowness vector estimates for the initial P arrivals from the events in Fig. 1 selected for manual review (panels c and d). Whilst

the apparent velocity, v_{app}, and the backazimuth, BAZ, are the most commonly used parameters in event location, it is usual to use the horizontal slowness parameters $s_x = \sin(BAZ)/v_{app}$ and $s_y = \cos(BAZ)/v_{app}$ for calculating the delay times in the $f-k$ analysis. Each of the panels in Fig. 3 indicates the values of s_x and s_y for which the optimal beam gain was achieved for the arrivals displayed. To assist in the interpretation of these plots, contours of equal apparent velocity are drawn in the $s_x - s_y$ parameter space as are lines of constant backazimuth from each of the sites considered. For display purposes only, a map for site identification is drawn at the center of the slowness plots.

The exact nature of the first P arrival varies somewhat for the different sites labelled in Fig. 1 (see, for example, STORCHAK et al., 2003, for a description of regional phases and nomenclature). The one-dimensional velocity model which provides the best fit for the region is the BAREY model (HICKS et al., 2004), and this model is used to predict travel times when forming the site-specific templates. The Pn phase is the first predicted arrival at ARCES for most of the sites considered. For the Finnish explosion site and the mines at Zapoljarni, the situation may be somewhat more complicated; a short time-window starting at the signal onset time is likely to include both the predicted Pn and Pg phases.

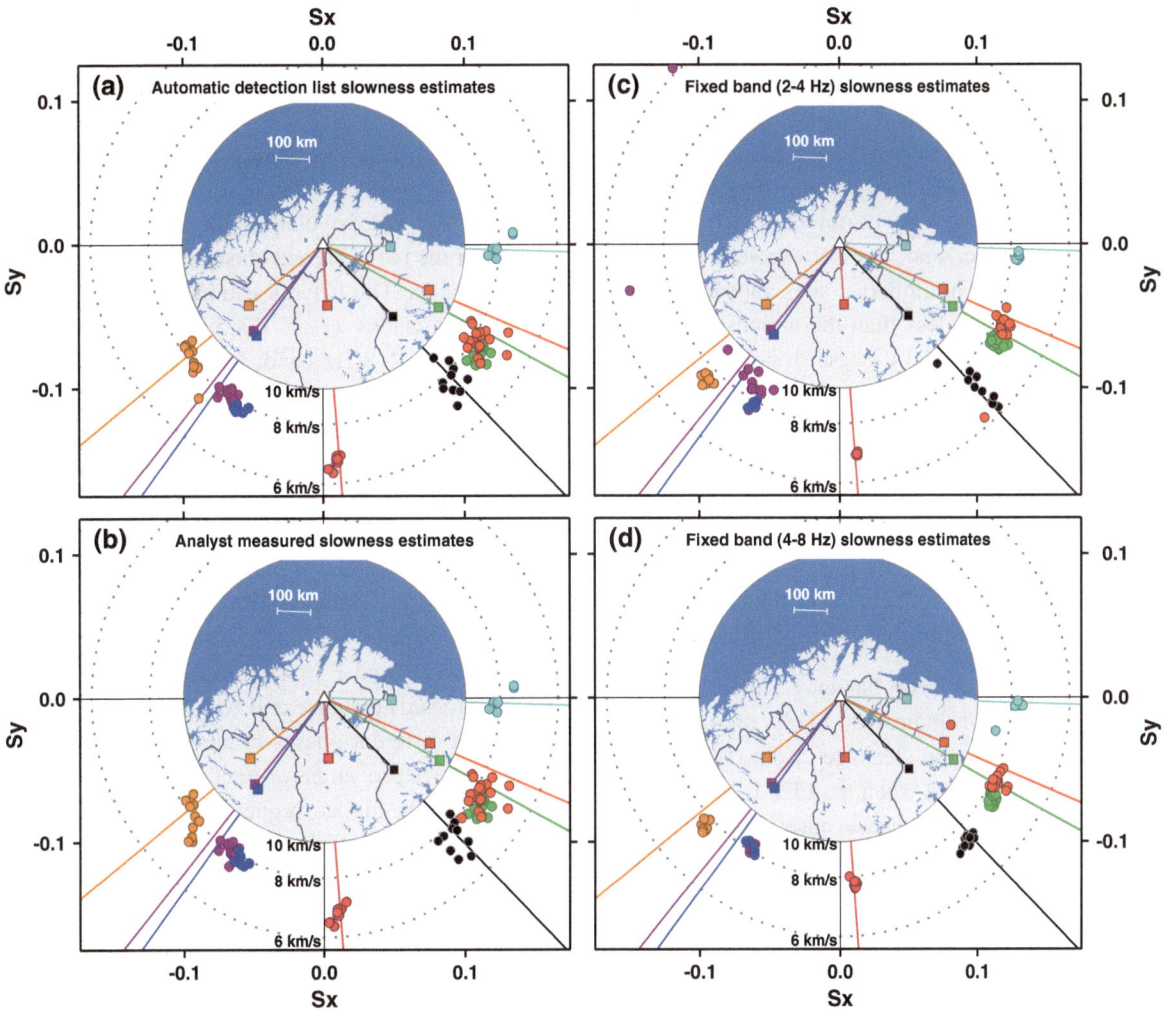

Figure 3

Slowness vector estimates for initial P arrivals at ARCES from each of the events selected for manual relocation in Fig. 1(c,d). Details of the different estimates plotted in panels **a**, **b**, **c**, and **d** are provided in the text. The *dotted-line circles* indicate the regions of slowness space corresponding to the apparent velocities as labelled. Within the 10 km/s circle is drawn a map of the region to allow a straightforward association of the slowness estimate symbols (*circles*) with the corresponding source locations as labelled in Fig. 1a (*squares*). The maps are drawn in a Gnomonic azimuthal projection which preserves directions with respect to the array at the center of the map. The projection is not distance-preserving. In these plots, *common symbols* are used for the different reporting mines in each of the clusters Zapoljarni, Khibiny and Olenegorsk

Comparing panels (a) and (b) in Fig. 3 reveals that there are no qualitative differences between the slowness estimates obtained from the fully automatic detection/estimation procedure and those resulting from the analyst review. The slowness estimates in panel (a) are made with respect to an automatically determined onset time, in a frequency band resulting from an SNR-optimizing algorithm, using in-house developed software. The estimates in panel (b) are made with respect to an analyst-adjusted arrival time,

using a frequency band chosen from a menu, using the Analyst Review Station (ARS) software developed by Science Applications International Corporation (SAIC). Without making direct comparisons between the processing parameters selected by the automatic algorithms and those chosen by the analyst, it appears that the two procedures result in similar distributions of slowness estimates for each of the sites displayed here.

All of the apparent velocity estimates for these initial P arrivals fall between approximately 7.0 and

8.5 kms^{-1}, with the exception of the Finnish explosion site for which the estimates fall between 6.0 and 7.0 kms^{-1} (more typical of Pg-type phases). A number of consistent biases in the backazimuth estimates are clear in these panels. P-type arrivals from the Olenegorsk and Khibiny mining regions always appear to arrive from a backazimuth greater than the geographical value, and the corresponding phases from the Aitik and Malmberget sites appear to come from a backazimuth less than the anticipated value. The populations of initial P-phase slowness estimates for mines in the Olenegorsk and Khibiny clusters overlap considerably. The slowness estimate populations for the closely spaced Aitik and Malmberget sites are surprisingly well separated.

Panels (c) and (d) in Fig. 3 show the same sets of phase arrivals processed in the 2–4 and 4–8 Hz frequency bands, respectively. In obtaining these estimates, the following steps were taken:

(i) The waveforms were bandpass filtered in the frequency band indicated. (This mitigates the danger of spectral leakage from frequencies outside of this band when the FFT is performed.)

(ii) The analyst-revised phase onset-time pick was selected as the reference time, t_{ref}.

(iii) A time window $[t_{ref} - 0.2{:}t_{ref} + 2.7]$ was selected as the analysis window.

(iv) The data were multiplied by a tapering function that preserves all but the first and final 0.2 s long segments (which were tapered smoothly to zero, c.f., HARRIS, 1978). Whilst less powerful than the multitaper methods (THOMSON, 1982), this simple tapering procedure is also expected to mitigate the problems of spectral leakage when estimating slowness over short data segments in relatively narrow frequency bands.

(v) Broadband f–k analysis was performed on the tapered waveforms between the specified frequencies (KVÆRNA and DOORNBOS, 1986). KENNETT (2002) provides a more detailed summary of the formulation.

At a backazimuth of approximately 90°, the 2–4 Hz fixed-band processing for the (admittedly few) selected Zapoljarni events appears to indicate an improvement (i.e., a decrease) in the variability of the slowness estimates. A small bias seems to be present in all the estimates which may contribute to the location bias visible in Fig. 1d). In the 4–8 Hz band (Fig. 3d), one Zapoljarni slowness estimate appears to be an outlier.

The fixed-band estimates for the Khibiny and Olenegorsk events still show a significant bias but, significantly, the populations of the slowness estimates for the two different clusters are almost distinct in both bands. The Kovdor estimates (also studied in detail by GIBBONS et al., 2005) are arguably more stable for the 4–8 Hz band. In the 2–4 Hz band, the backazimuths obtained appear relatively stable whilst the apparent velocities obtained vary greatly.

The estimates for the different events at the Finnish explosion site are almost identical in the 2–4 Hz band. In the 4–8 Hz band the variability has increased slightly but, more significantly, the apparent velocity has jumped from about 6.8 to about 8.0 kms^{-1}. We recall that, according to the BAREY model, the Pn and Pg phases arrive almost simultaneously for this distance and so we may primarily be observing different phases in the different frequency bands. The steep angle-of-incidence Pn phase may dominate in the higher frequency band and the upper-crust propagating Pg phase may dominate at the lower frequency.

The Aitik event slowness estimates show good stability (and a significant bias) in the 2–4 Hz band. The Malmberget event estimates are highly variable between 2 and 4 Hz, possibly a result of low energy in this frequency band. In the 4–8 Hz band, both Aitik and Malmberget estimates display good stability but occupy essentially the same region of slowness space. The Kiruna estimates are relatively stable in both frequency bands although the bias for the 2–4 Hz band is visibly larger than that for the 4–8 Hz band.

Even from the limited sets of events displayed in Fig. 3, it is clear that the azimuth and slowness estimates vary in a non-trivial way with respect to the frequency band applied. Given that only events with a preliminary magnitude estimate greater than M = 2 are usually considered for manual analyst review, it follows that most of the events included in Fig. 3 were probably quite well observed with a reasonable SNR in many frequency bands. As the event magnitude decreases, the range of frequencies over which an acceptable SNR is observed will also decrease. This is of course the very reason that variable frequency bands are used in the automatic processing. In the following section, we will consider phases from all of the mining Ground Truth events in the period October 2001 to September 2002 from two mines in the Zapoljarni group (ZP1 and ZP2, separated by approximately 2.5km) and two from the Khibiny group (KH1 and KH3, separated by approximately 6.3km). As indicated in Table 1, the backazimuth from ARCES to ZP1 is almost identical to that from ARCES to ZP2 and, in theory, this array should not be able to distinguish between events from these two

sites. The same is true for the two sites KH1 and KH3.

Figure 4 shows the fully automatic (variable frequency band) slowness estimates for first P arrivals from Zapoljarni events (top left panel, hereafter simply referred to as *variable band* estimates) together with fixed-band estimates in many different, overlapping frequency bands as labelled. Each waveform was examined manually, with an arrival-time picked on a display of array beams filtered in numerous frequency bands. The fixed-band estimates were then obtained using tapered data windows as described above. Of the 42 Ground Truth events

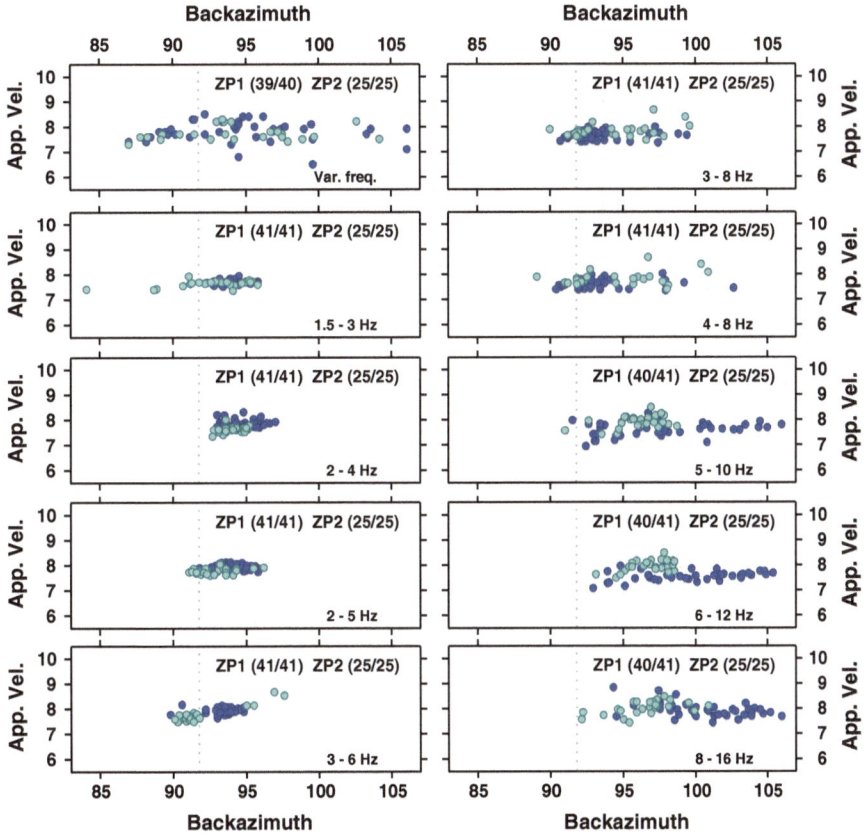

Figure 4

Slowness estimates at ARCES for initial P arrivals for events at the two mines near Zapoljarni (ZP1, *dark symbols*, and ZP2, *light symbols*) between October 2001 and September 2002. The *dotted line* indicates the true great circle backazimuth. The *top left panel* displays the variable frequency estimates from the fully automatic detection lists and the remaining panels display fixed frequency band estimates as indicated. Note that the outermost nine sites of the array were not included for the 5–10, 6–12, and 8–16 Hz frequency bands. This is due to the diminished coherence of the high frequency signals over these greater distances (c.f. Fig. 5c of GIBBONS *et al.*, 2005). Forty-two events are ascribed to the ZP1 mine during this period and 25 events to the ZP2 mine. Some qualitatively incorrect slowness estimates fall outside of the region of parameter space displayed. The *numbers in brackets* indicate how many estimates from the total populations fall within the region plotted

associated with the ZP1 mine during the period shown, 40 GBF events were identified for which the origin time fell within 10 s of the true origin time. The label ZP1 (39/40) in the top left panel indicates that one of the first P phase slowness estimates for the ZP1 mine GBF events falls out of range on the plot (the missing event had v_{app} = 7.9 kms^{-1}BAZ = 116.1°). The label ZP2 (25/25) in the top left panel indicates that all of the 25 confirmed ZP2 events in this period correspond to an automatic GBF solution within 10 s of the true origin time, and that all of the corresponding variable band estimates for the initial P arrival fall within the range plotted. The top left panel indicates a spread of at least 20° in the variable band estimates for Zapoljarni event backazimuths.

The first observation with regards to the fixed-band estimate panels in Fig. 4 is that almost all frequency bands return initial P-phase slowness vector estimates which fall within the range plotted. (The only exceptions are for the highest frequency bands, for which a single ZP1 event estimate falls out of range.) Each of the fixed-band plots in the left panel of Fig. 4 (i.e., 1.5–3.0, 2–4, 2–5, and 3–6 Hz) shows a significant reduction in the spread (variance) of estimates compared with the variable band estimates in the top-left panel. However, in spite of the significant overlap in the frequency bands displayed, the corresponding slowness vector distributions do look significantly different. The two bands with the greatest stability in the backazimuth estimates are the 2–4 and 2–5 Hz bands, which both display a range of approximately 5° with a bias that is approximately 3° greater than the true geographic backazimuth of 191.8°. The most surprising of these panels is for the 3–6 Hz frequency band where it appears that the slowness vector estimates from the ZP1 and ZP2 mines form almost distinct clusters in the parameter space. Aside from 3 or 4 outliers for each population, the ZP1 and ZP2 slowness estimates form quite compact clusters centered on approximately BAZ = 91° and BAZ = 94°, respectively. It is unclear as to how one would attempt to calculate an SASC for events from the Zapoljarni region in this frequency band.

The plots in the right-hand panels of Fig. 4 show that, for the higher frequency bands, the spread in slowness vector estimates is significantly greater than for the lower frequency bands. The spread in back-azimuth estimates ranges from approximately 10° for the 3–8 Hz band to almost 20° (i.e., approaching the same spread as the variable band estimates). The populations of estimates from the ZP1 and ZP2 mines are never entirely distinct but, in each of the plots, there are clearly regions of parameter space which are dominated by either the light or the dark symbols. In the global network association and location problem, it is essentially the backazimuth estimate which is the most important for regional phases (the apparent velocity estimate often does little more than identify the most likely candidate for a phase identification). From the plots in Fig. 4, one would probably advocate the use of the 2–4 Hz fixed band for stable parameter estimates for Zapoljarni P phases. It should be noted that whilst the slowness estimates from the different events are highly clustered for many frequency bands, the clusters in the slowness-azimuth parameter space are often skewed and biased such that a mean and standard deviation value alone may not provide a very useful description of the observed variation.

Figure 5 shows slowness estimates for the confirmed events at mines KH1 and KH3 on the Khibiny Massif for the same time period as shown in Fig. 4. Apart from the data sets used, the only difference between Figs. 4 and 5 is the range of backazimuth values displayed. The majority of Khibiny events result in automatic detections for which the apparent velocity and azimuth fall within the range displayed (top left panel of Fig. 5). Almost all measurements overestimate the backazimuth by between 2 and 10°. The distributions of slowness estimates obtained from analysis in the different frequency bands varies greatly. A side-by-side comparison of Figs. 4 and 5 shows that, whilst the variabilities of slowness estimates in the optimal frequency band are comparable, the frequency bands for which the minimum variability is obtained are very different for the two mining regions. The best band for the Zapoljarni events (2–4 Hz) is relatively poor for the Khibiny events and it is worth noting that approximately 13% of estimates (28 out of 208 arrivals) do not even fall within the range displayed. (The 1.5–3.0 Hz range is even poorer with fewer than half of the events resulting in even a qualitatively correct backazimuth estimate.) The 3–6, 3–8 and 4–8 Hz frequency bands appear to provide the best stability for the initial P arrivals from Khibiny events. On the basis of the distributions shown in Fig. 4, none of these bands would be chosen for accurate measurement of signals from Zapoljarni events.

Whilst Figs. 4 and 5 demonstrate how much poorer the stability of the variable band estimates are compared with the estimates from the optimal fixed bands, the plots demonstrate why it is necessary to generate direction estimates in variable frequency bands for the automatic detection lists. There is no single frequency band which results in good stability for all sites and, for any given source region, it is impossible to know which band would provide

optimal processing in the absence of extensive Ground Truth information. One solution would of course be to perform fixed-band analysis in a number of fixed bands for every given detection, but such multiple band lists would be of little use without a calibrated post-processing algorithm to interpret them.

Part of the discrepancy between Figs. 4 and 5 can be attributed to SNR issues. The background noise at ARCES is greatest below 2 Hz, dominated by ocean-generated microseisms. The SNR from the Khibiny events, in particular the lower-yield shots, is low at frequencies below 4 Hz and increases at higher frequencies due to the decreasing background noise. The 4–8 Hz frequency range clearly provides some kind

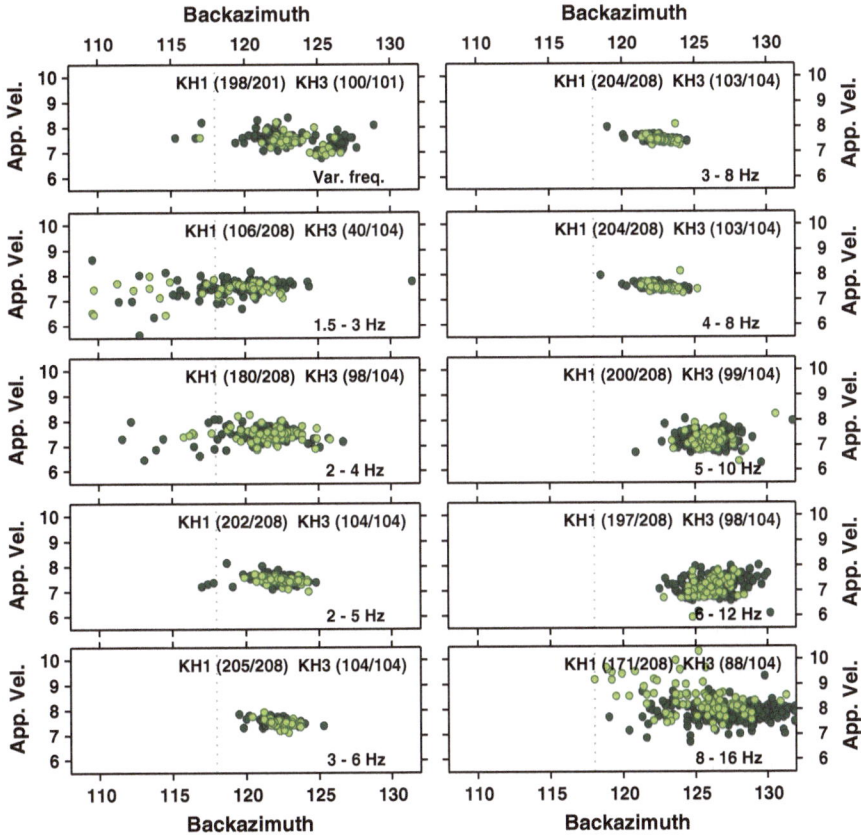

Figure 5

Slowness estimates at ARCES for initial P arrivals for events at two of the mines on the Khibiny massif (KH1, *dark symbols*, and KH3, *light symbols*) between October 2001 and September 2002. The *dotted line* indicates the true great circle backazimuth. The *top left panel* displays the variable frequency estimates from the fully automatic detection lists and the remaining panels display fixed frequency band estimates as indicated. Note that the outermost nine sites of the array were not included for the 5–10, 6–12, and 8–6 Hz frequency bands. 208 events are ascribed to the KH1 mine during this period and 104 events to the KH3 mine. Some qualitatively incorrect slowness estimates fall outside of the region of parameter space displayed. The *numbers in brackets* indicate how many estimates from the total populations fall within the region plotted

of optimal trade-off between SNR and signal incoherence at the even higher frequencies. The events from the closer Zapoljarni sites have a better SNR across the full spectrum of frequencies recorded by ARCES. This can explain why the stability in the 2–4 Hz band is better than that for the Khibiny events, but does not explain why the stability in the 4–8 Hz band is so much worse.

We may attempt to form an explanatory hypothesis in terms of the likely propagation paths resulting from events in these two source regions. We recall that the initial P arrival from the Khibiny events is likely to be a Pn-type phase, which has propagated primarily through a relatively homogeneous lower crust/uppermost mantle path. We recall also that the Pn-phases from Zapoljarni events are likely to compete with the possibly more highly scattered Pg phases, which have propagated through a potentially more heterogeneous upper crust. The stability in the 2–4 Hz band for the Zapoljarni events may simply be because the larger wavelength parts of the signal have a good SNR (in contrast to the Khibiny events in this frequency band). The poor stability in the 4–8 Hz band for the Zapoljarni events may be the result of competition between wavelets of shorter wavelength arriving over a broader range of paths. The apparent slowness vector chosen will be a function of the strength of the wavefield along any incident ray, and different events may preferentially excite different frequencies and different source-to-receiver paths. Even if our hypothesis is correct, it does not provide any kind of prescription for selecting a frequency band for a given phase from a given site.

All of the slowness estimates displayed so far have been for initial P arrivals. The detection and estimation of secondary phases is crucial for events at regional distances for which very few observations are available. RINGDAL and HUSEBYE (1982) showed how the coherence of the Lg phase diminished far more rapidly as a function of sensor separation than the Pg phase from the same event, and it can be anticipated that this will have direct consequences for the stability of direction estimates. In addition, the weak S phases from many of the mining events considered here are compromised by interference from strong coda and/or repeating events (see, for example, Fig. 5(e) of GIBBONS et al., 2005).

Considering P arrivals from all of the sites listed in Table 1, the 2–5 Hz band was deemed to provide the most stable estimates overall (even if it was rarely the best frequency band for one specific site). Figure 6 shows confidence ellipses containing 95% of all slowness vector estimates for both P and S phases for the four mining clusters at Zapoljarni, Olenegorsk, Khibiny and Kovdor. The spread in backazimuth estimates is without exception far greater for the secondary phases than for the primary phases, and this should be taken into account in location procedures.

4. Case Study: Ammunition Destruction Explosions in Northern Finland

One source of repeating man-made seismicity in this region has proved highly illustrative due to the simple source-time functions of these events. Most of the mining events considered are ripple-fired and often with highly variable geometries of explosive sources. These are factors which may contribute to the variation of slowness and azimuth estimates in a way which is not easy to quantify, and it is useful to have the opportunity to study a source for which these factors do not need to be considered. The free-surface explosions in Finland are performed with the purpose of disposing of old ammunition and the high similarity of the signals indicates a very repeatable, single-event source. This eliminates the spurious association of seismic phases from consecutive events as a reason for location error. However, the large spread in the red diamonds due south of the ARCES array in Fig. 1b) indicates that significant errors still occur, probably due to variability of arrival time and direction estimates, compounded by the association of coda detections as incorrectly identified seismic phases. 142 events at this site between 2001 and 2006 were identified using the correlation procedure of GIBBONS and RINGDAL (2006). Due to the well-defined maxima of the correlation coefficient traces, this procedure also provides consistent origin time estimates (i.e., the error in the origin time is identical for each event).

The repeatability of the waveforms themselves make this data set an interesting test of the stability of

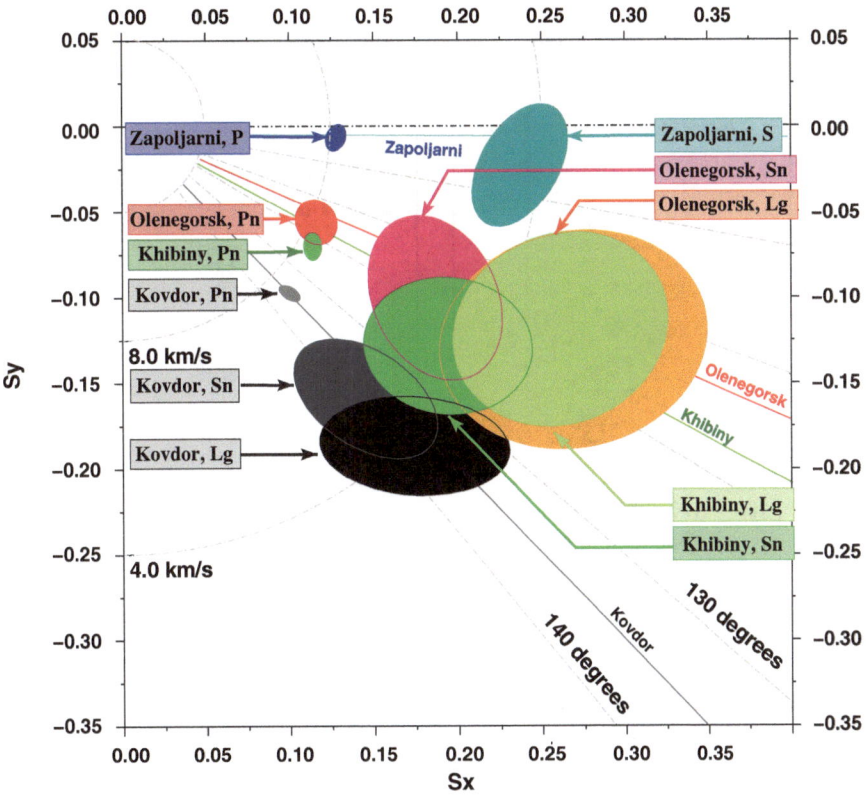

Figure 6

Schematic overview of the variations in slowness estimates for the main seismic phases recorded from mining clusters on the Kola Peninsula. Instead of plotting individual slowness values as was done in Fig. 3, we show estimated 95% confidence ellipses based on the events analyzed for each phase/mine combination. All slowness estimates were made in the fixed frequency band 2.0–5.0 Hz, and the event set comprises all events confirmed to have occurred at the mines between October 2001 and September 2002. Events from different mines within the Zapoljarni, Khibiny and Olenegorsk groups are considered together. As discussed in the text, it is likely that the first few seconds of the wavetrain resulting from Zapoljarni events contains both Pn and Pg phases, and likewise for Sn and Sg. We therefore simply use P and S for this site

slowness estimates. The left-hand panel of Fig. 7 shows the spread of the slowness estimates for the initial P arrivals for the 142 events obtained from the automatic detection lists. The variability is somewhat surprising considering the similarity of the waveforms that the estimates are based upon. The right-hand panel of Fig. 7 shows fixed band estimates for the same set of arrivals. It should also be noted that the variable and fixed-band estimates may be taken using a slightly different time-window. The time-windows for the fully automatic detections are set relative to an automatic onset-time estimate on an event-by-event basis. The time-windows for the fixed-band calculations are set relative to a reference time which was picked manually for a single (master) event and then inferred for the remaining events on

the basis of cross correlation with the master event waveform. The cross correlation determined onset-time and the fully automatic single event onset-time estimates are likely to vary by up to approximately one second. A fixed window length of 3.0 s was used for the fixed-frequency band calculations. The window length applied in the automatic processing is also approximately 3.0 s long, although the exact length may vary somewhat according to different processing parameters.

Two observations are evident from a comparison of the two panels in Fig. 7. Firstly, the spread of the variable band slowness estimates is far greater than would have been assumed on the basis of the variability between the different fixed-band estimates. Secondly, the fixed-band estimates in the frequency

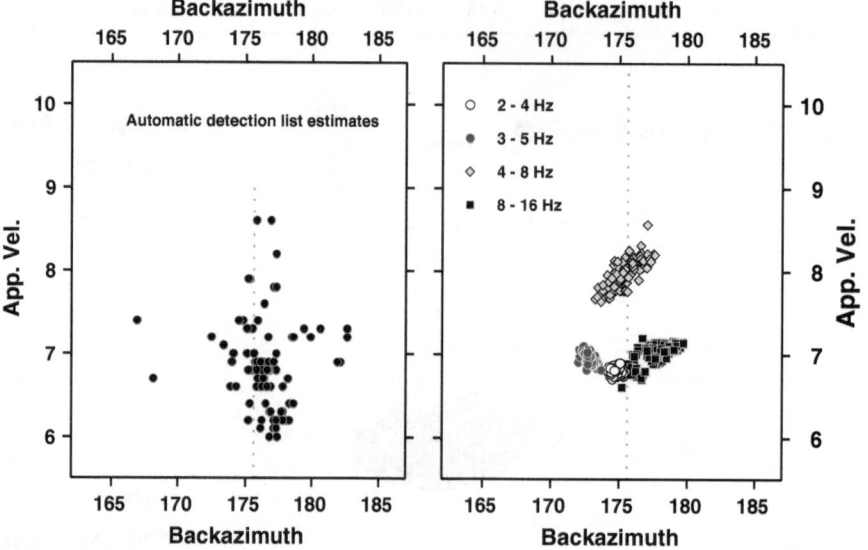

Figure 7

Slowness estimates at ARCES for initial P-arrivals for 142 events at the Finnish explosion site (FES) between 2001 and 2006. The *left panel* displays fully automatic (variable frequency band) estimates and the *right panel* displays fixed frequency band estimates as indicated

bands as indicated result in very compact populations which are almost entirely distinct from each other. The frequency band displaying the best (smallest) variability is the 2–4 Hz band. The remaining frequency bands all provide a better SNR, but poorer stability. Most of the bands indicate an apparent velocity of approximately $v_{app} = 7$ kms^{-1}, whereas the 4–8 Hz band indicates estimates closer to $v_{app} = 8$ kms^{-1}. This is consistent with the subset of observations displayed in Fig. 3 and, as previously discussed, may be related to the competing Pn and Pg phases. That the slowness has reverted to a more typical Pg velocity for the 8–16 Hz band is likely to be at least partly due to a loss of signal coherence (see, for example, Fig. 5(c) of GIBBONS *et al.*, 2005).

The analysis of the slowness estimates allows us to formulate a template location algorithm analogous to that described for events at the Kovdor mine by GIBBONS *et al.* (2005). Given that only a single P and a single S phase appears to be observed, the template is somewhat simpler than that for the Kovdor events. However, a similar procedure is followed. For the P phase, an autoregressive arrival-time estimate, t_a, is made (c.f., Fig. 7 of GIBBONS *et al.*, 2005) and the fixed frequency band $f-k$ calculation for P made in a time window relative to this reference time. The fixed-band $f-k$ estimate for the S-phase is made in a

time-window beginning 21.2s following t_a and then an autoregressive arrival time is re-estimated with $t_a + 21.2$ as the initial guess. As a result of the stability investigation, the 2–4 Hz fixed frequency band was selected for all $f-k$ calculations. The location estimates were made using the HYPOSAT program (SCHWEITZER, 2001a) and are displayed in Fig. 8 together with location estimates made by three different procedures.

Panel (a) of Fig. 8 shows the GBF fully automatic locations as displayed in Fig. 1. Figure 8b) shows 10 of the events which were selected for analyst review. (Given that these events are almost always below magnitude 2, they are very seldom selected for manual relocation.) The variability of the location estimates in Fig. 8b) is clearly far better than that shown in panel (a). The reviewed location estimates from the Finnish national network are displayed in panel (c). These are better constrained that the NORSAR solutions in a North-South direction and more poorly constrained than the NORSAR solutions in an East-West direction. This is clearly a result of the geometry of the very sparse seismic network in the far North of the country. (Additional stations have been in operation since 2006 which is likely to improve the variability of the location estimates for subsequent events.) Panel (d) displays the template-

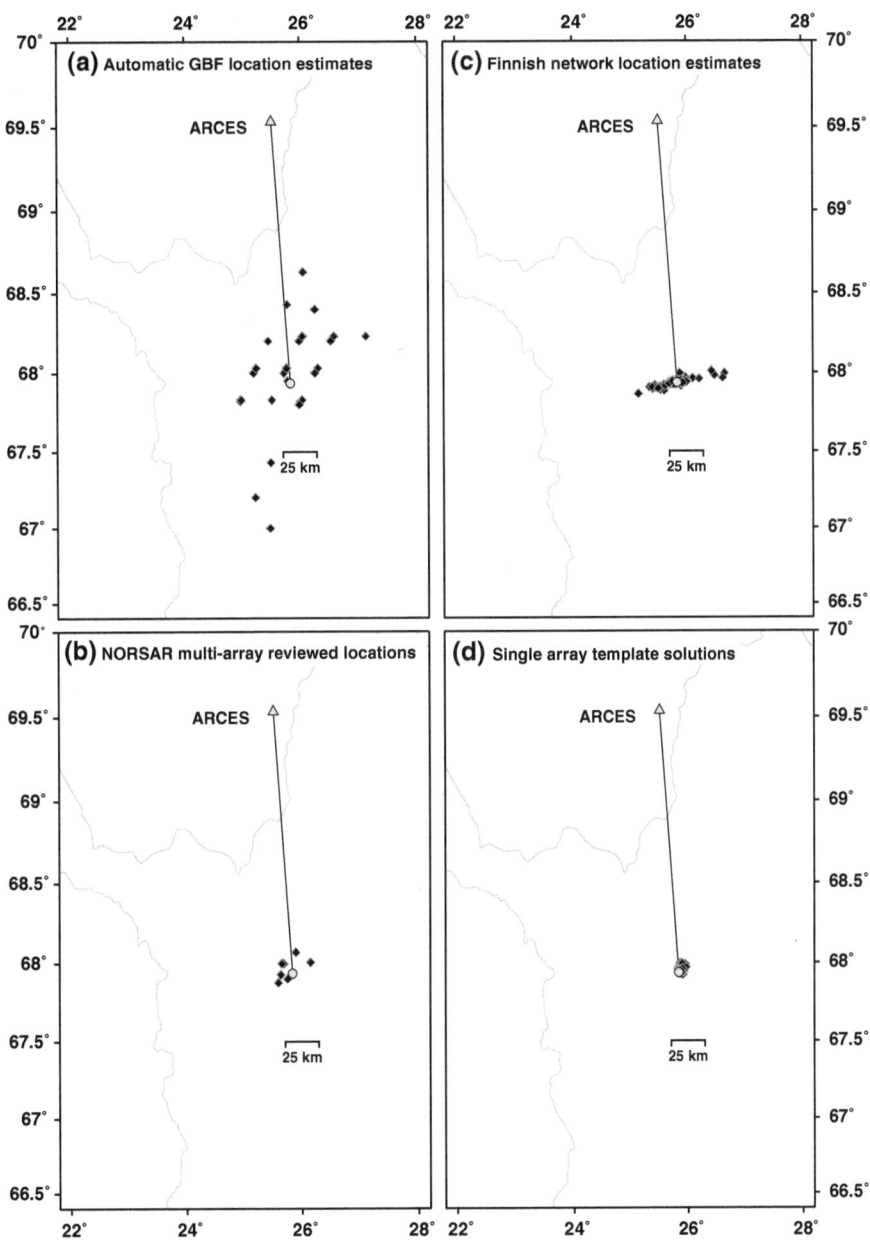

Figure 8
Location estimates of events at the Finnish explosion site between 2001 and 2006. Details of the methods employed are provided in the text. A total of 142 events are included in this database and a location estimate for all these events is included in panels **a** and **d**. The panels which display analyst reviewed location are subject to omissions; panels **b** and **c** contain 10 and 89 event locations, respectively

based (single-array) solutions which exhibit clearly a far smaller variance. (Note that there is a small bias in the population of event locations; no attempt has been made to correct for the backazimuth obtained in the 2–4 Hz frequency band, see Fig. 7.) It is interesting to note how much more compact the template

solutions are than even the NORSAR multi-array analyst solutions. The stability of the estimates in panel (d) is largely due to the fact that the direction estimate is almost entirely dictated by the variability of the corresponding slowness estimates. Each of the events located in panel (b) is based upon a frequency band chosen by the analyst on a case by case decision. The right-hand panel of Fig. 7 suggests that this may result in a large variability in location estimates. The analyst can improve the SNR significantly by choosing a different frequency band than that offering the most stable direction estimates, and is unlikely to pick the 2–4 Hz band for analysis.

5. Conclusions and Discussion

Small-aperture seismic arrays are very powerful tools with which to detect and locate low-magnitude seismic events. Under current observatory procedures, it is however almost impossible to associate with a satisfactory level of confidence a seismic signal with a known source of repeating seismicity without manual analysis. Waveform correlation methods can be highly successful at identifying explosions from repeating sources, but the effectiveness is often diminished due to differences between complicated temporal and spatial source functions. Using only traditional array processing (whereby seismic arrivals are treated as planar wave fronts) can produce very accurate event location estimates, but differences between predictions and observations must be taken account of and corrected for. The need to calibrate seismic arrays has been appreciated essentially since their first deployment (see, for example, BERTEUSSEN, 1976, and references therein). We have demonstrated that the situation for small-aperture regional arrays can be even more complicated with a significant dependence upon frequency which cannot be predicted *a priori*.

Even assuming perfect signal coherence over a seismic array, there is an unavoidable trade-off between resolution and uniqueness for wavefronts of different frequencies and slowness. At low frequencies, whereby the wavelength is long compared with the sensor separation, satisfactory resolution is only achieved for (possibly unphysical) low apparent velocities. Higher frequency wavefronts will, in theory, result in better resolution at realistically low slowness values at the expense of ambiguity between main and side lobes (KENNETT *et al.*, 2003, provide an ingenious solution for the sparse array problem but at considerable computational expense and without

guaranteed convergence to the correct slowness vector). ROST and THOMAS (2002) and SCHWEITZER *et al.* (2002) both provide an excellent discussion of array response functions and their consequences for the observation of real seismic phases on various array geometries.

The reality of parameter estimation is far worse than predicted due to (a) the presence of noise in otherwise optimal frequency bands and (b) lack of coherence (due to diffraction and scattering) at higher frequencies with optimal SNR (e.g., BUNGUM and HUSEBYE, 1971). This is very much the case for regional phases at the ARCES array where the ocean-generated microseismic noise dominates the waveforms below 2 Hz and where, even over this small array aperture, the coherence diminishes rapidly for higher frequencies. Ideally, one fixed frequency band could be used for parameter estimation in order to reduce the slowness estimate variability associated with the variable band estimates. Unfortunately, the band resulting in the most stable estimates varies from site to site (compare Figs. 4 and 5) and small changes in the limits of the frequency band used can often have significant consequences for the variability and bias in the slowness and azimuth estimates. Furthermore, without access to a large number of Ground Truth calibration events, it is impossible to know which frequency band(s) would provide the greatest stability for any given site. The band providing the most repeatable direction estimates is not necessarily that associated with the best SNR, or the best beam gain, and may therefore seldom be chosen by an observatory analyst. This is why the fully automatic template location estimates for the almost co-located Finnish explosions (Fig. 8d) are far less scattered than the corresponding analyst estimates (Fig. 8b). An analyst, examining each new event without reference to historical events, has good reasons not to choose the frequency band which actually results in the most repeatable measurements.

The calculation of SASCs is valuable in reducing location error in event monitoring. However, if the full capability of the array is to be exploited, these corrections must contain a reference to the frequency band employed. For example, the orange circles in Fig. 3 represent slowness estimates for initial P arrivals for events at the Kiruna mine (true backazimuth from ARCES 231.4°). The mean back-azimuth estimate in the 2–4 Hz frequency band is 225.1° compared with a value of 227.2° in the 4–8 Hz frequency band with almost no overlap between the two populations. The difference between the necessary corrections is small but enough to lead to a significant mislocation. An even more significant problem is arguably the large spread in estimates resulting from very wide frequency bands. This is illustrated in Fig. 7. Each of the estimates in relatively narrow fixed frequency bands (right panel) covers a small region of slowness space. The algorithm in the automatic detection process constructs the broadest possible frequency band which maintains an acceptable SNR and the resulting spread (illustrated in the left panel) is far worse than can be predicted from observing the fixed band estimates. Different frequency components contribute energy to different regions of slowness space such that the location of the relative beam-power maximum is a poorly determined function of the spectral content of the signal. The instability of the narrow-band estimates identified by KVÆRNA and RINGDAL (1986) is probably the result of a combination of aliasing problems and spectral leakage (compounded by the use of untapered waveforms). Both of these difficulties are mitigated by widening the frequency band used, resulting in a greater stability of slowness estimates. However, each part of the usable spectrum appears to have apparent directional properties under the plane wavefront formulation such that a very wide band is likely to result in a less predictable direction estimate according to the relative excitation of different frequencies.

In many of the fixed band estimates, populations of slowness estimates from very closely spaced sources separate clearly (see in particular Figs. 4 and 5). This is likely to be the result of scattering along the path, and may be dominated by effects in the source region (McLAUGHLIN et al., 2004, demonstrate that the radiation patterns from quarry blasts can vary greatly according to the location and orientation of explosives within the pit). This result is negative in that it makes no sense to apply a large SASC to a given slowness estimate if a different source at essentially the same location (to within the theoretical resolution of the array) will result in a very different backazimuth estimate. The correction vector between theoretical and observed slowness will be a very discontinuous function of the relevant parameter space and may actually lead to more unstable processing. The result is positive in that it confirms that there are characteristics of the wavefield, associated with a given source, that are highly repeatable on subsequent measurements. In the same way that the matched filter detectors (e.g., GIBBONS and RINGDAL 2006) recognize the source location specific wavetrain at a given receiver, a narrow frequency band calibration procedure may capture spatial characteristics of the wavefield from a wider source region without reference to the plane wavefront formulation (HARRIS and KVÆRNA, 2008). Such a procedure would actually exploit the scattering to provide a detection/classification system which was less inhibited by variations in the source-time function than the correlation detectors, albeit providing a less sensitive detector.

All the work presented in this paper is based upon observations on a single seismic array, ARCES. The same principles are likely to apply to many other arrays but possibly with very different consequences given different tectonic settings, conditions and array configuration. It would be, for example, pointless examining frequencies greater than 2 Hz on the BRTR array in Turkey due to the strong incoherence of high-frequency signals over this aperture.

It is important to note that the GBF solutions presented in Fig. 1b are multi-station solutions and, paradoxically, the contribution of additional, more distant, stations can lead to a substantial worsening of the preliminary event location estimate. If only phase determinations from a single array are associated, the receiver-to-source backazimuth estimate will fall within the range of the backazimuth estimates for the individual phases recorded at the array. The receiver-to-source distance estimate will depend upon the times separating the associated phases and the identification of those phases obtained from the slowness estimates. Unrelated phases from a different array, detected in the same time-window, are likely to contribute to the GBF event location estimate if their arrival times and slowness estimates fall within the permitted ranges. This can result in a qualitatively incorrect solution which simply provides a better fit to multi-array criteria than the two event model provides for the single-array criteria. The likelihood of this scenario can be reduced dramatically by narrowing the permitted ranges of slowness estimates, which in turn requires calibration of the kind advocated in this paper.

It may be desirable to complement global association type algorithms with single or two-array solutions targeted at certain source regions. Much of the spread in the fully-automatic solutions displayed in Fig. 1b) is due to the contribution of less accurate arrival-time and direction estimates from more distant stations. It is not the case that the more observations one has, the more accurate the location estimate will be. In any location estimate inversion, all phase estimates should be weighted according to the quality of the available measurement. In cases where the quality of the information provided by one highly sensitive station, close to the source, vastly exceeds that available from more distant stations, it may be that a weighting which truly reflects the quality of the measurements would effectively lead to the exclusion of arrival information from all other stations. This will almost certainly be the case if accurate, calibrated bias-corrections are available for the best station.

Acknowledgments

This work was funded by the United States National Nuclear Security Administration under contract number DE-FC52-03NA99517. This report was prepared as an account of work sponsored by an agency of the United States Government. Neither the United States Government nor any agency thereof, nor any of their employees, make any warranty, express or implied, or assumes any legal liability or responsibility for the accuracy, completeness, or usefulness of any information, apparatus, product, or process disclosed, or represents that its use would not infringe privately owned rights. Reference herein to any specific commercial product, process, or service by trade name, trademark, manufacturer, or otherwise does not necessarily constitute or imply its endorsement, recommendation, or favoring by the United States Government or any agency thereof. The views and opinions of authors expressed herein do not necessarily state or reflect those of the United States Government or any agency thereof. We gratefully acknowledge the work of Vladimir Asming and colleagues at the Kola Regional Seismological Center (KRSC) in Apatity, Russia, for collection of the Ground-Truth information on explosions at the Russian mines, and also to the mine operators for supplying the information. We thank Mats Stålnacke of LKAB, Sweden, for confirming the source location of events at the Kiruna and Malmberget ore mines, and the Seismology Research Group at the University of Helsinki, Finland, for providing their Seismic Event Bulletin and for providing the coordinates of the military explosion site. We are grateful to Berit Paulsen of NORSAR for the reviewed event locations. Maps were created using GMT software (WESSEL and SMITH, 1995).

REFERENCES

BAME, D. A., WALCK, M. C., and HIEBERT-DODD, K. L. (1990), *Azimuth estimation capabilities of the NORESS Regional Seismic Array*, Bull. Seism. Soc. Am. *80*, 1999–2015.

BEN HORIN, Y., KOCH, K., and BARTAL, Y. (2004), *Use of GSETT-3 gamma data in the slowness-azimuth calibration of IMS primary arrays at regional distances*, J. Seismol. *8*, 129–142.

BERTEUSSEN, K. A. (1976), *The origin of slowness and azimuth anomalies at large arrays*, Bull. Seism. Soc. Am. *66*, 719–741.

BONDÁR, I. and NORTH, R. G. (1999), *Development of calibration techniques for the Comprehensive Nuclear-Test-Ban Treaty (CTBT) international monitoring system*, Phys. Earth Planet. Inter. *113*, 11–24.

BONDÁR, I., NORTH, R. G., and BEALL, G. (1999), *Teleseismic slowness-azimuth station corrections for the International Monitoring System seismic network*, Bull. Seism. Soc. Am. *89*, 989–1003.

BUNGUM, H. and HUSEBYE, E. S. (1971), *Errors in time delay measurements*, Pure Appl. Geophys. *91*, 56–70.

CAPON, J. (1969), *High-resolution frequency-wavenumber spectrum analysis*, Proc. IEEE *57*, 1408–1418.

CARR, D. B. (1993), *Azimuth estimation capabilities of the ARCESS regional seismic array*, Bull. Seism. Soc. Am. *83*, 1213–1231.

DOUGLAS, A., Seismometer arrays—their use in earthquake and Test Ban seismology. In *International Handbook of Earthquake and Engineering Seismology* (eds. W. H. K. Lee, H. Kanamori, P. C. Jennings, and C. Kisslinger) (Academic Press 2002), pp. 357–367.

GELLER, R. J. and MUELLER, C. S. (1980), *Four similar earthquakes in Central California*, Geophys. Res. Lett. *7*, 821–824.

GIBBONS, S. J. and RINGDAL, F. (2006), *The detection of low magnitude seismic events using array-based waveform correlation*, Geophys. J. Int. *165*, 149–166.

GIBBONS, S. J., KVÆRNA, T., and RINGDAL, F. (2005), *Monitoring of seismic events from a specific source region using a single regional array: A case study*, J. Seismol. *9*, 277–294.

GIBBONS, S. J., RINGDAL, F., and KVÆRNA, T. (2007), *Joint seismic-infrasonic processing of recordings from a repeating source of atmospheric explosions*, J. Acoust. Soc. Am. *122*, EL158–EL164.

GUTENBERG, B. and RICHTER, C. F., *Seismicity of the Earth and Associated Phenomena* (Princeton University Press 1954).

HARRIS, D. B. (1991), *A waveform correlation method for identifying quarry explosions*, Bull. Seism. Soc. Am. *81*, 2395–2418.

HARRIS, D. B. and KVÆRNA, T. (2009), *Precise mapping of seismic events with empirical matched field processing*. Manuscript under review.

HARRIS, D. B., RINGDAL, F., KREMENETSKAYA, E. O., MYKKELTVEIT, S., SCHWEITZER, J., HAUK, T. F., ASMING, V. E., ROCK, D. W., and LEWIS, J. P., Ground-Truth collection for mining explosions in Northern Fennoscandia and Russia. In *Proc. 25th Seismic Research Review, Tucson, Arizona, September 23–25, 2003. Nuclear Explosion Monitoring Building the Knowledge Base*, LA-UR-03-6029 (2003), pp. 54–63.

HARRIS, F. J. (1978), *On the use of windows for harmonic analysis with the discrete Fourier transform*, Proc. IEEE *66*, 51–83.

HICKS, E. C., KVÆRNA, T., MYKKELTVEIT, S., SCHWEITZER, J., and RINGDAL, F. (2004), *Travel-times and attenuation relations for regional phases in the Barents Sea region*, Pure Appl. Geophys. *161*, 1–19.

KENNETT, B. L. N., *The Seismic Wavefield. Volume II: Interpretation of Seismograms on Regional and Global Scales* (Cambridge University Press 2002).

KENNETT, B. L. N., BROWN, D. J., SAMBRIDGE, M., and TARLOWSKI, C. (2003), *Signal Parameter Estimation for Sparse Arrays*, Bull. Seism. Soc. Am. *93*, 1765–1772.

KVÆRNA, T. and DOORNBOS, D. J. (1986), *An integrated approach to slowness analysis with arrays and three-component stations*, NORSAR Scientific Report: Semiannual Technical Summary No. 2-1985/1986, NORSAR, Kjeller, Norway, pp. 60–69.

KVÆRNA, T. and DOORNBOS, D. J. (1991), *Scattering of regional Pn by Moho topography*, Geophys. Res. Lett. *18*, 1273–1276.

KVÆRNA, T. and RINGDAL, F. (1986), *Stability of various f-k estimation techniques*, NORSAR Scientific Report: Semiannual Technical Summary No. 1-1986/1987, NORSAR, Kjeller, Norway, pp. 29–40.

KVÆRNA, T. and RINGDAL, F. (1994), *Intelligent post-processing of seismic events*, Annali di Geofisica *37*, 309–322.

KVÆRNA, T., GIBBONS, S. J., RINGDAL, F., and HARRIS, D. B. (2007), *Integrated seismic event detection and location by advanced array processing*, Tech. Rep. UCRL-SR-228092, Lawrence Livermore National Laboratory, Livermore, CA, United States. Available to the public via the http://www.osti.gov/bridge/product.biblio.jsp?osti_id=902233.

MACCARTHY, J., HARTSE, H., GREENE, M., and ROWE, C. (2008), *Using waveform cross correlation and satellite imagery to identify repeating mine blasts in Eastern Kazakhstan*, Seism. Res. Lett. *79*, 393–399.

MASSA, M., FERRETTI, G., SPALLAROSSA, D., and EVA, C. (2006), *Improving automatic location procedure by waveform similarity analysis: An application in the South Western Alps (Italy)*, Phys. Earth Planet. Inter. *154*, 18–29.

MCLAUGHLIN, K. L., BONNER, J. L., and BARKER, T. (2004), *Seismic source mechanisms for quarry blasts: Modelling observed Rayleigh and Love wave patterns from a Texas quarry*, Geophys. J. Int. *156*, 79–93.

MYKKELTVEIT, S. and BUNGUM, H. (1984), *Processing of regional seismic events using data from small-aperture arrays*, Bull. Seism. Soc. Am. *74*, 2313–2333.

MYKKELTVEIT, S., ÅSTEBOL, K., DOORNBOS, D., and HUSEBYE, E. (1983), *Seismic array configuration optimization, Bull.* Seism. Soc. Am. *73*, 173–186.

MYKKELTVEIT, S., RINGDAL, F., KVÆRNA, T., and ALEWINE, R. W. (1990), *Application of regional arrays in seismic verification research*, Bull. Seism. Soc. Am. *80*, 1777–1800.

RINGDAL, F. and HUSEBYE, E. S. (1982), *Application of arrays in the detection, location, and identification of seismic events*, Bull. Seism. Soc. Am. *72*, S201–S224.

RINGDAL, F. and KVÆRNA, T. (1989), *A multi-channel processing approach to real time network detection, phase association, and threshold monitoring*, Bull. Seism. Soc. Am. *79*, 1927–1940.

ROST, S. and THOMAS, C. (2002), *Array seismology: Methods and applications*, Rev. Geophys. *40*. 1008, doi:10.1029/2000RG000100.

SCHULTE-THEIS, H. and JOSWIG, M. (1993), *Clustering and location of mining induced seismicity in the Ruhr Basin by automated Master Event Comparison based on Dynamic Waveform Matching (DWM)*, Comput. Geosci. *19*, 233–241.

SCHWEITZER, J. (2001a), *HYPOSAT—An enhanced routine to locate seismic events*, Pure Appl. Geophys. *158*, 277–289.

SCHWEITZER, J. (2001b), *Slowness corrections—One way to improve IDC products*, Pure Appl. Geophys. *158*, 375–396.

SCHWEITZER, J., FYEN, J., MYKKELTVEIT, S., and KVÆRNA, T., Chapter 9: Seismic Arrays. In *IASPEI New Manual of Seismological Observatory Practice* (ed. P. Bormann) (GeoForschungsZentrum, Potsdam 2002), 52 pp.

STORCHAK, D. A., SCHWEITZER, J., and BORMANN, P. (2003), *The IASPEI standard seismic phase list*, Seism. Res. Lett. *74*, 761–772.

THOMSON, D. J. (1982), *Spectrum estimation and harmonic analysis*, Proc. IEEE *70*, 1055–1096.

WESSEL, P. and SMITH, W. H. F. (1995), *New version of the Generic Mapping Tools*, EOS Trans. Am. Geophys. Union *76*, 329.

(Received August 25, 2008, revised December 18, 2008, accepted January 12, 2009, Published online December 12, 2009)

Pure Appl. Geophys. 167 (2010), 401–412
© 2010 Birkhäuser Verlag, Basel/Switzerland
DOI 10.1007/s00024-009-0031-2

Analysis of Signals from an Unique Ground-Truth Infrasound Source Observed at IMS Station IS26 in Southern Germany

KARL KOCH[1,2]

Abstract—Quantitative modeling of infrasound signals and development and verification of the corresponding atmospheric propagation models requires the use of well-calibrated sources. Numerous sources have been detected by the currently installed network of about 40 of the final 60 IMS infrasound stations. Besides non-nuclear explosions such as mining and quarry blasts and atmospheric phenomena like auroras, these sources include meteorites, volcanic eruptions and supersonic aircraft including re-entering spacecraft and rocket launches. All these sources of infrasound have one feature in common, in that their source parameters are not precisely known and the quantitative interpretation of the corresponding signals is therefore somewhat ambiguous. A source considered well-calibrated has been identified producing repeated infrasound signals at the IMS infrasound station IS26 in the Bavarian forest. The source results from propulsion tests of the ARIANE-5 rocket's main engine at a testing facility near Heilbronn, southern Germany. The test facility is at a range of 320 km and a backazimuth of ~280° from IS26. Ground-truth information was obtained for nearly 100 tests conducted in a 5-year period. Review of the available data for IS26 revealed that at least 28 of these tests show signals above the background noise level. These signals are verified based on the consistency of various signal parameters, e.g., arrival times, durations, and estimates of propagation characteristics (backazimuth, apparent velocity). Signal levels observed are a factor of 2–8 above the noise and reach values of up to 250 mPa for peak amplitudes, and a factor of 2–3 less for RMS measurements. Furthermore, only tests conducted during the months from October to April produce observable signals, indicating a significant change in infrasound propagation conditions between summer and winter months.

Key words: Infrasound, atmospheric propagation, sound source, ground-truth, explosion monitoring.

[1] CTBTO Preparatory Commission, Provisional Technical Secretariat, Vienna International Centre, P.O. Box 1200, 1400 Vienna, Austria.
[2] *Present Address:*
Federal Institute for Geosciences and Natural Resources (BGR), Stilleweg 2, 30655 Hannover, Germany. E-mail: karl.koch@bgr.de

1. Introduction

Monitoring of the Comprehensive Nuclear-Test-Ban Treaty (CTBT) for atmospheric nuclear explosions is mainly based on two technologies: infrasound and radionuclide monitoring. While radionuclide monitoring captures relevant air particles and noble gases generated during the explosion, the sonic signature of such explosions will be detected by infrasound monitoring. This technology has been developed strongly during the 1950s and 1960s (DONN and EWING, 1962; WEXLER and HASS, 1962; DONN and SHAW, 1967; PIERCE and POSEY, 1971; LISZKA, 1974) due to the large number of atmospheric tests at that time, but declined afterwards (c.f. BROWN et al., 2002). It is only now reemerging since establishing the CTBT in 1996.

Though no atmospheric nuclear tests have been conducted since the early 1980s, there are numerous sources that can be detected by the currently installed IMS infrasound network of about 40 of the final 60 stations. These sources include meteorites, supersonic aircraft including re-entering spacecraft and rocket launches, volcanic eruptions as well as non-nuclear explosions, most of them controlled like mining and quarry blasts, occasionally, however, accidental and well recorded (c.f., CAMPUS, 2004). Specifically it should be mentioned that there are reports of infrasound observations related to rocket launches, including ARIANE-5 starts from the European Space Agency (ESA) launch pad in Kourou, Guyana, which have been recorded at distances of several thousand kilometers, e.g., for IS09 (Brazil) at ~2,400 km from Kourou, or for IS34 at ~3,300 km from Baikonur (Kazakhstan) (N. Brachet, pers. communication).

All these infrasonic sources, including the rocket launches, have one feature in common, in that their source history may not be well established and the quantitative interpretation of the corresponding signals is therefore somewhat ambiguous. For example, the rocket launch infrasound observations have a couple of unknowns associated with them, although the launch process itself may be well controlled: (1) The exact location of the infrasound source, since the infrasound excitation may occur close to the ground or at considerable altitude, even at stratospheric or thermospheric heights; (2) the departure direction, since any difference may produce strong directivity effects in terms of infrasound generation; and (3) differences in the states of the atmosphere could produce significant effects and these states may not be well represented for large altitudes.

However, for quantitative modeling of infrasound signals it is mandatory that well-calibrated sources are available that can be used to establish infrasound propagation models. These propagation models in turn have to be used to infer the strength of any detected infrasound source for any CTBT-related infrasound monitoring activity. For this reason it is highly desirable during the establishment of such infrasound propagation models to use calibration sources which are greatly repeatable during any atmospheric conditions (DONN and RIND, 1971; GARCES et al., 1998; DROB et al., 2003; LE PICHON and DROB, 2004).

Sources for which all or many source parameters are exactly known are called Ground-Truth (GT) events. It is mandatory to use such ground-truth events in order to quantitatively model infrasound propagation to infer both dynamic and kinematic characteristics of the wave field. A source, fulfilling these requirements in a superb manner, has recently been identified. The infrasound source is related to engine tests of ESA's ARIANE-5 rocket with known source coordinates, excitation times and well-controlled source parameters. This source repeatedly produced infrasound recordings at the IMS infrasound station IS26, co-located with the seismic GERESS array in the Bavarian forest, as is described below. The scope of this study is the description of the ground-truth data set and the

initial analysis of the recorded infrasound signals (KOCH, 2005).

2. Description of Ground-Truth Data Set

In early 2004 a personal report was obtained that suggested that signals from main engine tests of the ARIANE-5 rocket, conducted at the testing facilities of the Space Propulsion Institute of the German Aerospace Center (Deutsches Zentrum für Luft- und Raumfahrt/DLR) near Heilbronn in southern Germany, may be detectable at IMS infrasound station IS26. The geographic coordinates for this facility are latitude 49.3°N and longitude 9.4°E. Based on this location (Fig. 1), the test facility is thus at a range of 320 km and at a backazimuth of 280° from IS26. After establishing contact with the DLR staff in charge of the engine tests, ground-truth information was obtained for nearly 100 of these tests conducted during five years starting in 2000 (see Table 1) (R. Hupertz, pers. communication). This ground-truth information included the dates and local times of all tests as well as their duration times. The local times were taken from the computer system which is controlling the tests and acquiring data from various measurements, and are expected to be within seconds of standard time. The test facility exists since 1990 and has been used during development and acceptance testing of the main engines of ESA's launch vehicle ARIANE-5 (Fig. 2). In the years 1990–1996 the first version of the VULCAIN engine underwent development testing, followed by acceptance tests for flight engines from 1996 to 1998. Testing activity continued with developmental testing of the VULCAIN-2 engine from 1999 onward, while in recent years the focus has shifted again to acceptance testing (DLR, 2004).

Subsequent search in the available data archives, at both the IDC and BGR, for IS26 data for the period since its installation in fall 1999 revealed that some 30 of these tests showed signals at IS26 above the background noise level. The existence of these signals is verified by the consistency of different signal parameters: arrival time of about 17–19 min after test start, matching signal and test duration (as provided by the ground-truth data) as well as appropriate signal

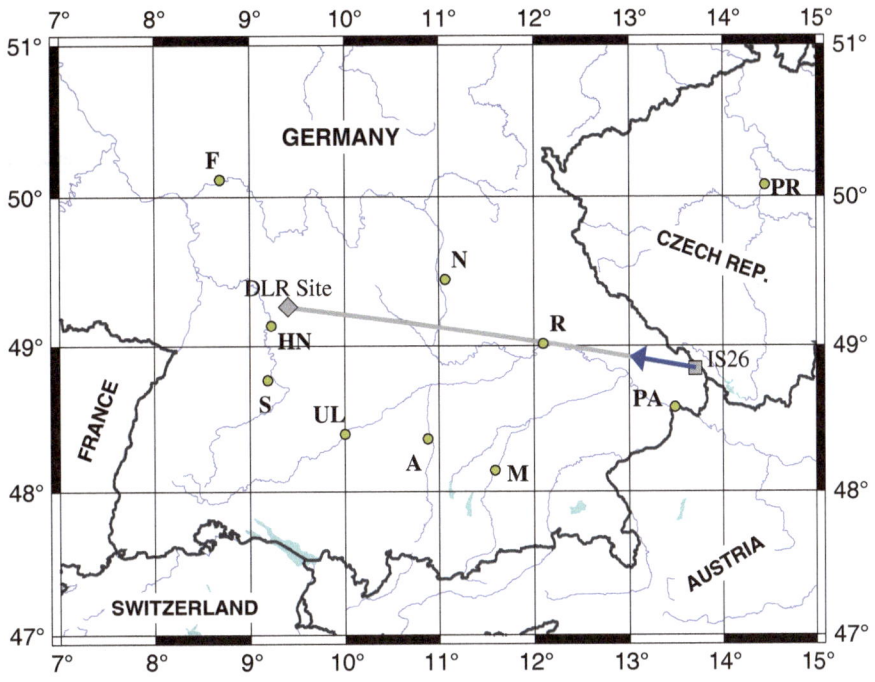

Figure 1

Geographical location of IMS infrasound array station IS26 and the ARIANE-5 testing facility at DLR Lampoldshausen near Heilbronn, Germany. The distance of Lampoldshausen from IS26 is about 320 km at a backazimuth of 280° (*solid line* is for great circle path and *arrow* for direction). The locations of the following cities are given as: *F*—Frankfurt, *S*—Stuttgart, *M*—Munich, *N*—Nuremberg, *R*—Regensburg, *PR*—Praha, *HN*—Heilbronn, *UL*—Ulm, *A*—Augsburg

characteristics such as backazimuth and apparent velocity.

3. Infrasound Observations and Data Processing

Infrasound signals commonly detected at IMS infrasound stations are observed in the frequency band between a fraction of one Hz and several Hz (CAMPUS, 2004). Therefore, the initial search for the signals of the ARIANE-5 tests for frequencies of 0.5–4 Hz did not reveal the presence of any infrasound energy at IS26. Only after application of high-pass filtering to enhance frequencies above 3 Hz did the signals emerge from the background noise.

Figure 3 displays the data processing results for an event on 11 March 2004, for which first ground-truth parameters were obtained after initial contact to DLR. The test was conducted at 15:00 UTC (16:00 local time) and the engine test lasted some 135 s. The raw data were filtered with a Butterworth high-pass filter with a corner frequency of 3 Hz, as used

throughout the data analysis. All five channels of the IS26 infrasound array recorded the signal with signal-to-noise ratios on the order of 3–5. Therefore, these signals represent one of the tests that were most easily identified. Other tests showed signals that were more emergent or showed more variability among the array elements. The initial onset of the signal occurred nearly 19 min after test start. The window of 165 s used for the ensuing frequency–wavenumber (F–K) analysis, chosen to embrace the whole signal train, clearly indicates that the observed signal correlates well with the test duration. An F–K analysis was carried out for frequencies between 1 and 4 Hz (see Fig. 3) since the software used in this case provided otherwise unstable results when including higher-frequency contributions. With a maximum slowness range of 500 s/deg and a discretization with 125 points, the analysis yields parameters consistent with the ground-truth location of the source: A slowness of 340 s/deg corresponding to an apparent velocity near the sound speed and an azimuth of 277° for the direction of the incoming wavefield. Although

31

Table 1

Listing of ground-truth information pertaining to the ARIANE-5 engine tests described in this study

Event no.	Origin time (dd/mm/yy hhmm)	Duration (s)	Observation.
1	**14/01/00 16:00:00**	**595.328**	**Yes**
2	21/01/00 15:50:00	498.711	–
3	27/01/00 15:25:00	4.612	–
4	09/03/00 14:45:00	400.004	–
5	20/03/00 15:30:00	629.199	–
6	19/04/00 15:00:00	92.959	–
7	27/04/00 15:20:00	650.052	–
8	09/05/00 13:45:00	627.248	–
9	16/05/00 13:45:00	600.030	–
10	22/05/00 14:50:00	676.084	–
11	30/05/00 13:00:00	2.657	–
12	14/06/00 13:00:00	449.374	–
13	20/06/00 13:30:00	690.718	–
14	28/06/00 13:00:00	4.632	–
15	30/06/00 12:30:00	630.006	–
16	12/07/00 13:30:00	640.890	–
17	19/07/00 13:05:00	367.628	–
18	28/07/00 14:00:00	405.821	–
19	03/08/00 13:35:00	535.986	–
20	11/08/00 12:15:00	279.049	–
21	17/08/00 12:15:00	364.243	–
22	27/09/00 15:30:00	261.565	No data
23	09/10/00 14:00:00	600.012	–
24	13/10/00 13:30:00	55.386	–
25	25/10/00 14:05:00	610.014	–
26	**02/11/00 15:30:00**	**675.526**	**Yes**
27	**23/11/00 15:10:00**	**600.013**	**Yes**
28	**30/11/00 14:45:00**	**600.019**	**Yes**
29	07/12/00 14:30:00	630.013	–
30	14/12/00 14:00:00	600.027	–
31	**19/12/00 13:45:00**	**600.020**	**Yes**
32	**08/02/01 14:26:00**	**600.011**	**Yes**
33	**14/02/01 14:15:00**	**164.902**	**Yes**
34	21/02/01 14:15:00	739.294	–
35	07/03/01 14:15:00	757.542	–
36	**13/03/01 15:30:00**	**728.735**	**Yes**
37	05/04/01 13:20:00	15.127	–
38	**09/04/01 13:19:59**	**297.314**	**Yes**
39	**23/04/01 13:10:00**	**840.129**	**Yes**
40	**27/04/01 13:40:00**	**847.363**	**Yes**
41	04/05/01 12:19:59	785.761	–
42	18/06/01 15:05:00	3.722	–
43	25/06/01 13:00:00	735.358	–
44	03/07/01 13:45:00	742.267	–
45	09/07/01 13:25:00	754.776	–
46	12/07/01 12:40:00	542.824	–
47	09/08/01 13:45:00	600.018	–
48	15/08/01 13:35:00	600.016	–
49	23/08/01 12:50:00	600.018	–
50	30/08/01 12:45:00	600.019	–
51	05/09/01 13:15:00	600.010	–
52	25/10/01 13:00:00	380.016	–
53	27/11/01 15:21:00	7.126	–

Table 1 continued

Event no.	Origin time (dd/mm/yy hhmm)	Duration (s)	Observation.
54	**03/12/01 14:00:00**	**600.022**	**Yes**
55	**20/12/01 14:30:00**	**600.014**	**Yes**
56	**14/01/02 15:30:00**	**600.826**	**Yes**
57	**17/01/02 14:15:00**	**646.954**	**Yes**
58	31/01/02 13:50:00	245.025	–
59	**13/02/02 13:50:00**	**215.920**	**Yes**
60	**06/03/02 14:00:00**	**659.430**	**Yes**
61	**11/03/02 14:00:00**	**645.070**	**Yes**
62	15/03/02 14:00:00	511.825	–
63	**25/03/02 14:00:00**	**665.174**	**Yes**
64	04/04/02 12:45:00	600.013	–
65	**15/04/02 12:45:00**	**400.017**	**Yes**
66	26/04/02 14:20:00	540.022	–
67	03/05/02 12:46:00	5.012	–
68	22/05/02 12:50:00	190.024	–
69	13/06/02 13:30:00	714.655	–
70	25/06/02 13:15:00	160.301	–
71	31/07/02 14:10:00	440.010	–
72	07/08/02 14:00:00	5.671	–
73	13/08/02 13:50:00	190.007	–
74	**15/11/02 14:25:00**	**440.001**	**Yes**
75	**05/12/02 14:20:00**	**708.504**	**Yes**
76	16/07/03 14:10:00	200.000	–
77	23/07/03 14:10:00	539.993	–
78	25/07/03 13:00:00	179.003	–
79	01/08/03 13:00:00	178.997	–
80	20/08/03 13:10:00	190.004	–
81	12/09/03 14:00:00	540.010	–
82	**13/10/03 16:15:00**	**540.012**	**Yes**
83	**16/10/03 13:40:00**	**540.008**	**Yes**
84	03/11/03 15:13:00	0.000	No test?
85	**11/12/03 14:15:00**	**240.005**	**Yes**
86	**05/03/04 14:45:00**	**11.344**	**Yes**
87	**11/03/04 15:00:00**	**135.071**	**Yes**
88	16.04.04 13:15:00	105.020	–
89	10.05.04 13:35:00	389.988	–
90	28.05.04 14:55:00	693.089	–
91	15.06.04 13:30:00	561.575	–
92	09.07.04 13:05:00	746.927	–
93	21.07.04 13:10:00	749.503	–
94	30.07.04 12:50:00	680.183	–
95	06.09.04 14:45:00	440.007	–
96	**29.10.04 13:05:00**	**190.001**	**Yes**

Times given in UTC. Bold font is used for events with associated infrasound signals at IS26

the determined parameters are reasonable, the figure also shows quite large secondary peaks. It is suspected that these peaks are related to spatial aliasing when using very high frequencies. This also seems to be the reason for obtaining unstable results when including frequencies up to the Nyquist frequency.

Figure 2
Pictures of the testing facility P5 and the ARIANE-5 main engine (VULCAIN-2) during a test. For acceptance tests the engine generates a thrust of up to 1,000 kN during a minimum propulsion duration of 600 s, corresponding to a successful rocket launch. Development tests, however, may be shorter

Figure 4 shows a second example of the infrasound signals from a test carried out on 14 January 2002, which had an initiation time at 15:30 UTC and a source duration of 600 s corresponding to launch conditions. The signal is clearly visible, however, it is also evident that element I26H4 has distinctly smaller signal amplitudes, and hence a much poorer signal-to-noise ratio. Since the data were high-pass filtered, it suggests a strong reduction of measured signal energies at I26H4 above 3 Hz. The signal deficiency for I26H4 was later found (T. Hoffmann, pers. communication) to be related to the fact that this element had been equipped with experimental impedance reducers to suppress spurious spectral contributions from the spatial noise reduction filter used at IS26. These impedance reducers were removed in fall 2003, and hence, Fig. 3 does not show a comparable signal difference to other array channels.

The result from the F–K analysis is less convincing than the one shown in Fig. 3. This may certainly be related to the signal deficiency for one array element, which certainly contributes to the spatial aliasing effects from the high frequencies used. Parameters are again consistent in terms of azimuth and slowness, although slowness is somewhat larger than before. The travel time for the initial signal arrival is close to the travel time observed for the event shown in Fig. 3.

In order to carry out an initial quantitative study, signal and noise amplitude were measured in the frequency band between 3 and 10 Hz, i.e., up to the Nyquist frequency of the IS26 system. Figure 5 shows two examples of signal measurements. As Table 1 indicates, only 28 events were found to show observable signals from the ARIANE-5 engine tests, while for the rest only noise measurements could be collected, which will give a maximum likelihood threshold of the signal strength not to be exceeded. For the identified signals appropriate windows were selected being representative for the signal to be measured. These windows were set to include, if possible, the entire signal, but in other cases to exclude spikes or noise bursts. Usually the noise signal was measured in the immediate vicinity of the signal window. For absent signals the signal window was selected such that the time span of the expected signal arrivals was covered. Two amplitude measurements were taken on each channel: the maximum amplitude and the root-mean-squares (RMS)

Reprinted from the journal

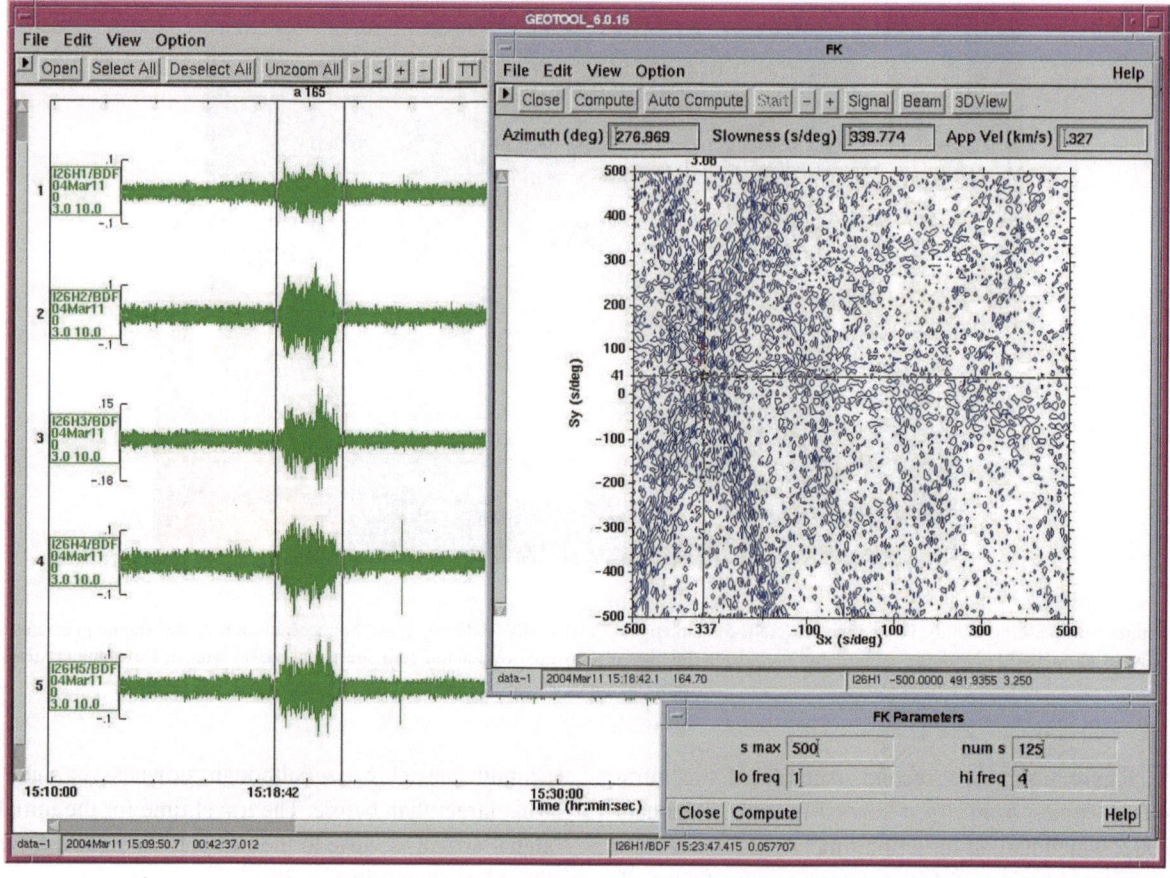

Figure 3
Data examples for an engine test in March 2004 with a duration of 135 s. The signals are observed only at frequencies higher than 3 Hz. The frequency–wavenumber analysis shows consistent values in terms of the measured slowness of 340 s/deg and an incidence direction of 277°

amplitude. The latter was intended to obtain a more stable amplitude estimate based on the observation that it was sometimes difficult to decide whether individual peaks had been affected by incidental noise bursts.

4. Data Analysis Results

Results from peak-to-peak measurements for the raw maximum amplitudes are displayed in Fig. 6 for all five elements of IS26. Depending on signal and noise they range between a few mPa to 400 mPa. It is also obvious that there is considerable scatter among the different elements, on the order of 50–100%. Furthermore, starting in fall 2001 and through fall 2003, element I26H4 always showed much lower signals and noise amplitude estimates than all other

elements. This is a clear indication that this element's sensitivity was significantly different in the relevant frequency band. Signal amplitudes corresponding to tests that could not be identified in the IS26 records were set to zero and are thus displayed on the x axis. From the figure it is clearly seen that valid signal amplitudes are observed for dates corresponding to winter months, while zero amplitudes group within summer months. Equivalently, RMS amplitudes determined for the array elements display the same pattern.

When averaging the raw measurements at the individual elements to obtain a more stable estimate for both noise and signals (Fig. 7), we obtain a result closely matching the features of Fig. 6. These averaged amplitudes range between 10 and 80 mPa for the noise and are up to 6–8 times higher for the signals reaching maximum values of 250 mPa. While

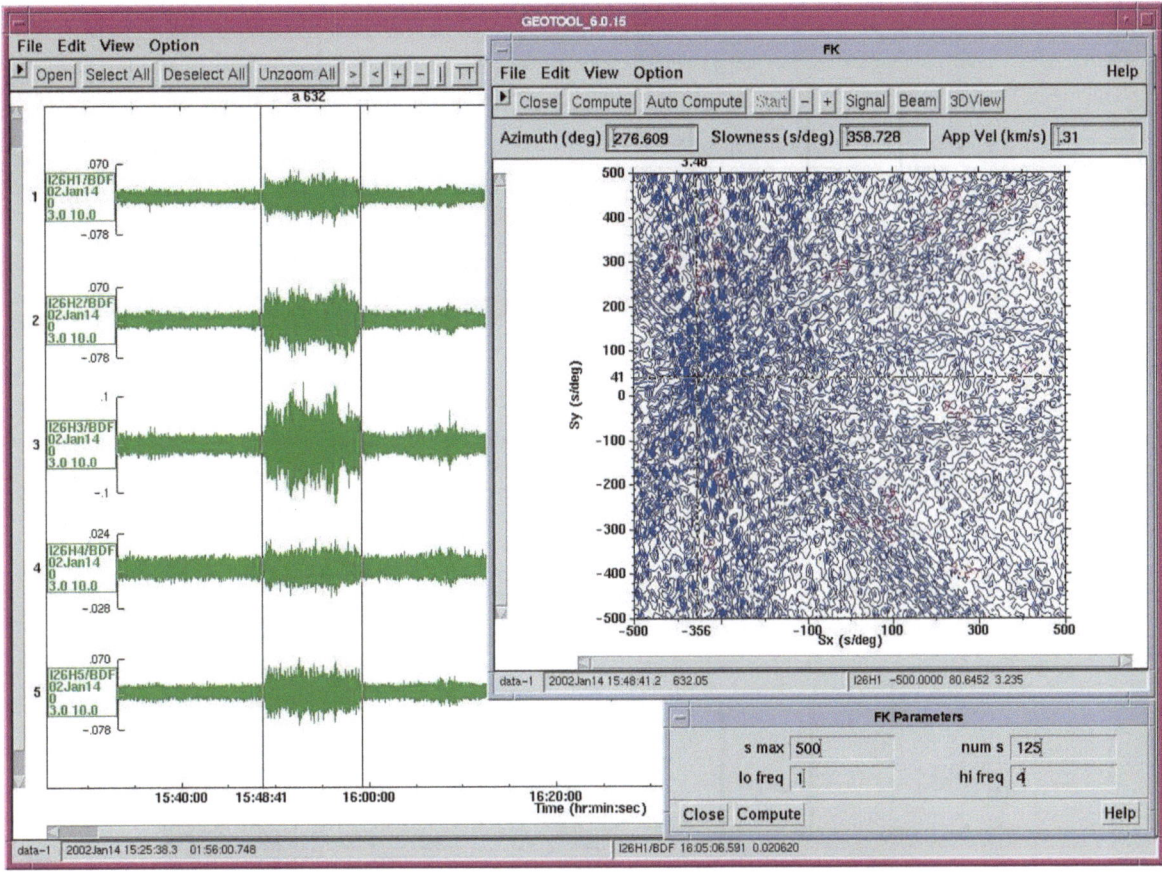

Figure 4

Data examples for an engine test in January 2002 with a duration of 600 s as used in actual launches. The frequency–wavenumber analysis shows consistent values by the measured values in slowness of 359 s/deg and incidence direction of 276°

the observation of signals is clearly bound to the winter season, i.e., the months from October to April, there is no trend in the noise data that would argue for higher noise levels during these winter months. QUASCHNING (2000) refers to meteorological observations throughout Germany with the monthly average wind speed being reduced during the summer months. From this, seasonal trends in the noise level could be expected.

Equivalently, array-averaged RMS amplitudes range between 5 and 20 mPa for the noise and reach maximum values of 50–60 mPa for the signals (Fig. 8). Even the RMS values considered substantially more robust estimates than the maximum amplitude estimates do not show a seasonal trend for the background noise. This result means that a strong relationship between seasonal wind conditions and observed background noise does not exist. Figure 9 shows the locally measured wind speed against the background noise amplitudes (both maximum as well as RMS). Although there is a clear trend, and hence a relationship between these quantities, the correlation is not very high, as expressed by the correlation coefficient of 0.2758 (for maximum noise amplitudes) and 0.1785 (for RMS noise amplitudes).

For all observed signals a preliminary travel-time analysis was carried out by picking the initial onset of the identified waveform package. By dividing the source–receiver distance with the travel-time estimate, an average propagation speed, called celerity, was determined. The corresponding results are shown in Fig. 10, showing a scatter in the travel time between 1,020 and 1,185 s, around an average value of 1,100 s, and corresponding celerity values between 270 and 313 m/s, with an average value of 292 m/s. The scatter in the observed travel times seems

35

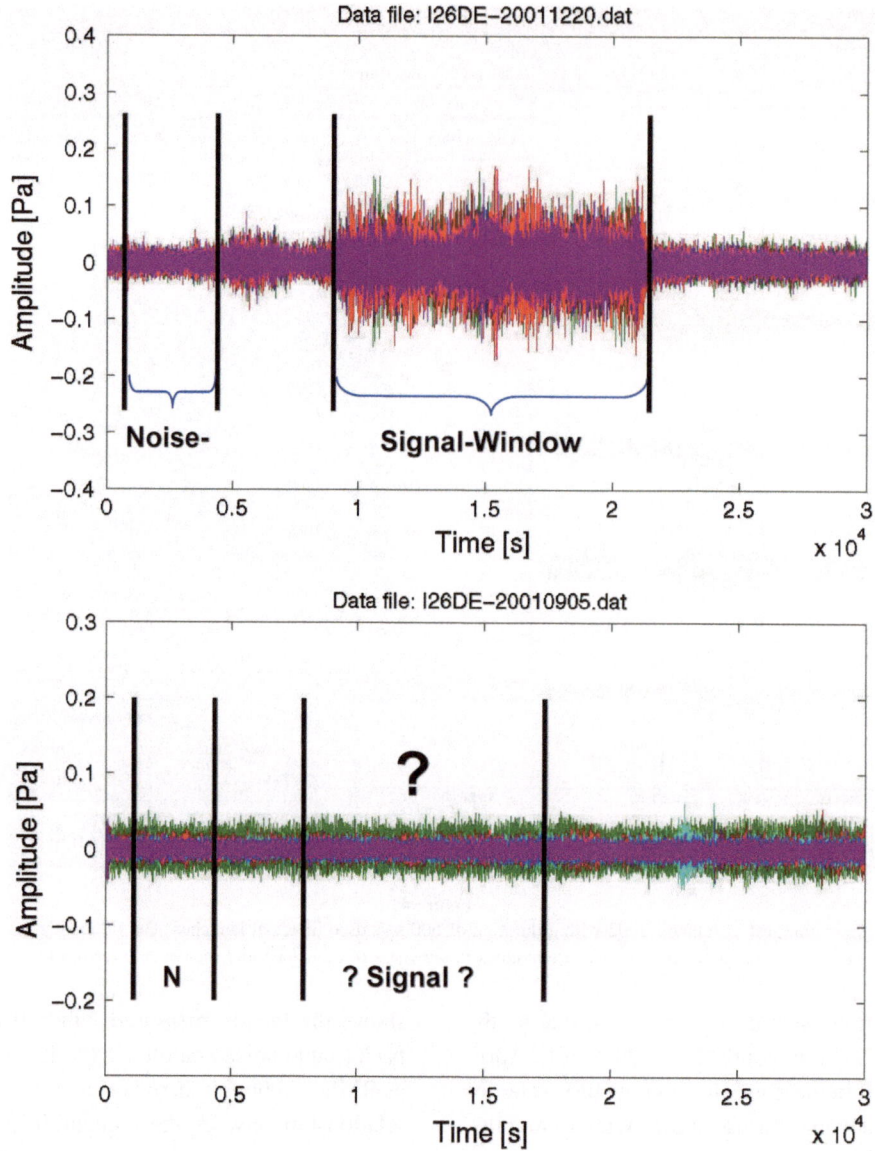

Figure 5
Measurement of amplitudes by selecting appropriate windows for the signal and the noise. Measurements usually encompass the entire signal. Noise windows were set close to the signal window, where possible when no noise bursts interfered. For missing signals, windows were set according to theoretical predictions, but actually without measuring a signal

incidental and not obeying seasonal changes. The celerity values also indicate a clear trend for these values to be lower than the normal sound speed of 330 m/s, hence being indicative of stratospheric propagation paths.

The initial data processing carried out indicated a potential lack of convergence of the F–K analysis for these high-frequency signals. PMCC processing (LE PICHON and CANSI, 2003) was not available for this study. Alternatively it was found that the F–K analysis using the Seismic Handler software (STAMMLER, 1993) provided for a viable alternative in producing fairly stable F–K estimates, even for the case of incorporating frequencies as high as 8 Hz. The slowness estimates obtained in the F–K analysis were converted to apparent velocities of the infrasound waves propagating across the IS26 array and are shown in Fig. 11. The apparent velocity values span the range

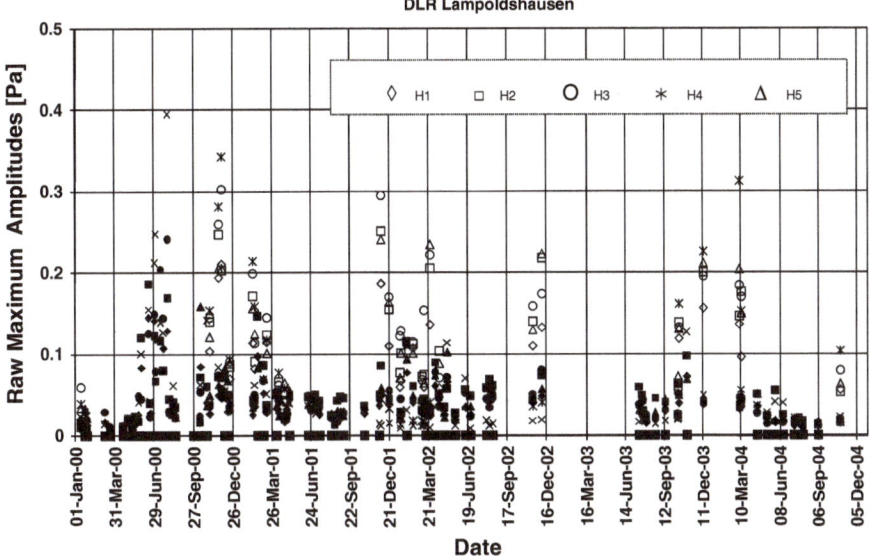

Figure 6
Results of maximum amplitude measurement for all elements at IS26. Signal amplitudes are represented by *open symbols* and noise measurements by the same type *closed symbols* (I26H4 shown as *star* and *cross*, respectively). A significant scatter between measurements on different elements is obvious. Note the lower signal and noise amplitudes for array element I26H4 between fall 2001 and fall 2003. Missing signals are indicated by zero amplitude values

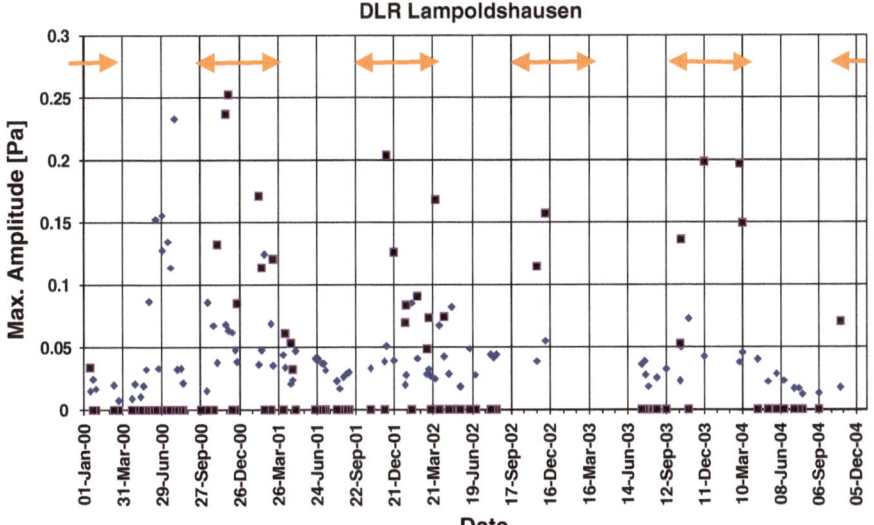

Figure 7
Array averaged maximum amplitudes to obtain more stable estimates (see also Fig. 6). Note the fairly uniform noise level and the varying signal-to-noise ratio. Noise amplitudes (*diamonds*) are between 20 and 80 mPa, while signal amplitudes (*squares*) reach 250 mPa. *Note*: *arrows* on the top indicate winter months; no signals were observed during summer months

between 311 and 373 m/s and are on average 343 m/s, thus being fairly close to sound speed. These values are distinctly higher than those obtained for celerity, which are reproduced for comparison, and at higher

resolution, in Fig. 11. The second parameter estimated during F–K-analysis, azimuth, has an average value of 282° for the 28 signals analysed, scattering between 275° and 290°, i.e., within about ±7.5°.

37

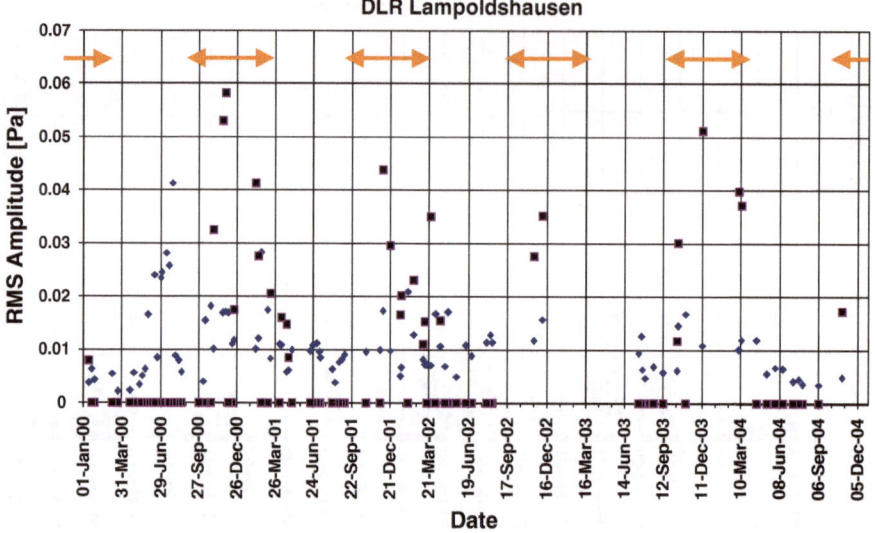

Figure 8
Array averaged root-mean-squares (RMS) amplitudes of signals (*squares*) and noise (*diamonds*). Noise RMS amplitudes are between 5 and 20 mPa, while signal RMS amplitudes are less than 60 mPa. *Note*: *arrows* on the top indicate winter months; no signals were observed during summer months

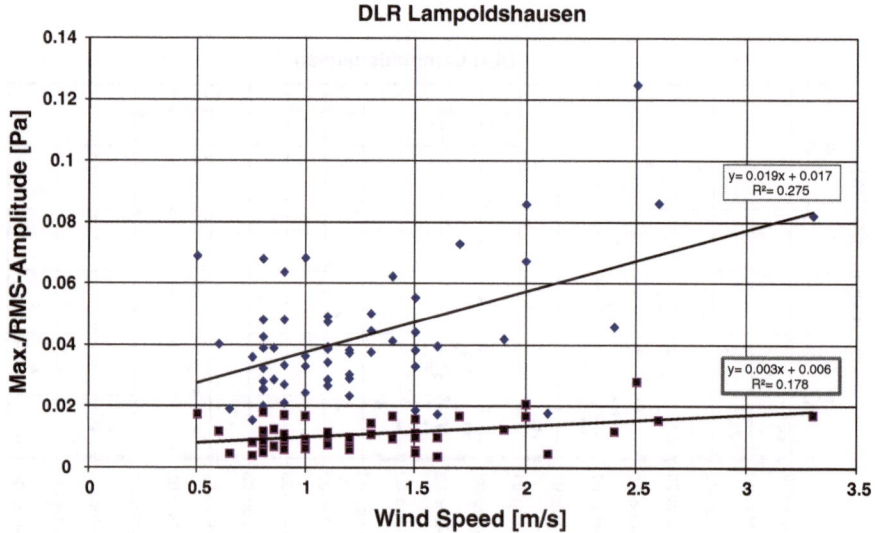

Figure 9
Correlation between measured noise amplitudes (maximum—*diamonds*; RMS—*squares*) and local wind speed. No strong correlation between wind velocity and infrasound background noise exists, as indicated by small correlation coefficients of about 0.2–0.3

5. Summary and Conclusions

Our data analysis reveals that there is a subtle pattern of the signals from repetitive ARIANE-5 engine tests observed at IS26 to be recorded above background noise only for tests conducted from October to April. The 28 detected signals at IS26 are all from some 50 engine tests conducted within these winter months. Although nearly equal in number, no signal was detected from any of the tests conducted in summer months. This finding hints at a significant change in the atmospheric layers in which the signal is propagating from Lampoldshausen to IS26, with significantly better transmission conditions during winter months.

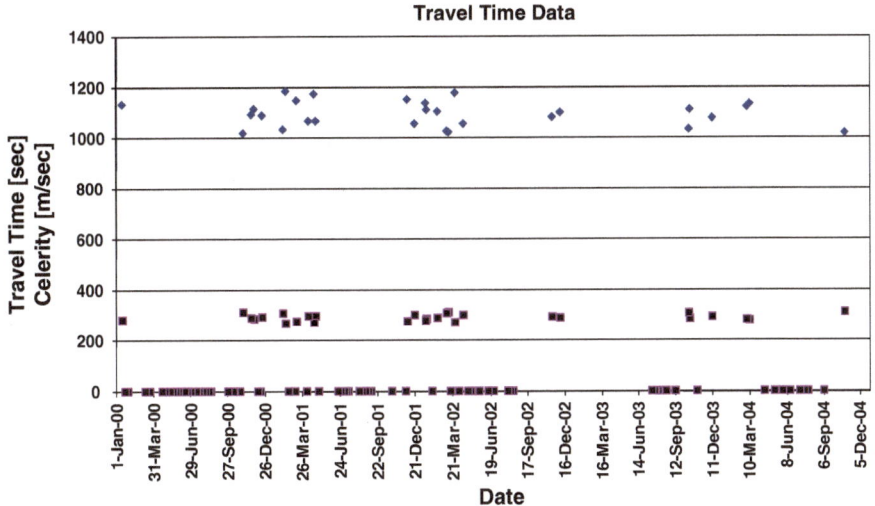

Figure 10
Results of the travel-time analysis obtained by picking signal onset times. Observed travel times (*diamonds*) range between 1,000 and 1,200 s. The average signal speed (*squares*), determined by dividing distance through travel time and called celerity, gives values between 270 and 313 m/s with a mean value of 292 m/s

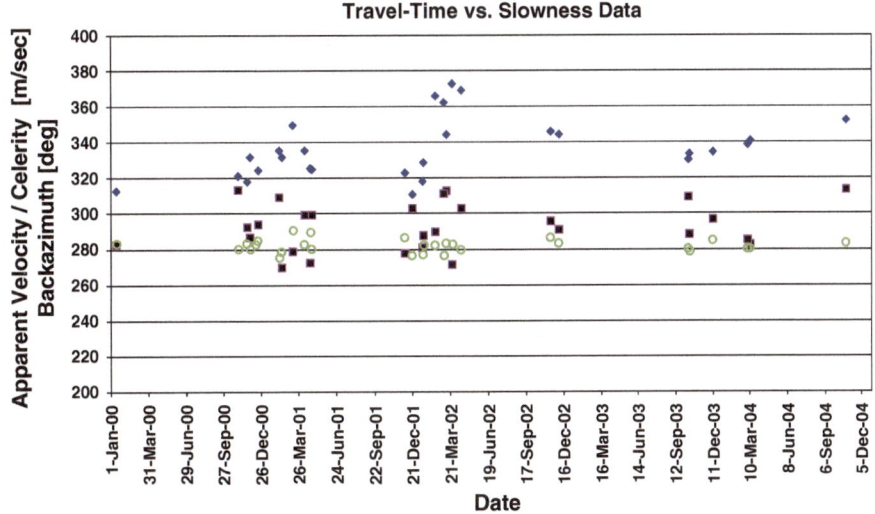

Figure 11
Results of the frequency–wavenumber (F–K) analysis by array processing of signals. Observed apparent velocities (*diamonds*), obtained from the estimated wave slowness, range between 311 and 373 m/s with a mean value of 343 m/s. The celerity values (*squares*) (reproduced from Fig. 10 for comparison) are significantly different from apparent velocity. The backazimuth estimates (*open circles*) confirm expected incidence directions, with differences indicative of changes in stratospheric wind conditions

These results conform to the detection capability analysis by LE PICHON et al. (2008) which predicts lower detection thresholds for events west of IS26 during winter, due to eastward blowing stratospheric winds compared to westward winds in summer.

That atmospheric propagation for the present source–receiver path is indeed within the stratospheric waveguide is indicated by the measured travel time exceeding the approximately 16 min for direct sound propagation along the Earth's surface. This result is further supported by a comparison of the effective velocity between source and receiver, called celerity and calculated from distance and travel time, and the apparent velocity of the signals from F–K-analysis.

While the celerity ranges between 270 and 310 m/s, the apparent velocity measured at the infrasound array indicates values between 310 and 370 m/sec. The signals' estimated backazimuths scatter between 275° and 285°. This scatter may in part be related to the high-frequency contents of the signals and some difficulties encountered in F–K estimation. The study of signal and noise amplitudes showed that the signal-to-noise ratio is generally between 2 and 8, with signal amplitudes extending to about 200 mPa for maximum amplitudes and about a factor of two less for RMS estimates. The identified signals are therefore potentially useful for the kinematic and dynamic validation of infrasound propagation models in Central Europe.

Acknowledgments

This research was triggered by a family visit and an embedded coffee afternoon. Only by a factual report fortuitously received from my sister D. Sahm was this study initiated. I am deeply indebted to the Institute for Propulsion Studies of DLR (German Aerospace Agency) at Lampoldshausen for providing me with ground-truth information of the ARIANE-5 engine tests for the years 2000–2004. In particular, I gratefully acknowledge the support by Prof. W. Koschel and Dipl.-Ing. R. Hupertz through stimulating discussions and a guided tour to the testing facility. The data used for this study are openly available from the Federal Institute for Geosciences and Natural Resources (Bundesanstalt für Geowissenschaften und Rohstoffe-BGR, Hannover) by AutoDRM.

REFERENCES

BROWN, D.J., KATZ, C.N., LeBRAS, R., FLANAGAN, M.P., WANG, J., and GAULT, A.K. (2002), *Infrasonic signal detection and source location at the Prototype International Data Centre*, Pure Appl. Geophys. *159*, 1081–1125.

CAMPUS, P. (2004), *The IMS infrasound network and its potential for detecting events: Examples of a variety of signals recorded around the World*, Inframatics Newsletter *06*(Jun), 13–22 (http://www.inframatics.org/newsletter.html).

DLR (2004), http://www.la.dlr.de/ra/.

DONN, J.W. and EWING, M. (1962), *Atmospheric waves from nuclear explosions*, J. Geophys. Res. *67*, 1855–1866.

DONN, J.W. and RIND, D. (1971), *Natural infrasound as atmospheric probe*, Geophys. J. R. astr. Soc. *26*, 111–133.

DONN, J.W. and SHAW, D.M. (1967), *Exploring the atmosphere with nuclear explosions*, Rev. Geophys. *5*, 53–82.

DROB, D. P., PICONE, J. M., and GARCÉS, M. A. (2003), *The global morphology of infrasound propagation*, J. Geophys. Res. *108*(D21), 4680, doi:10.1029/2002JD003307.

GARCES, M.A, HANSEN, R.A., and LINDQUIST, K.G. (1998), *Travel-times for infrasonic waves propagating in a stratified atmosphere*, Geophys. J. Int. *135*, 255–263.

KOCH, K. (2005), *Data analysis of infrasound recordings at IS26 from ARIANE-5 engine tests* (abstract), 65th Annual Meeting of the *German Geophysical Society*, Graz, Austria, 21–25 February, 2005.

LE PICHON, A. and CANSI, Y. (2003), *PMCC for infrasound data processing*, Inframatics Newsletter *02*(Jun), 1–9 (http://www.inframatics.org/newsletter.html).

LE PICHON, A. and DROB, D.P. (2004), *Probing high-altitude winds using infrasound from volcanoes*, Inframatics Newsletter *08*(Dec), 1–17 (http://www.inframatics.org/newsletter.html).

LE PICHON, A., VERGOZ, J., HERRY, P., and CERANNA, L. (2008), *Analyzing the detection capability of infrasound arrays in Central Europe*, J. Geophys. Res. *113*, D12115, doi:10.1029/2007JD009509.

LISZKA, L. (1974), *Long-distance propagation of infrasound from artificial sources*, J. Acoust. Soc. Am. *56*, 1383–1388.

PIERCE, A.D. and POSEY, J.W. (1971), *Theory of the excitation and propagation of Lamb's atmospheric edge mode from nuclear explosions*, Geophys. J. R. astr. Soc. *26*, 341–368.

QUASCHNING, V. (2000), *Systemtechnik einer klimaverträglichen Elektrizitätsversorgung in Deutschland für das 21. Jahrhundert*, Fortschr.-Ber. VDI Reihe 6 *437*, Düsseldorf, VDI Verlag.

STAMMLER, K. (1993), *SeismicHandler—Programmable multi-channel data handler for interactive and automatic processing of seismological analyses*, Computers and Geosci. *19*, 135–140.

WEXLER, H. and HASS, W.A. (1962), *Global atmospheric pressure effects of the October 30, 1961 explosion*, J. Geophys. Res. *67*, 3875–3887.

(Received October 1, 2008, revised March 20, 2009, accepted April 14, 2009, Published online January 12, 2010)

Pure Appl. Geophys. 167 (2010), 413–436
© 2009 Birkhäuser Verlag, Basel/Switzerland
DOI 10.1007/s00024-009-0038-8

Pure and Applied Geophysics

Detection and Analysis of Near-Surface Explosions on the Kola Peninsula

STEVEN J. GIBBONS[1] and FRODE RINGDAL[1]

Abstract—Seismic and infrasonic observations of signals from a sequence of near-surface explosions at a site on the Kola Peninsula have been analyzed. NORSAR's automatic network processing of these events shows a significant scatter in the location estimates and, to improve the automatic classification of the events, we have performed full waveform cross-correlation on the data set. Although the signals from the different events share many characteristics, the waveforms do not exhibit a ripple-for-ripple correspondence and cross-correlation does not result in the classic delta-function indicative of repeating signals. Using recordings from the ARCES seismic array (250 km W of the events), we find that a correlation detector on a single channel or three-component station would not be able to detect subsequent events from this source without an unacceptable false alarm rate. However, performing the correlation on each channel of the full ARCES array, and stacking the resulting traces, generates a correlation detection statistic with a suppressed background level which is exceeded by many times its standard deviation on only very few occasions. Performing *f-k* analysis on the individual correlation coefficient traces, and rejecting detections indicating a non-zero slowness vector, results in a detection list with essentially no false alarms. Applying the algorithm to 8 years of continuous ARCES data identified over 350 events which we confidently assign to this sequence. The large event population provides additional confidence in relative travel-time estimates and this, together with the occurrence of many events between 2002 and 2004 when a temporary network was deployed in the region, reduces the variability in location estimates. The best seismic location estimate, incorporating phase information for many hundreds of events, is consistent with backazimuth measurements for infrasound arrivals at several stations at regional distances. At Lycksele, 800 km SW of the events, as well as at ARCES, infrasound is detected for most of the events in the summer and for few in the winter. At Apatity, some 230 km S of the estimated source location, infrasound is detected for most events. As a first step to providing a Ground Truth database for this useful source of infrasound, we provide the times of explosions for over 50 events spanning 1 year.

Key words: Seismic arrays, explosion monitoring, correlation detectors, event location, infrasound.

1. Introduction

A recurring challenge in the field of nuclear explosion monitoring is the detection and classification of seismic events from a repeating source. Often these events are routine mining explosions and the need to identify the source is primarily for screening purposes (see, for example, SCHULTE-THEIS and JOSWIG, 1993; GIBBONS *et al.*, 2005). That is to say that we wish to attribute the observed signals, completely automatically and with a very high level of confidence, to a known industrial source such that they can be eliminated from the list of events which need to be examined manually. Signals are also generated by military explosions and, in such cases, event detection and classification may be desirable to monitor the level of, or changes in the pattern of, the activity at a given site. The waveforms recorded at any given receiver are like a fingerprint for events within a very specific source region and, as a result, signal cross-correlation has become an increasingly common means for the detection and identification of seismic events from a given source (e.g., ISRAELSSON, 1990; HARRIS, 1991; MACCARTHY *et al.*, 2008). In this paper we examine the application of a correlation detector on seismic array data to detect signals from presumed near-surface explosions in order to build up an event database.

In March 2005, residents of the Varanger Peninsula in the far north of Norway reported the observation of loud audible bangs to the local authorities, who in turn approached NORSAR with the aim of identifying a source. A number of seismic events had been detected automatically and located using the generalized beamforming (GBF) method (RINGDAL and KVÆRNA, 1989) which, due to the times

[1] NORSAR, P.O. Box 53, 2027 Kjeller, Norway. E-mail: steven@norsar.no

and approximate location estimates, were considered to be candidate sources of the noises. Waveforms recorded at the ARCES seismic array near Karasjok, northern Norway, are displayed in Fig. 1. For each of the events, the regional seismic P and S arrivals were followed, some 12 to 14 min later, by high amplitude signals which were demonstrated to be sound waves arriving from essentially the same direction. The directions of all incoming wavefronts are determined using broadband *f-k* analysis (e.g., KVÆRNA and RINGDAL, 1986; KENNETT, 2002). While the direction estimates of all phases indicate a source to the east of the array, a relatively high variance in the backazimuth estimates is observed which provides a challenge to accurate event location using only classical array processing. This variability is not unexpected given studies of slowness estimates for phases from repeating sources in this region (see GIBBONS et al., 2009) which showed the stability and bias to be very frequency-dependent, and with different frequency bands providing optimal stability for different source locations.

The fully automatic location estimates for the events in Fig. 1 are displayed in panel (a) of Fig. 2. The spread in these network estimates is considerable. Panel (b) of Fig. 2 shows the corresponding location estimates for quarry explosions confirmed to have taken place at the open-cast mines at Zapoljarni (labelled Z1, Z2 in panel d). The spread and geographical coverage of the Zapoljarni event location estimates is very similar to that in panel (a), indicating that a labour-intensive analyst location procedure would be required to separate these populations. A third source of seismic signals in the region is a site of ongoing earthquake swarms near the town of Kirkenes (panel c). ARCES has detected many hundreds of events between magnitudes 0 and 3 from this region and, based upon waveform similarity, the earthquakes are assumed to be separated by no more than 1 or 2 km. The pattern of swarm activity is similar to that at many locations on the western coast of Norway (e.g., BUNGUM et al., 1979; GIBBONS et al., 2007a) and the best location estimate is labelled KE in panel (d) of Fig. 2. While the exact nature of the sources detected in March 2005 is not known, the seismic and infrasonic signals are similar to those generated by an annual series of ammunition

destruction explosions in northern Finland [labelled FE in panel (d) of Fig. 2: see GIBBONS et al., 2007b] which suggests that the Kola events may be of a similar nature. While weak infrasonic signals are occasionally observed at ARCES from quarry explosions, they are essentially never of as high amplitude as those generated by the ordnance demolition events.

An initial analysis of the seismic and infrasound signals from the March 2005 events was presented by RINGDAL et al. (2005). This resulted in the location estimate labelled RE in panel (d) of Fig. 2, approximately 250 km to the east of ARCES, on the northern coast of the Kola Peninsula near Polustrov Ribachy (the Fisherman's Peninsula). This region is over 100 km from the settlements in the far north of Norway where the sounds were initially heard, suggesting that very unusual atmospheric sound propagation conditions led to the exceptional observations. At the time, it was not known whether or not these events were related to an exceptional, one-off, activity or whether (like the Finnish explosions) they were a routine and fairly regular occurrence.

The seismic signals from the Finnish events are almost identical from explosion to explosion which makes them amenable to detection using a matched filter detector. Multichannel waveform correlation detectors have previously been applied at NORSAR with great success for the detection of chemical explosions (STEVENS et al., 2006), natural seismicity (GIBBONS et al., 2007a), and mining-induced seismicity (GIBBONS, 2006), in situations where the events have been known to take place within a very limited source region. Although the signals displayed in Fig. 1 share many characteristics, there is no ripple-for-ripple correspondence as observed for the Finnish explosions. Performing cross-correlation between a single waveform from one explosion and the corresponding waveform from a different event often resulted in a slight increase of the correlation coefficient value. However, the maximum correlation coefficient was never sufficiently greater than the background level for the procedure to be used as a detector for identifying new occurrences of signals from events at this site. The increased correlation coefficient values always occurred over an extended time-window of up to several seconds, and did not

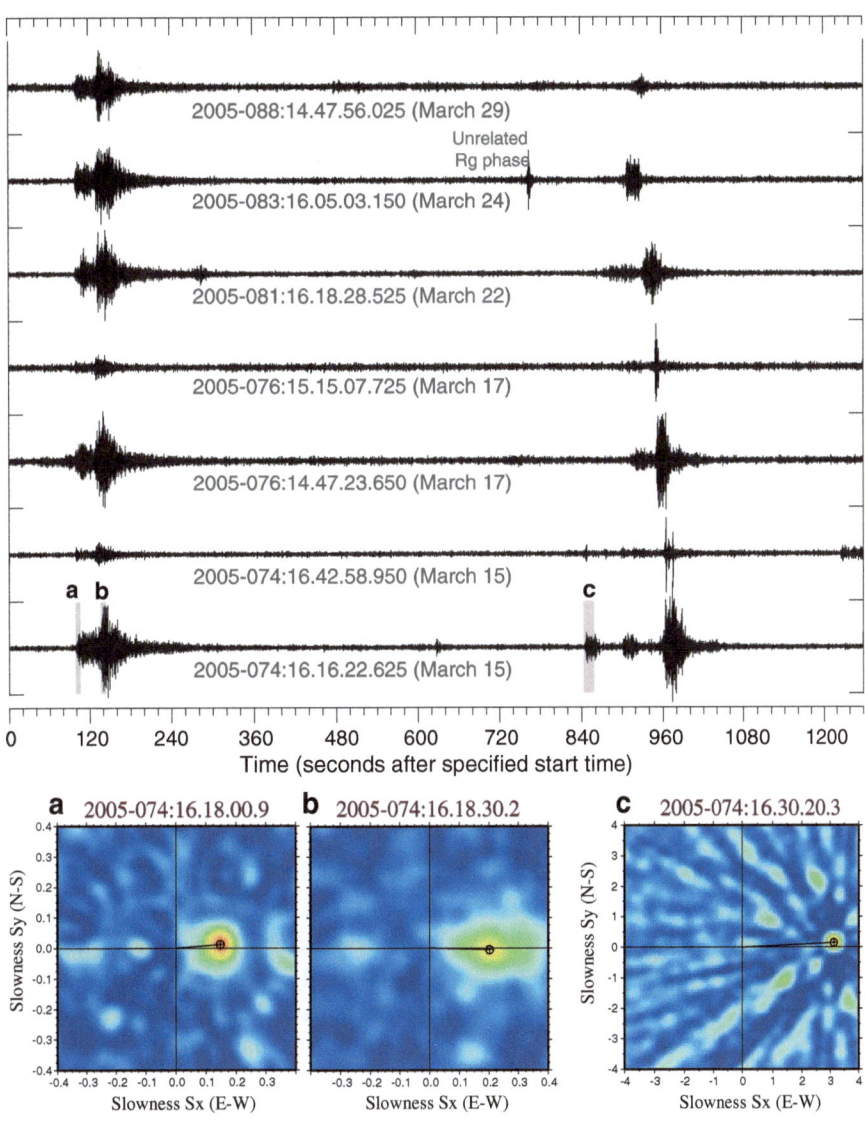

Figure 1

Segments of waveform data from the ARA0_sz sensor of the ARCES array, bandpass filtered 2.0–7.0 Hz at the times as labelled. All traces are plotted to the same scale. Seismic P and S phases arrive at approximately 100 and 130 s respectively and most of the arrivals after 800 s correspond to signals propagating with an air-sound speed from approximately the same direction. Slowness estimates (**a**), (**b**) and (**c**) are made using broadband *f-k* analysis at the times indicated. Sensors in the outer two rings of the array are not used for estimate (**c**). S_x and S_y are always specified in s km^{-1}. Note that *f-k* grid (**c**) has a different scale to grids (**a**) and (**b**)

resemble a delta-function at a single instant, as is characteristic of correlations between almost identical signals.

GIBBONS and RINGDAL (2006) demonstrated that if marginal correlation detections on a small-aperture array are caused by wavefronts approaching from the same direction, then performing *f-k* analysis on the single-channel correlation coefficient traces will result in the zero slowness vector, indicating a zero-delay alignment. If however, the correlation is the result of two wavefronts which approach from different directions, the *f-k* analysis will result in a non-zero slowness vector which is approximately equal to the difference between slownesses of the correlating wavefronts in the detected and master event signals. Performing *f-k* analysis on the correlation coefficient

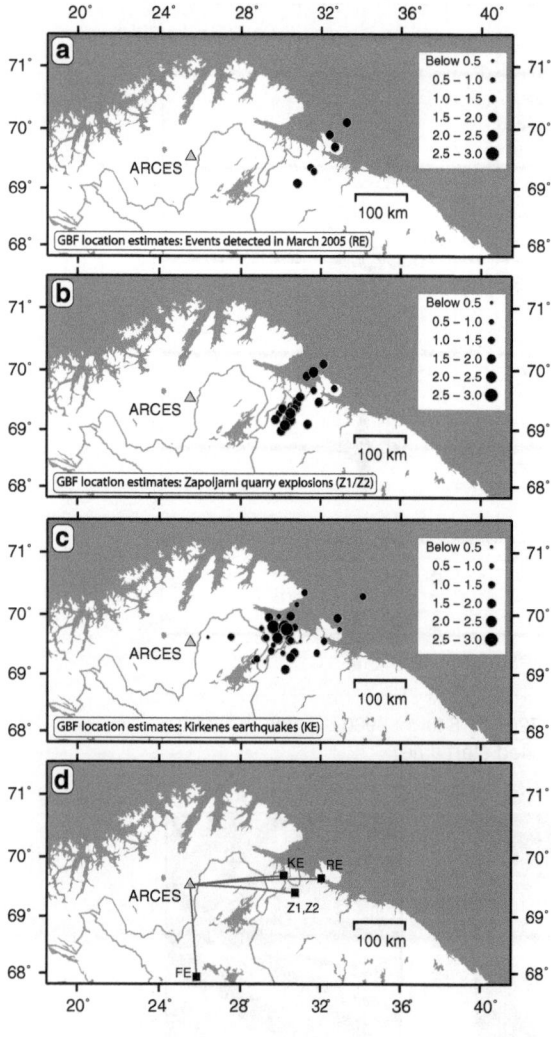

Figure 2
The uppermost three panels show fully automatic GBF location estimates for the March 2005 events (**a**), quarry explosions confirmed to have taken place at Zapoljarni (**b**), and earthquakes in the vicinity of Kirkenes in Norway (**c**). The locations of Zapoljarni (Z1, Z2) and the Kirkenes swarm (KE) are shown in panel (**d**) together with the site of the military explosions in Finland (FE, GIBBONS et al., 2007b) and the preliminary location estimate for the March 2005 events (RE, RINGDAL et al., 2005)

traces at the times of local maxima of the correlation coefficient stack, when the events in Fig. 1 were correlated against each other, resulted invariably in an unambiguous zero slowness vector which suggests that the wavefronts (despite the waveform mismatch) approach from exactly the same direction. This in turn suggests that a correlation detection procedure can be run, at a very low detection threshold, with an alignment-testing postprocessing system which will

eliminate the false alarms and leave only the detections which are truly indicative of co-located events.

2. Detection Using Multi-Channel Waveform Correlation

The process followed in the multichannel correlation detection procedure is essentially that described by GIBBONS and RINGDAL (2006). The process follows a number of steps:

1. A frequency band is identified which provides an optimal signal-to-noise ratio (SNR) for the signals.
2. A waveform template is extracted from a master event, with signals bandpass-filtered in the frequency band selected. As a general rule, the longest signal template possible should be selected. A template with the greatest time-bandwidth product will provide the most sensitive matched filter. As many channels as are available should be selected. Waveform similarity between channels is not necessary. We can therefore use, for example, both vertical and horizontal seismograms in the same analysis.
3. A segment of incoming multichannel data is read in and bandpass-filtered in the frequency band selected.
4. A fully normalized cross-correlation trace is calculated between each channel of the waveform template and the corresponding channel of the incoming data.
5. Each of the single-channel cross-correlation traces is stacked with zero time-delays. (If the waveform templates for the different channels do not start at the same time, then we need to take the time differences into account when stacking the correlation coefficient traces.)

Once we have the correlation coefficient stack (or beam) we need to determine whether or not we have a detection. There are a number of ways to detect significant values of the resulting detection statistic (see, for example, SCHAFF, 2008, for a discussion of different detection statistics). We have followed an idea similar to that outlined by SHELLY et al. (2007). Statistics of the correlation coefficient stack are calculated and values which exceed a specified multiple of the standard deviation are considered to be outliers and are returned as detections. A detection threshold of 10 times the standard deviation was typically set in the current study.

At the detection thresholds required by the poorly correlating master event candidates, a relatively large number of detections was generated by each of the event templates. For each of these detections, the correlation coefficient trace f-k analysis post-processing described by GIBBONS and RINGDAL (2006) was performed and all detections which did not result in an almost zero slowness vector were discarded. The number of detections for which the post-processing is required can be reduced by calculating the zero-lag cross-correlation (ZLCC), in a moving window, as a function of the individual single-channel cross-correlation traces only. This measure provides an indication of how well aligned the single-channel cross-correlation traces are at any given time. If $C_i(t)$ is the value of the fully normalized correlation coefficient trace for channel i, and if we have a total of N channels, then we define the ZLCC $Z(t, \Delta t)$ as

$$Z(t, \Delta t) = \frac{2}{N(N-1)} \sum_{i=1}^{N-1} \sum_{j=i+1}^{N} C_{ij}(t, \Delta t) \qquad (1)$$

where $C_{ij}(t, \Delta t)$ is the fully normalized correlation coefficient between the traces $C_i(t)$ and $C_j(t)$ over the interval $[t - \Delta t, t + \Delta t]$. This is, in other words, a pairwise average correlation coefficient between all correlation coefficient traces. In all examples presented in this paper, the ZLCC is calculated using $\Delta t = 0.5$ s. It was noted that for all occasions on which the f-k analysis test was passed, the value of $Z(t, \Delta t)$ was large compared with the background level (the actual value of Z never exceeds unity). Therefore, this function was evaluated continuously and the post-processing was only performed for times at which significant values of both the correlation coefficient stack and the ZLCC were obtained. It should be noted that the f-k analysis performed upon the single-channel cross-correlation traces is only really meaningful over small aperture arrays. If we are using a very large array or network, the ZLCC trace could probably replace the f-k post-processing.

A typical detection of an event on the ARCES array using this procedure is displayed in Fig. 3. It is clear that the values of the single-channel correlation coefficient traces at the time of the detected signal do not exceed the background level by a significant factor. The stacking operation suppresses the background level greatly, and the level close to the start of the detected signal far less. This has the effect that the local maxima of the correlation coefficient beam at this time exceed the background level by a far higher multiple, allowing a detection threshold to be set which is likely to be exceeded far less often. The value of the ZLCC trace at this time is also far higher than the background level. If two signals (with a reasonable time-bandwidth product) are truly similar, then cross-correlation will result in a trace which approximates a delta function at the time of greatest similarity. The correlation coefficient stack displayed here exceeds the background level over a duration exceeding 10 s. We assume that this is due to the signals being the result of a complicated source-time function as is observed in ripple-fired mining events.

The detection recipe identifies local maxima of the correlation coefficient stack which exceed the set threshold and then ignores all subsequent values within a short time interval of these detections; a window length of 1 or 2 s is usually used. This is to prevent multiple detections which can result from side-lobes in the correlation coefficient traces. For situations like that displayed in Fig. 3, this procedure will inevitably result in multiple detections simply because of the long time-interval containing significant values of the correlation coefficient stack. It is undesirable to simply extend the window for detection reduction since we would still seek to resolve relatively distinct detections which are separated by a few seconds. For this reason, it had to be accepted that a correlation of compound events will lead to multiple detections. The list of detections was reduced manually by inspecting the appropriate waveforms at ARCES.

The signals from these events are also well recorded by the three-component KEV station at Kevo in northern Finland, from which data have been available for many years through the IRIS data management center. It is noted that the SNR is almost always better at KEV than on single sensors of AR-CES, although ARCES has the advantage of multiple channels with which to improve the SNR through beamforming. The improved SNR is partly to be expected due to the shorter distance to the source,

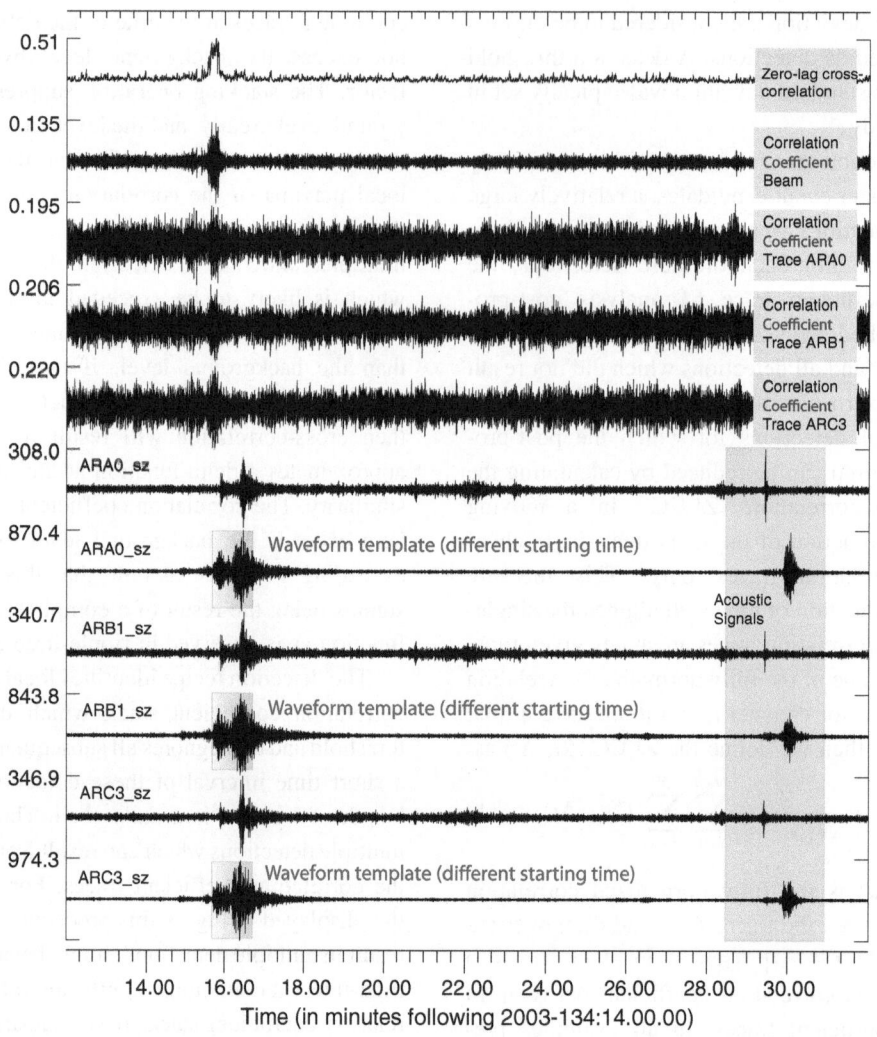

Figure 3
A positive identification from a multichannel waveform detection procedure on ARCES array data using a 60-s long signal template beginning at a time 2005-076:14.49.00.000. All waveforms are bandpass-filtered 4.0–8.0 Hz prior to calculating the correlation coefficient traces. In addition to the array correlation coefficient beam, we have calculated a trace, $Z(t, \Delta t)$, from the single channel correlation coefficient traces (see Eq. 1). The data segment that includes the waveform template is displayed for three channels, aligned according to the maximum of the Correlation Coefficient Beam at a time 14.15.39.908 on 14 May 2003. The wavetrains following both events are characterized by acoustic signals (see the *shaded box*)

although focusing effects may also be significant since the signals frequently appear to be more impulsive at KEV with arrival times that are far easier to pick. However, as is clear from Fig. 4, the correlation procedure has simply not worked using only this station. The single-channel cross-correlation traces are comparable to those at ARCES but, given only three channels with which to perform the stacking, the maxima fail to exceed the background level by a significant factor. The detection would

almost certainly be missed unless the triggering threshold was set so low that the false alarm rate would be unacceptable.

Templates from several different master events were used in the matched filter detection procedure and all templates detected essentially the same set of events. Some events, presumably those with the simplest of source-time functions, resulted in more significant values of the correlation coefficient beam and ZLCC measures. However, the lists of "safe"

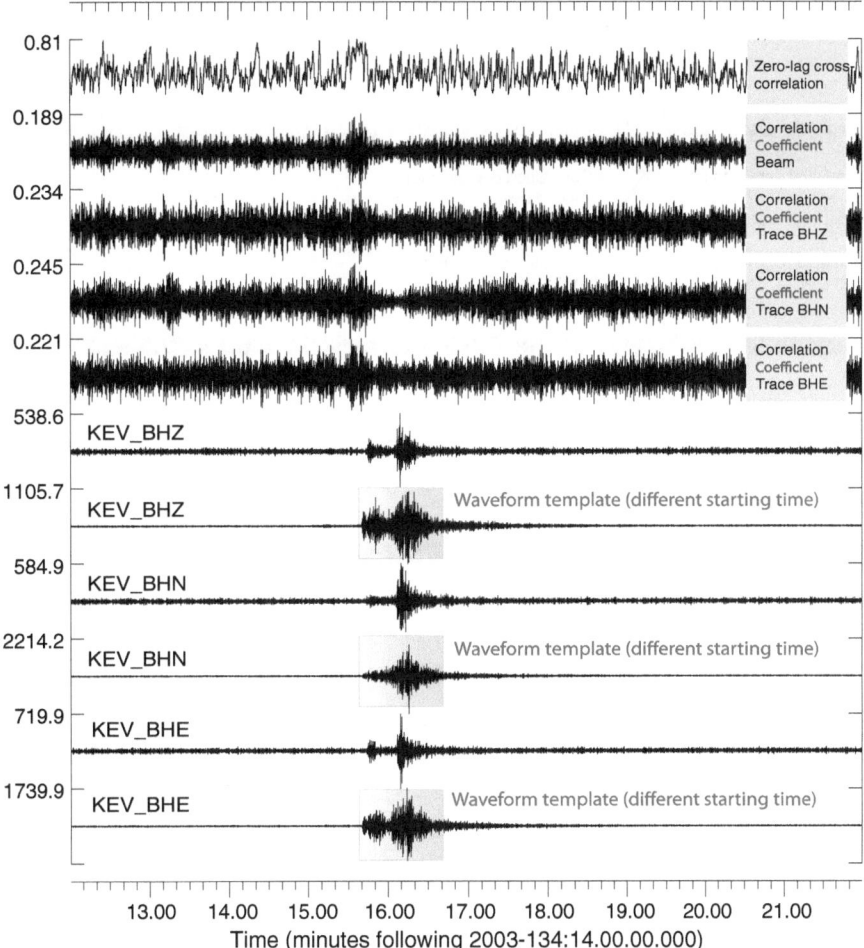

Figure 4

An attempt to detect the 14 May 2003, event using the 17 March 2005, signal as a template on the IRIS three-component station KEV in northern Finland. The waveforms are bandpass-filtered 4.0–8.0 Hz and the 60-s long signal template begins at a time 2005-076:14.48.54.01. The maximum of the cross-correlation stack does lead to an approximate alignment of the P and S phases of the master and detected signals but the local maxima of the correlation coefficient stack and corresponding ZLCC function are not significantly greater than the background levels. The signal-to-noise ratio on single channels for these events is usually better at KEV than at ARCES

detections were repeatable for each of the templates chosen. (The template displayed in Fig. 3 appears to be one of the more complicated waveforms.) With the exception of the very weakest signals, there was no clear relationship between the SNR of the master signal and the effectiveness of the waveform template. SHELLY *et al.* (2007), for example, describe a seismic matched filter monitoring situation in which many of the master event waveform templates are very weak.

In the preliminary description of this event detection project, GIBBONS and RINGDAL (2007) provide a list of the detections which were clearly false alarms. In most cases, these corresponded to correlations with high amplitude Rg-type local phases coming from a backazimuth of approximately 89°. All of these detections were subsequently eliminated by tightening the constraints applied to *f-k* analysis post-processing of the cross-correlation trace channels as described by GIBBONS and RINGDAL (2006).

Most of the correlation detections which passed the post-processing diagnostics corresponded to signals visible in the waveform data at ARCES with appropriate arrival times. There were a number of exceptions, the majority of which occurred shortly after other events which were visible in ARCES data. Such an example is shown in Fig. 5. An initial

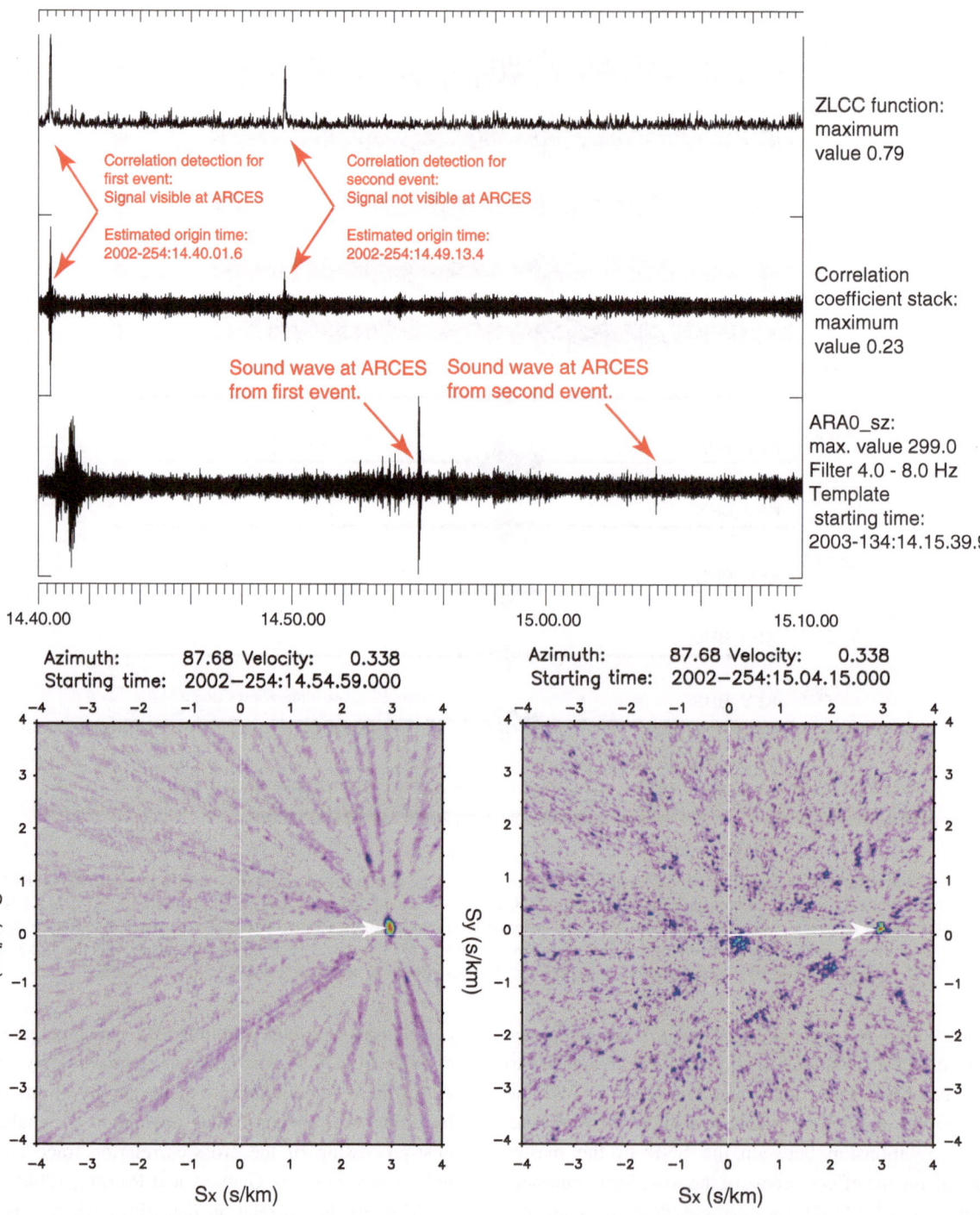

Figure 5

An example of a correlation detection for which no signal is visible in the waveform data. Two detections occur within ten minutes of each other on 11 September 2002. The first corresponds to a visible signal which is followed by the characteristic acoustic phase approximately 14 min later (see slowness grid in lower *left panel*). Some 9 min after the first detection a second trigger of high ZLCC and correlation beam value occurs without a clear corresponding seismic signal. However, there is a second acoustic signal matching the ZLCC delay (see slowness grid in lower *right panel*)

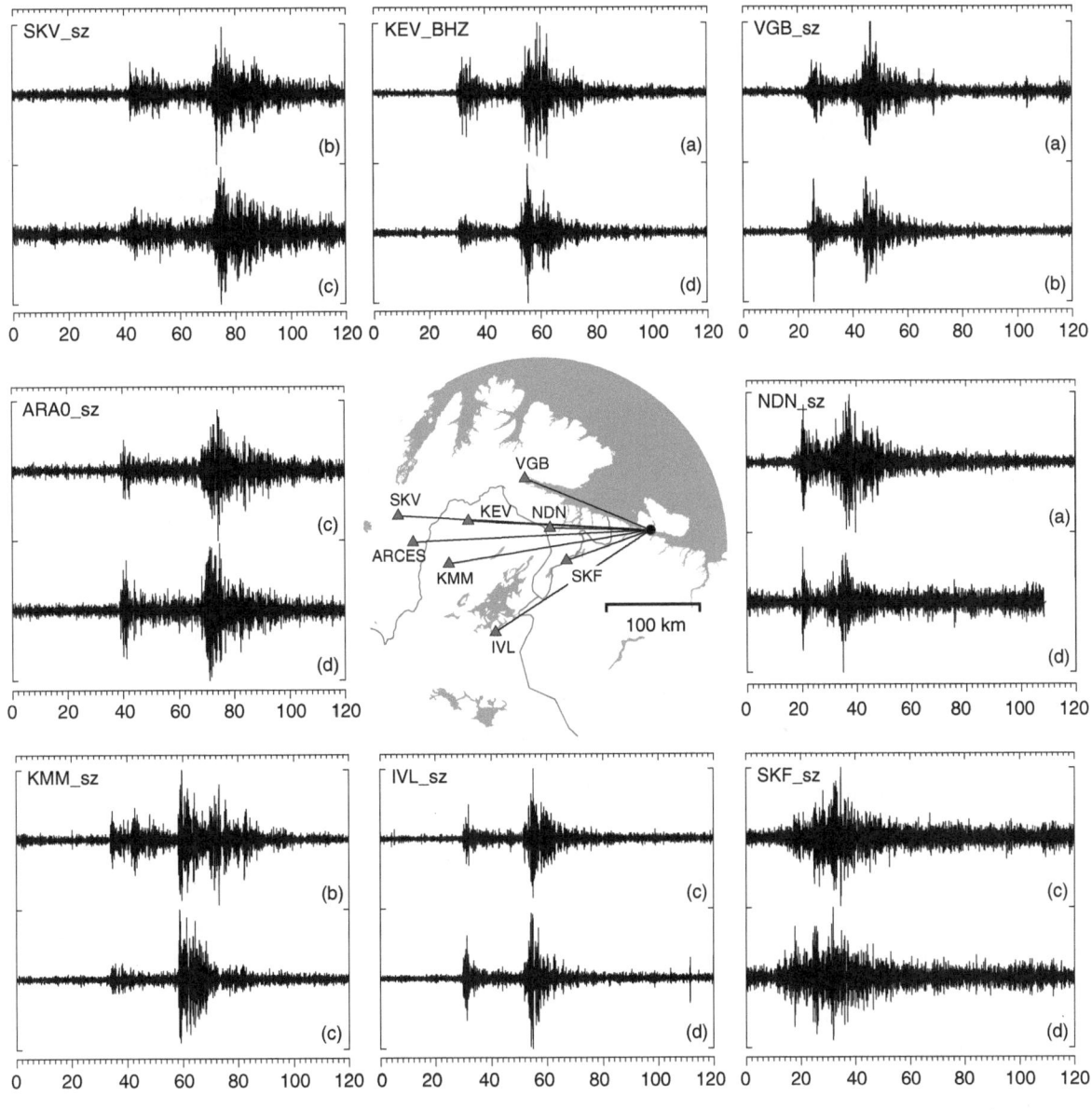

Figure 6
Selected waveforms from the stations listed in Table 1, all bandpass-filtered 3–8 Hz. The time axis in all plots indicates the number of seconds following estimated origin time. The letters *a*, *b*, *c* and *d* identify events with the origin times *a* August 30 2002-242:13.52.36, *b* April 25 2003-115:12.42.47, *c* May 14 2003-134:14.15.14 and *d* May 14 2003-134:14.58.09

correlation detection corresponds to a clear signal with a characteristic sound wave observed on the array approximately 14 min later. A second correlation detection occurs some nine minutes following the first. There is no seismic signal at ARCES which can be measured using any means other than the matched filter detector. However, a small signal is detected approximately 14 min after the presumed origin time which propagates with the same sound velocity observed for the first infrasound signal. The higher frequency band 4.0–8.0 Hz was necessary to see both the acoustic signals over the background seismic noise. The observation of this weak infrasonic signal at this time is significant evidence for the validity of the correlation detection. It is worth noting (see, for example, the March 15 and March 17 events

49

Table 1

Seismic stations used in the event location estimates

Station	Location	Latitude (°N)	Longitude (°E)	Elev. (m)	Notes
ARCES	Karasjok	69.53486	25.50578	403	25 site array
KEV	Kevo	69.75530	27.00670	81	IRIS 3-C station
IVL	Ivalo	68.71523	27.80159	120	Field deployment 2002–2004
SKF	Skogfoss	69.39041	29.70506	60	Field deployment 2002–2004
NDN	Neiden	69.69850	29.26333	90	Field deployment 2002–2004
VGB	Varangerbotn	70.16445	28.57550	6	Field deployment 2002–2004
SKV	Skoganvarre	69.77436	25.04407	220	Field deployment 2002–2004
KMM	Kaamasmukka	69.34849	26.51432	220	Field deployment 2002–2004

in Fig. 1) that weak seismic signals are frequently observed a short time following relatively large signals.

The detection and identification of events was performed using seismic data from the ARCES array alone. Data from additional stations, such as KEV, could have been added to the correlation detector system although this would have only added three channels to the 33 available on the full ARCES array and it is not very likely that this would have improved the detection capability significantly. (We recall that the *f-k* analysis performed on the correlation coefficient traces is only meaningful for a small aperture array and data from additional stations would have to be excluded from this part of the process.)

3. Event Characteristics

The exact nature of this sequence of events is currently not known. The generation of both seismic and infrasound signals indicates that the events probably take place close to the surface. Seismic signals are recorded by numerous stations and, in Sect. 3.1, we display waveforms recorded at regional distances and compare spectral properties of the signals with those of signals generated by other known sources of repeating seismicity in the region.

The generation of infrasound signals is in many ways the primary motivation for compiling a data base of these events. Whereas the travel time of a seismic phase from a given source to a given receiver will remain unchanged from event to event, the observation of an infrasound signal at a station will depend upon the properties of a propagation medium which is in continuous change. It is therefore essential for our understanding of atmospheric sound propagation that we have large numbers of Ground Truth events with tight constraints upon the origin time and source location. The observations on the seismic sensors at ARCES are clearly of interest and, in addition, we have obtained segments of infrasound data from the three-element arrays at Lycksele and Apatity (see Table 2). The detection or non-detection of infrasound at these station is discussed in Sect. 3.2.

3.1. Analysis of Seismic Signals

Figure 6 displays waveforms from selected events on the permanent stations ARA0 (central element of the ARCES array) and KEV, together with six stations of a temporary network deployed between 2002 and 2004 (see Table 1). At all stations, energy at the lower frequencies is dominated by ocean-generated microseismic noise, and the 3–8 Hz band typically provided the best SNR for both P and S phases. The IVL station was the best of the temporary deployment with very little down-time and very little environmental noise. The IVL signals displayed in Fig. 6 are very typical with very clear P and S onsets. The station VGB also recorded many events well, but was often subject to quite high noise levels. The closest stations to the site are SKF and NDN (at distances of around 100 km). Although signals from these events were often visible in the data, the noise at high frequencies usually made accurate phase readings impossible. Both SKV and KMM had considerable down-time (see MAERCKLIN *et al.*, 2005) and were of somewhat limited use for event location given their proximity to ARCES and KEV.

However, very clear S phases were recorded at KMM.

Each of the panels in Fig. 7 displays amplitude density spectra for P arrivals at ARA0 from four different events from the specified source in Fig. 2 (panel d). The event source covering the greatest range of magnitudes is the earthquake swarm near Kirkenes (top left panel). Despite the differences in event magnitude (the largest event has $M_L \sim 3$), the spectral shapes are very similar. The most notable

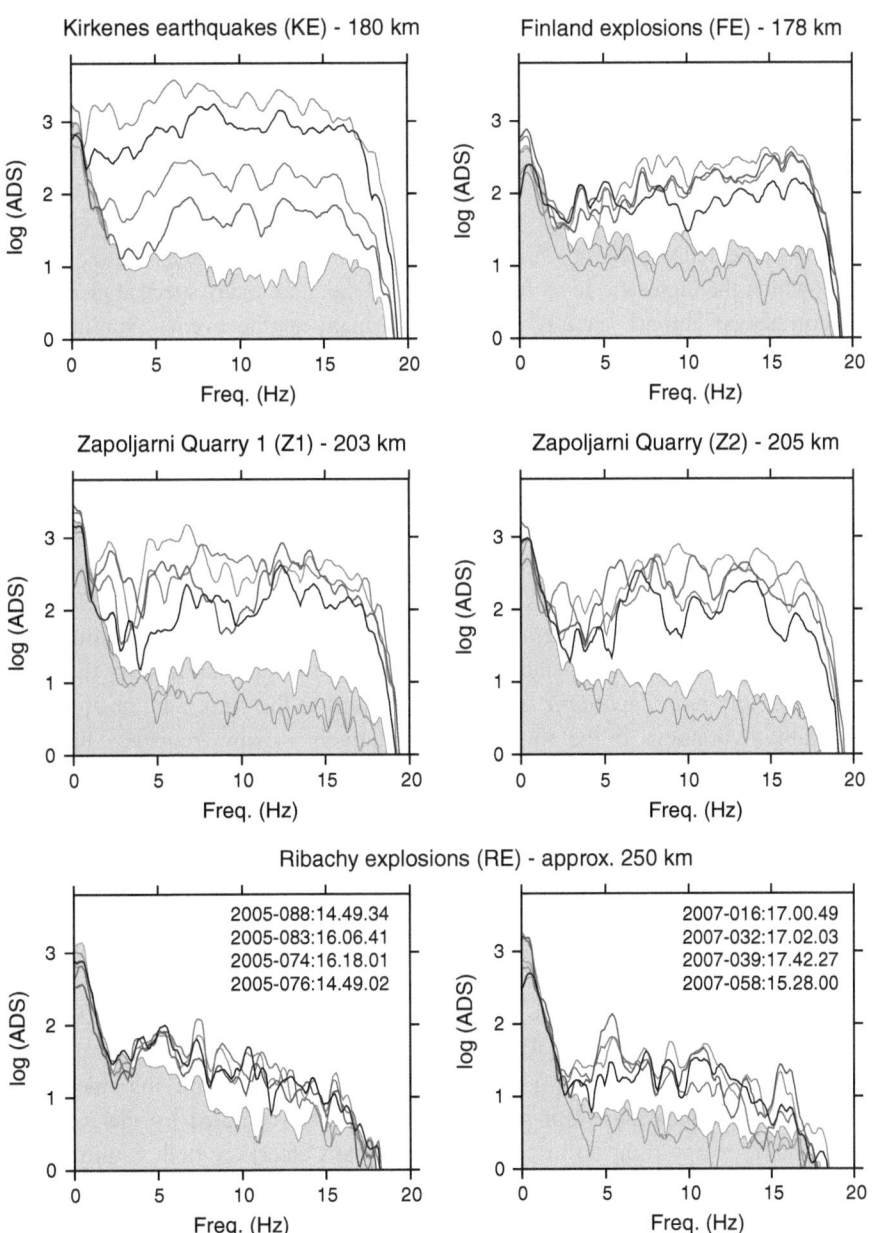

Figure 7

Estimates of amplitude density spectrum (ADS) for 10-s-long windows of short-period vertical data from the site ARA0 for four different events from each of the source locations indicated. The spectral estimate for each window starting with the P-arrival time t_P is displayed using a *solid line*, and the superimposed *grey shapes* indicate the corresponding noise spectra (calculated in the 10-s window ending at time $t_P - 3$). All spectra calculated using the multitaper method of Thomson (1982). The times displayed in the lower two panels are the arrival times at ARCES (not origin times)

feature is that the low-frequency energy diminishes more rapidly than the high-frequency energy as the event magnitude decreases. All of the Kirkenes events were detected using a correlation detector from a single master event, and so they show a significant degree of waveform similarity. The similarity of the spectra is therefore not unexpected. It is worth noting that the largest event has a high SNR right down to 2 Hz, whereas the smallest event shown here is essentially not visible above the background noise level below 5 Hz.

The top right panel shows spectra for four of the Finnish explosions discussed by GIBBONS et al. (2007b). Again, the documented waveform similarity results in spectra in which the highs and lows for the different events correspond almost exactly as a function of frequency. Note the very small variation in the spectral amplitudes; all of these explosions have a yield within about 10% of 20,000 kg. Whereas the earthquake spectra in the top left panel are relatively flat right up the anti-aliassing cut-off frequency, the spectra for the explosions in the top right panel appear to increase for increasing frequencies. (GIBBONS et al., 2009, show that, despite the 2–4 Hz band having the lowest SNR, slowness estimates at ARCES in this band are the most stable.)

The spectra for the different Zapoljarni quarry blasts vary considerably. Variations in the shape of the waveforms are typical given ripple-fired sources (e.g., HARRIS, 1991), and this is usually exacerbated by differences in the location within the mine and the orientation of the rock faces being excavated (e.g., BONNER et al., 2003; McLAUGHLIN et al., 2004). The superimposed signals will influence the spectra accordingly and spectral properties can be used to distinguish ripple-fired quarry blasting from other types of seismic sources (see, for example, ARROW-SMITH et al., 2007, and references therein). One feature common to all of the uppermost four panels is that the signals contain energy right up to cut-off frequency of the anti-aliassing filter. In other words, we would need to employ a higher sampling rate to capture all the energy present in these signals.

The lowermost two panels both show spectra from the Ribachy explosions. The lower-left panel shows four events from March 2005 and the lower-right panel shows four events from January and February 2007. It was decided to display both sets of events due to the different spectra of the background noise at these times. Whereas the spectral amplitudes for the events in both panels are very similar, the noise spectral amplitudes are considerably higher between 3 and 6 Hz for the March 2005 events, resulting in a lower SNR at these frequencies. The Ribachy explosions are the only events displayed for which the spectral amplitudes decrease with increasing frequency and the only events for which the signal spectral amplitudes have decreased to the background level at a frequency lower than the cut-off. The spectra for the different events are far more similar to each other than, for example, the Zapoljarni events and there are many spectral peaks which are common to many of the events. Similarity of the amplitude density spectra supports the hypothesis that the event locations are closely spaced.

In addition to the stations listed in Table 1, seismic signals from these events are also recorded by the IRIS three-component station LVZ at Lovozero and the seismic array at Apatity (phase detections from this array contribute to the automatic network location estimates). However, there appears to be a far greater attenuation of high frequency energy for the stations to the south than for the stations to the west, and the signals are confined to a relatively narrow frequency band (around 4–8 Hz). The SNR at Lovozero is exceptionally poor for most events.

3.2. Analysis of Infrasound Signals

Figure 8 shows a network of three stations which have recorded infrasound signals generated by these events back to 2002. The ARCES seismic array (which has been demonstrated to record many acoustic signals) and the microbarograph subarray at Apatity (operated by the Kola Regional Seismological Center) are both within 260 km of the site. A third station, at Lycksele in Sweden at a distance of approximately 800 km, is one of four 3-site microphone arrays operated by the Swedish Institute for Space Physics. This array was selected for inclusion here simply because it is the most distant of these infrasound stations and demonstrably records signals from a large number of these events at far greater

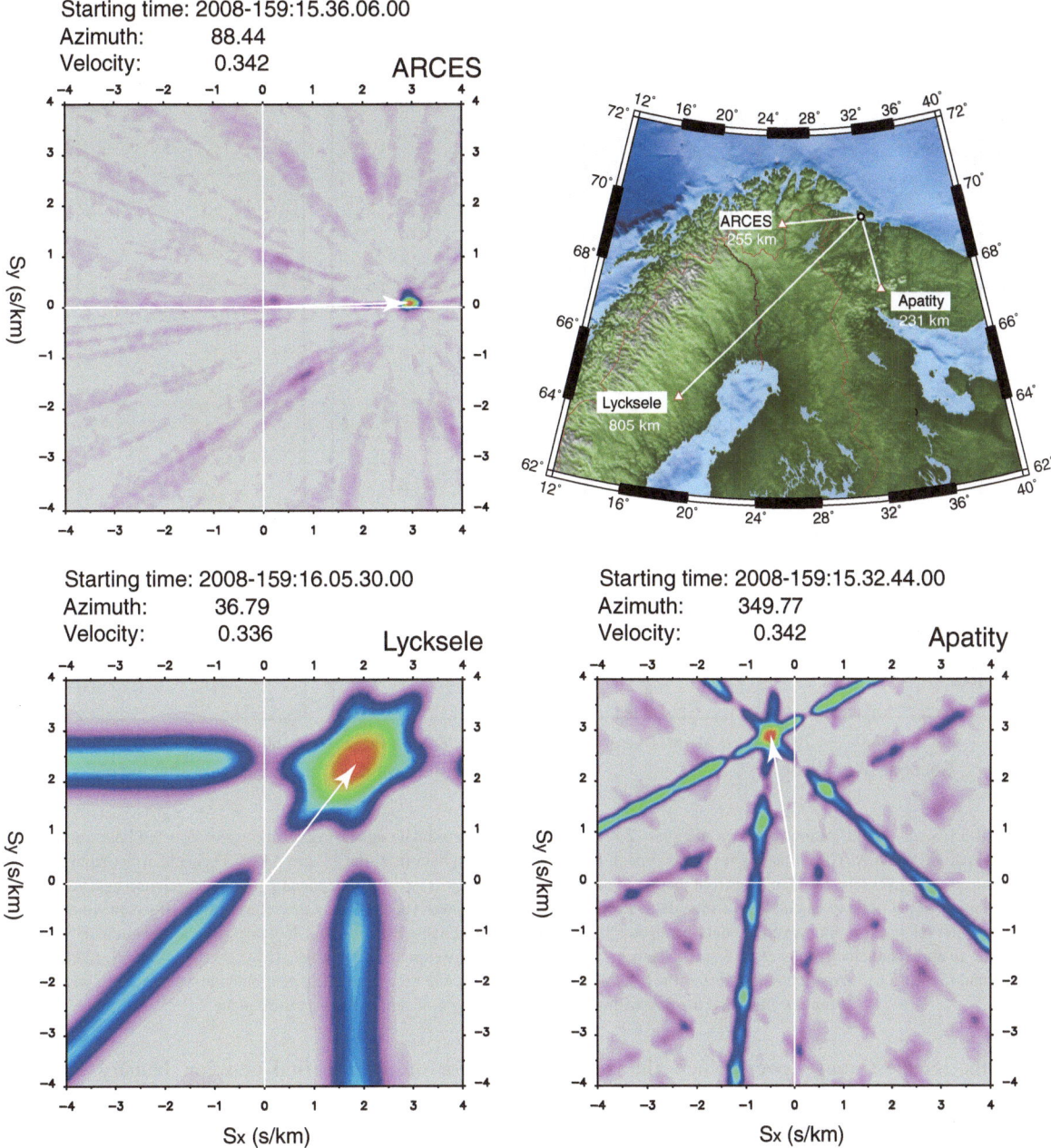

Figure 8
Map of three selected seismic and infrasound stations in Fennoscandia and Russia which recorded acoustic signals from the event on June 7, 2008 (estimated origin time 2008-159:15.22.04). A 10-s window was selected for the signal on each of the stations and a mean pairwise cross-correlation coefficient evaluated for time delays specified over the slowness grids shown (FRANKEL et al. 1991). All waveforms were bandpass-filtered 2–7 Hz prior to the correlation

distances than seismic stations are able to register the seismic phases. Figure 8 indicates that each of the stations records a backazimuth estimate consistent with the geographical location of the explosion site.

The resolution of slowness is clearly a function of the array aperture. For all of the panels displayed, the frequency band 2–7 Hz was used for processing of the sound waves. At ARCES, this frequency band

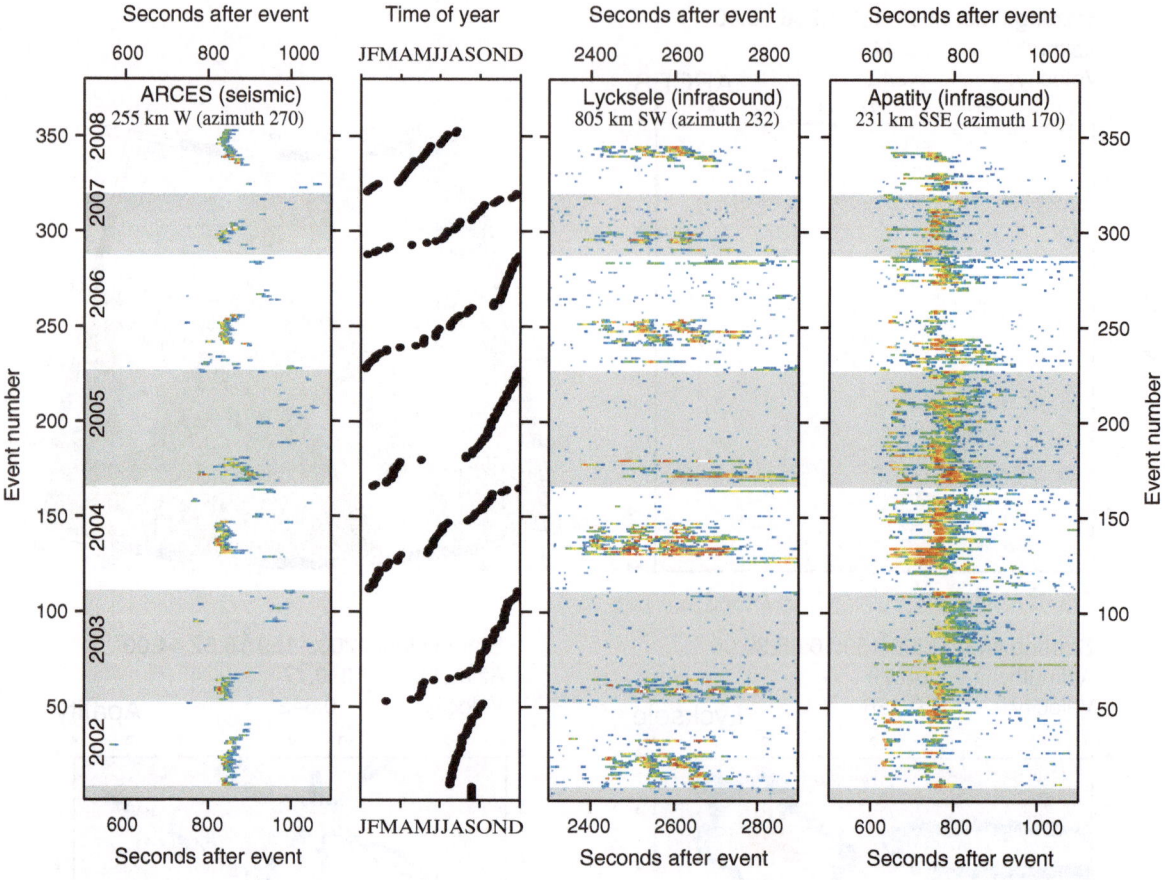

Figure 9

For each of 353 event identifications between the years 2001 and 2008, an approximate event origin time was estimated based upon the P-arrival times at KEV and ARCES. Starting from this time, a detector was run on data from the 25-site ARCES seismic array, from the 3-site microphone array at Lycksele, Sweden, and from the 3-site microbarograph array at Apatity. The mean pairwise correlation coefficient value Γ (see BROWN *et al.*, 2002) was calculated for time delays corresponding to slowness vectors over a dense grid using 10-s data windows, with one second interval spacing, for waveform data bandpass filtered 2.0–7.0 Hz. Each coloured pixel indicates that a maximum Γ value was obtained for an apparent velocity between 0.3 and 0.4 km s^{-1} and a backazimuth estimate in the ranges (75.0, 95.0) at ARCES, (25.0, 45.0) at Lycksele, and (345.0, 5.0) at Apatity. *Red* indicates the highest correlation coefficient between sensors and blue the weakest. The *second panel from left* indicates the day of the year on which each of the explosions took place

covers the greatest range over which the sound signals are observed. Ocean generated microseismic noise dominates at frequencies below 2 Hz and no acoustic signals from these events have been observed that have been visible at these frequencies. The 2–7 Hz band was also chosen for processing at the 3-site arrays in order to mitigate the effects of poor slowness resolution (at the lower frequencies) and aliassing (at the higher frequencies). The configuration of the microbarograph subarray at Apatity, with approximately 250 m between sensors, appears to provide a very good compromise for signals in this frequency range.

For the time period between January 2000 and July 2008, we have attributed 353 events to this source on the basis of the correlation detector on ARCES. Figure 9 displays the sound signals observed at the three stations for each of these events. An additional panel (second from left) indicates the time of year at which each shot occurred. It is clear from this panel that shots have taken place at essentially all times of the year. Sound signals at ARCES are observed for a high number of events, and it is clear that the events which do and do not generate sound signals on ARCES are clustered in time. Comparing the first and third panels of Fig. 9

indicates that the microphone array at Lycksele detects essentially the same events as ARCES. We note that both the travel times and durations of the signals at the more distant station are far greater. The coincidence of detected events at ARCES and Lycksele indicates that source characteristics or atmospheric sound propagation conditions, rather than local effects close to the receivers, are likely to be the principal factors in determining detection or non-detection. The fourth panel in Fig. 9 indicates that almost all events were detected at the micro-barograph mini-array at Apatity, 230 km SSE of the events. This makes it unlikely that the detection or otherwise at ARCES and Lycksele was a function of the source, since any variation of yield taking place does not seem to affect the detectability of infrasound at this third station. (We note that MCKENNA *et al.*, 2007, did not find evidence that explosion yield was closely related to the amplitude or detectability of infrasound signals.)

The second panel of Fig. 9 indicates that the events detected at ARCES and Lycksele take place mainly in the summer and that undetected events take place mainly in the winter. This observation is consistent with a hypothesis that it is the direction of the stratospheric winds which dominates the detectability at these stations to the West of the source, and is consistent with the findings of LE PICHON *et al.* (2009) who conclude that this factor dominates the global detectability of infrasound. As a preliminary resource for the evaluation of atmospheric propagation models, Table 3 provides the estimated origin times of 53 explosions attributed to this source in 2004, together with the times at which the highest amplitude infrasound phases were detected at ARCES. The somewhat speckled appearance of the panels for Lycksele and Apatity may be related to the fact that these mini-arrays have only three elements. Such an array has the disadvantage that there is no redundancy. If one element is subject to excessive noise, or is out of operation, no direction estimate can be obtained regardless of how well the remaining instruments perform. It should be noted that the apparently weaker detections at Apatity during 2003 were the result of a technical fault with one sensor of the array and should not be interpreted as being the result of weaker signals.

4. Event Location Estimates

In the absence of precise coordinates for these events, we need to exploit the seismic (and possibly also infrasound) data from all the events attributed to this site in order to obtain the best location estimate possible. All the available waveforms for every event were examined manually, in multiple frequency bands, to measure as accurately as possible both first P and first S arrival times. Seven of the eight stations listed in Table 1 are three-component stations and we were able to rotate the horizontal components in order to display radial and transverse components which maximize energy arriving for the P and S phases, respectively. It was possible to improve the SNR further for many P phases by applying the transformations specified by KIM *et al.* (1997) for isolating the P-wave energy. For the ARCES array, beamforming was performed in addition to improve the SNR for the different phase arrivals.

Figure 10 shows the manual arrival time measurements for the P arrivals at ARCES and IVL relative to the P-arrival time at KEV (always deemed to be the safest onset time estimate due to the high SNR and more impulsive nature of the arrival), and the S − P travel time difference estimates for the three stations KEV, ARCES, and IVL. The panels on the left indicate the times at which these events occurred, and hence an indication of the frequency and pattern of explosion occurrence. The panels on the right show the same travel-time difference estimates rearranged in order to indicate the consistency with which the arrival time estimates were made. Flatter curves indicate more consistent estimates (the P-time estimates are usually subject to less variation than the S-time estimates), although a bias resulting from a persistent misidentification of a phase would not show in these plots. Most of the S − P difference outliers indicate too large a time difference, and this is usually the case for the more complicated event sequences. The large number of events detected will provide a better overview of the uncertainty in arrival time measurements than would be possible given only the small number of events initially identified (Fig. 1).

The complexity of the event source-time functions precludes the application of the double-

Reprinted from the journal

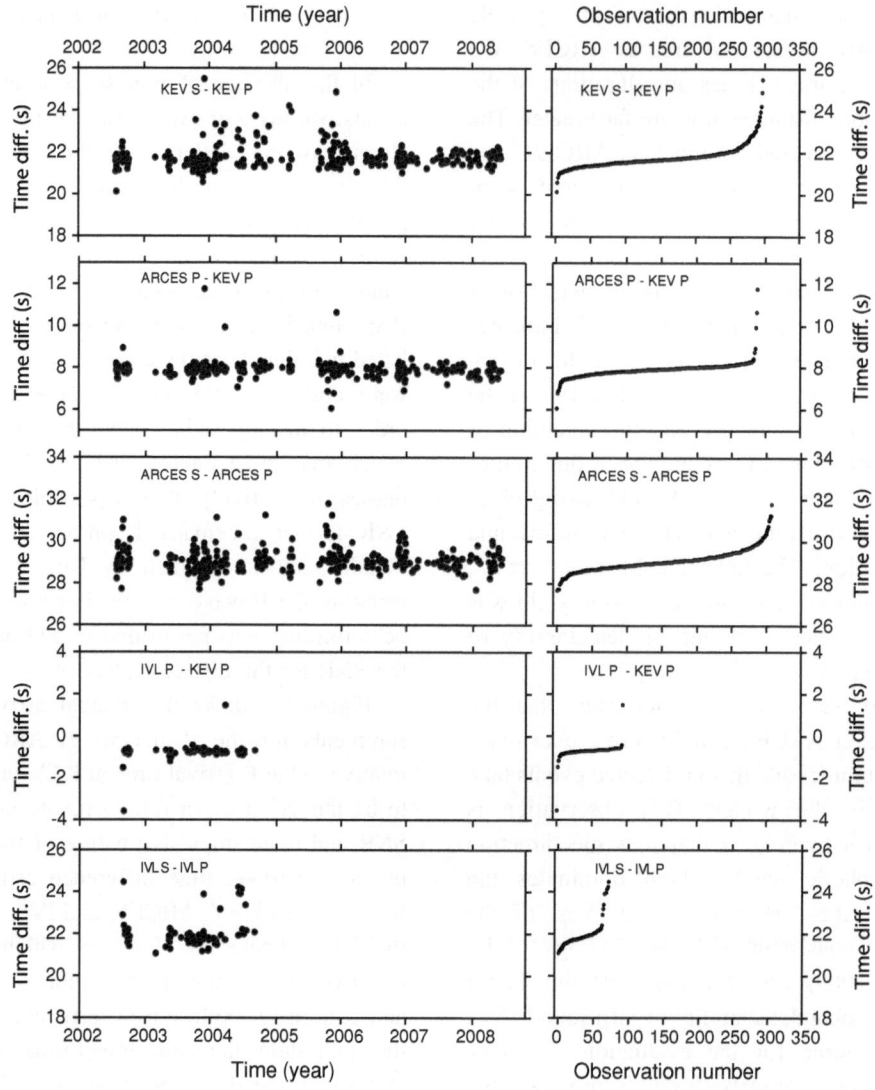

Figure 10
Time delays measured between different phase arrivals at the specified stations: On the left as a function of epoch time and on the right reordered according to increasing travel-time difference. Each event detected was subject to manual examination and, wherever possible, a P- and S- arrival time was picked for each observing station using appropriately rotated waveforms in many different filter bands. Beamforming was also applied for ARCES. In addition to the stations displayed, distributions were also calculated for the SKF, NDN, VGB, SKV, and KMM stations which had fewer satisfactory observations

difference methods to obtain relative location estimates (see RICHARDS et al., 2006, and references therein for an overview) since reliable measurements of relative travel times are not available. An assumption which we need to justify is the hypothesis that the events are "almost co-located". We are able to detect the signals from one event by correlating the incoming data stream with the signals from a different (master) event; this is to say that the detected

event is located within the correlation footprint of the master event. The spatial separation possible between two correlating events was suggested by GELLER and MUELLER (1980) to be no more than a quarter wavelength at the dominant frequency, which would correspond to no more than a few hundred meters at the high frequencies considered here. RINGDAL et al. (2008) examined the detectability of shots in a marine seismic profile in the Barents Sea using a matched filter procedure on ARCES array data. This experiment was particularly pertinent to the current study since (a) the source-to-receiver distance and (b) the frequency content of the signals were essentially the same as for the signals considered here. Using only a single channel, RINGDAL et al. (2008) found that a signal could be detected using correlation if the location was within 500–600 m of the master event. However, applying the stacking procedure over the full array, it was found that this distance was increased to over 2 km (this is consistent with the findings of HARRIS, 1991, who suggests a larger correlation footprint of approximately two wavelengths). For events over 3 km away from the master event, RINGDAL et al. (2008) were not able to detect the signals using the matched filter detector and we will assume that if the explosions in the current study had been separated by more than this distance then we would have been unable to detect them using the procedure described here. (For the sake of comparison, the ammunition destruction explosions described by GIBBONS et al., 2007b, are known to take place within a region of approximate aperture 350 m at 67.934°N, 25.832°E.) Even if the events are separated by the maximum distance which will still facilitate detection by waveform correlation, they can still be considered to be almost co-located with respect to the resolution possible applying classical location methods using a sparse seismic network at regional distances.

Very few of the events are recorded by a large number of stations, and many have only phase readings from ARCES and KEV. Attempting a solution for such events using, for example, HYPOSAT (SCHWEITZER, 2001) results in hypocenter estimates concentrated around the longitude of the Ribachy Peninsula, but which are spread over a large range of latitudes and are associated with error ellipses elongated in the North–South direction. Almost 100 events also included phase readings from the IVL field station (see Fig. 10) and these were generally located somewhat south of the Peninsula, to the north of the Titovka River. Considering the variation in the event location estimate which will result from the different station availability for each event, it makes more sense to examine how the location estimate varies with the estimates of travel-time differences

between each pair of phases. The assumption of a common source location implies that the true value of each of these travel-time differences is a constant which has been estimated for each of the events identified in the detection procedure. The travel-time differences for the most frequently measured pairs of phases are displayed in Fig. 10, although we also have available the corresponding measurements for the remaining stations in Table 1, albeit with significantly fewer observations in most cases.

The stations in Fig. 6 are between 50 and 300 km of all possible source locations and the exact nature of the first arriving regional P and S phases will depend upon source-receiver distance and the local crustal structure. It is rarely possible to identify an initial regional P arrival as a Pn or Pg phase (see STORCHAK et al., 2003) at these distances and so we exploit the option in HYPOSAT which allows an unidentified first regional P or S arrival to be labelled P1 or S1. HYPOSAT revises the phase identification attributed at each iteration, depending upon the source-receiver distances for the most recent location estimate.

The location procedure was run with many different velocity models in order to assess whether the choice of model made any significant difference to the event locations. We applied the global ak135 model (KENNETT et al., 1995), the Fennoscandian model (MYKKELTVEIT and RINGDAL, 1981), and the BAREY and BAREZ models (HICKS et al., 2004). For each velocity model, 10,000 different combinations of onset times for initial P and S arrivals at the stations listed in Table 1 were generated at random from the observed distributions as displayed in the right-hand panels of Fig. 10. Each phase arrival time estimate is weighted within HYPOSAT according to the variability of the measurements and the number of observations. Each combination consists of an arbitrarily chosen first P-arrival time at KEV (the clearest and most easily read arrival), a P-arrival time at the other stations specified with respect to the KEV P-onset time, and an S-arrival time set relative to the corresponding P time at all stations. Figure 11a displays the solutions obtained for each of the 10,000 arrival-time combinations, solved using the BAREY velocity model, with the colour of each symbol reflecting the number of arrival-time combinations

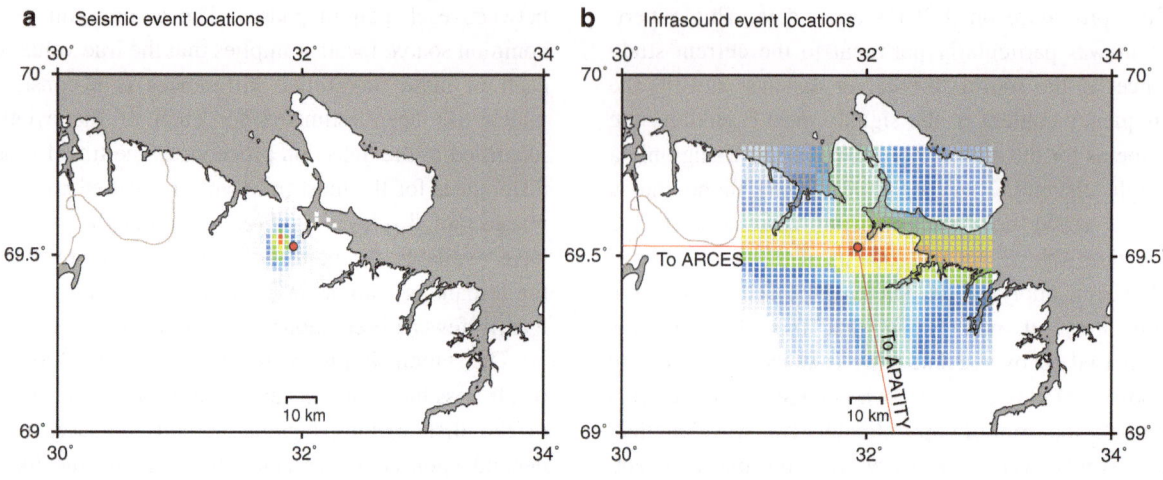

Figure 11

Distributions of location estimates **a** using only travel times of seismic P and S waves measured on the stations listed in Table 2 and **b** using only the backazimuths measured from the stations listed in Table 2 for each of the observations displayed in Fig. 9. The red circle marks 69.526°N, 31.930°E which is the approximate location indicated by SKORVE (1991) for an air defence site (based upon satellite imagery prior to 1989). The seismic location estimates are made using the BAREY velocity model (HICKS *et al.*, 2004)

which resulted in a location estimate within each of the cells displayed. The cell containing the greatest number of solutions is centered on 69.553°N, 31.821°E. The travel-time combination generator produced hypothetical observation sets that resulted in location estimates up to a distance of approximately 25 km from these coordinates (for these most distant location estimates, the arrival times of P phases at ARCES or IVL were chosen at the extremes of the distributions displayed). A single run of HYPOSAT, using the median traveltime differences as displayed in Fig. 10, results in a solution close to 69.55°N, 31.82°E.

We note that SKORVE (1991) marks the location of a presumed air defence unit on the northern bank of the Titovka River with the approximate coordinates 69.526°N, 31.930°E. We have no information to indicate whether or not this unit is still operational and, even if there is still a military installation at this site, there is no direct evidence to suggest that this site is the source of the explosions. However, in the absence of features visible elsewhere using, for example, Google Earth, we have adopted these coordinates for purposes such as the generation of waveform templates given the small travel-time residuals at the stations listed in Table 1. If, in the future, the true coordinates of this site should come to light, this will provide an excellent opportunity to

reevaluate the most appropriate regional velocity model.

As described in Sect. 3, almost all of these events generate infrasound signals observable at at least some stations at regional distances. All of the infrasound detections displayed in Fig. 9 are associated with backazimuth estimates which, as displayed in Fig. 8, correspond closely to the true receiver-to-source direction. If infrasound signals from an event are registered at two or more stations then a provisional event location can be estimated simply by finding the intersection of the lines (or a best-fit point of intersection in the case of more than two stations). Unlike seismic waves, which will take an identical path from source to receiver for every event at a given location, infrasound waves propagate along paths determined by the effective velocity in the atmosphere (see, for example, GARCÉS *et al.*, 1998) and substantial deviations from the true backazimuth can result from cross-winds.

Backazimuth estimates for infrasonic signals at the ARCES array range from 80 to 95°, with most estimates falling between 87 and 89°. ARCES is by far the largest of the three arrays considered and the large aperture provides excellent resolution in slowness (see the upper left panel of Fig. 8). This indicates that the variance in the backazimuth estimates is more the result of atmospheric conditions

Table 2

Stations used to process infrasound signals

Station	Latitude (°N)	Longitude (°E)	Notes
ARCES	69.53486	25.50578	Seismic array (25 elements, 3 km aperture)
Apatity	67.60330	32.99440	Microbarograph sub-array (3 elements, 150 m aperture)
Lycksele	64.61350	18.75070	Microphone array (3 elements, 75 m aperture)

Table 3

Estimated origin times of events detected during 2004

1	20 January	020:15.00.08	28	23 June	175:16.27.39 (836)
2	21 January	021:14.45.45	29	28 June	180:17.00.01 (843)
3	29 January	029:18.21.04	30	30 June	182:15.59.53 (830)
4	29 January	029:18.45.12	31	5 July	187:19.55.53 (846)
5	29 January	029:19.13.22	32	5 July	187:20.19.08 (828)
6	3 February	034:19.05.12	33	7 July	189:19.00.25 (825)
7	5 February	036:18.18.11	34	14 July	196:18.13.12 (826)
8	5 February	036:18.39.16	35	19 July	201:14.32.20 (837)
9	10 February	041:23.04.52	36	2 September	246:17.01.43
10	10 February	041:23.54.14	37	9 September	253:20.30.02
11	26 February	057:19.53.06	38	15 September	259:19.00.49 (918)
12	1 March	061:22.26.12	39	17 September	261:16.50.16 (923)
13	1 March	061:23.32.12	40	20 September	264:20.28.45
14	4 March	064:20.48.02	41	28 September	272:22.31.48
15	24 March	084:21.14.01	42	6 October	280:16.11.27
16	24 March	084:21.58.25	43	6 October	280:16.33.33
17	26 March	086:18.07.44 (953)	44	12 October	286:13.44.27
18	31 March	091:16.35.20 (978)	45	14 October	288:17.25.21
19	1 June	153:22.03.01 (839)	46	21 October	295:16.37.19 (771)
20	4 June	156:20.21.58 (848)	47	21 October	295:17.21.18 (768)
21	4 June	156:20.57.45 (837)	48	26 October	300:16.16.37
22	9 June	161:20.26.08 (837)	49	26 October	300:16.55.53
23	15 June	167:16.45.09 (830)	50	28 October	302:18.42.07
24	15 June	167:17.09.37 (824)	51	19 November	324:03.17.15 (952)
25	15 June	167:17.29.41 (824)	52	30 November	335:16.45.51 (923)
26	17 June	169:17.23.08 (824)	53	28 December	363:17.23.24
27	21 June	173:16.00.28 (836)			

If infrasound signals were detected at ARCES, the number in brackets is the number of seconds after origin of the greatest coherence for a phase with sound speed

than of uncertainty in azimuth estimation (see, for example, SZUBERLA and OLSON, 2004). Despite the somewhat poorer slowness resolution, the microbarograph subarray at Apatity provides backazimuth estimates for infrasound signals over a similar range to ARCES, from approximately 345–4°, with a median value of approximately 351°. The 3-site microphone array at Lycksele in Sweden produces backazimuth estimates which are most frequently between 30 and 40°. However, to compound the poorer slowness resolution, the considerable distance between the station and the source region means that the station is not able to resolve different source locations within the region of interest. Relative frequency histograms were constructed from the backazimuth estimates obtained at the three stations listed in Table 2 using bins of 1° width, The colours of the symbols in panel (b) of Fig. 11 indicate a scale proportional to the product of the three relative frequencies for the applicable receiver to source backazimuths. The region which is consistent with the greatest proportion of the backazimuth estimates

59

includes the area surrounding Novaya Titovka and the locations identified as the most probable sites from the seismic travel-time inversions.

The resolution indicated in Fig. 11b) is probably a considerable underestimate of what is possible using infrasound signals on the given station network (see, for example, ARROWSMITH *et al.*, 2008; CHE *et al.*, 2009). The location estimates are obtained from aggregates of backazimuth estimates from several hundred events over an interval of several years without any consideration of the atmospheric conditions at the time of each event. Modelling the propagation of sound for each event, with the atmospheric model for the appropriate time, may lead to reasonable predictions of the anticipated departure from the geographical backazimuth which can be accounted for when calculating the intersections. Such a study is beyond the scope of this paper.

5. Conclusions and Discussion

We have examined a series of explosions at an unknown site on the Kola Peninsula in North-West Russia. We were first alerted to the existence of these events in March 2005 since sound generated (either directly or indirectly) by the explosions was heard by residents on the Varanger Peninsula in the far north of Norway, at a distance of over 100 km. Examination of the seismic traces on the ARCES array indicates an event location approximately 250 km to the east of the array, however an accurate location is made difficult by the complexity of the waveforms observed for each event and the sparsity of the seismic network in the region. The events were detected automatically and located using the GBF algorithm (RINGDAL and KVÆRNA, 1989). Magnitude estimates ranging between 0.9 and 1.5 were attributed to the events. The seismic signals at ARCES from each of the initially observed events were followed about 14 min later by high amplitude infrasound signals from the same direction recorded on the seismic sensors.

A project was initiated to detect and identify as many signals from these events as possible using ARCES data. Multichannel waveform correlation is the simplest procedure to apply, but poor correlation

between the signals from the set of events in Fig. 1 suggests that a detection threshold would need to be set at a very low value. Performing *f-k* analysis on the correlation coefficient traces at the times of each of the local maxima indicates an excellent alignment of the correlation traces at these times, despite the modest values of the detection statistics obtained. This may suggest that the waveform dissimilarity between the signals is primarily the result of different source-time functions rather than source location. A continuous ZLCC function is calculated, providing a measure of the similarity and alignment between each of the single-channel correlation traces. A detector with conditions on the values of both the correlation coefficient stack and the corresponding ZLCC trace has far fewer triggers than a detector based on the values of the correlation stack alone and, applying the *f-k* analysis post-processing described by RINGDAL (2006), the number of false alarms is reduced greatly. This highlights the advantage of applying the procedure on seismic array data with a high number of channels. Although correlation detectors which exploit all three components of 3-C seismic data are more sensitive than single channel detectors (see SCHAFF, 2008), the constraints on correlation coefficient trace alignment for arrays and networks with more channels is superior. We demonstrate an event, presumed to be at the same site, which is detected clearly using array-processing of the correlation coefficient traces at ARCES, but which could not be detected at the closer KEV 3-component station using a matched filter detector without incurring an unacceptable false alarm rate. It is noted that the advantage of array processing in this case is not a SNR issue, but a question of an improved ability to detect signals which are related to the template signal by a common time-history at each site.

We are fortunate that the ARCES array is equipped with so many sensors (33 data channels at 25 different sites). Many newer array stations have only nine sites and a single three-component station. The array-gain on such a system would clearly be far smaller than at ARCES and it would need to be examined on a case-by-case basis as to whether the suppression of the background level of the detection statistic was high enough to produce a detector with a sufficiently low false alarm rate. Whilst the array used in this study is a small aperture array, the same principle could presumably be applied to a network of arbitrary configuration provided that the signals are reasonably well-recorded at most sites. (Reliable instrumental timing across the network is assumed.) The *f-k* analysis part of the post-processing is only meaningful for small aperture arrays, and alternative methods of measuring the significance of the alignment between the single-channel correlation traces would need to be applied given a network of arbitrary size. On the other hand, there are typically far fewer detections anyway on large aperture networks since the signals are usually less similar on the different sensors and it is more difficult to achieve a waveform match over the whole network in a given time-window.

We argue from the detection by waveform correlation that the sources are likely to be almost co-located. (Approximate co-location of events is also supported by the similarity of the signal spectra displayed in the lowermost two panels of Fig. 7.) The waveform recorded at any sensor from a point source at a given site is described by a Green's function, and a multiple event at the same site will result in the observation of a convolution of this Green's function with the appropriate source-time function (see, for example, STUMP and REINKE, 1988, for a discussion of signal superposition). When we correlate any two signals from these compound events with each other, the resulting correlation coefficient traces are a complicated function of the convolutions between the different source-time functions and, on single-channel correlation coefficient traces, may barely exceed the background level. However, since these source-time functions are identical for each sensor, the stacking operation will combine the traces constructively at the appropriate times as displayed in Figs. 3 and 4, with the greatest improvement for the largest number of sensors.

Between January 2000 and July 2008, a total of 353 events were identified using the waveform correlation procedure. No events were detected in 2000 and, in 2001, the only detections corresponded to a sequence of several events on September 13. Between 2002 and 2008, there have been approximately 60 detections per year. The events take place at all times of year and usually occur clustered in time

(see Fig. 9). The large number of events has enabled us to examine the variability in estimates of travel-time differences and so reduce the variability in location estimates. (This will not of course remove any location bias resulting from inaccuracies in the velocity model.) The ability to extrapolate the detection process back in time has enabled us to identify numerous events in a period between 2002 and 2004 when a temporary network was deployed in the region which has also contributed to a better location estimate. If the exact coordinates of the explosion site should become known at a later stage, the distributions of travel-time measurements obtained here will be able to be used to constrain regional velocity models.

The observation of infrasound signals associated with almost all of these explosions is reason enough to monitor and constrain as tightly as possible the source parameters for these events. This site, despite the uncertain location, can be added to the list of infrasonic sources on the Kola Peninsula compiled by LE PICHON *et al.* (2008) for the purposes of evaluating the detection capability of infrasound arrays. In addition, this source of repeating, well-constrained, seismo-acoustic events provides a superb set of calibration data against which atmospheric propagation models can be validated (see, for example, EVERS and HAAK, 2007). As a first step to the creation of a Ground Truth database for this ongoing sequence of events, we provide the estimated origin times for all events detected during 2004 (Table 3). This table also provides the number of seconds between the origin time and the greatest intensity of infrasound signals at ARCES, or an indication that no infrasound was detected. The explosions in northern Finland described by GIBBONS *et al.* (2007b) have the advantage that the seismic waveforms are essentially identical from event to event, indicating that the variation in source function and yield is very small. The Finnish events have the disadvantage that they are restricted to a one month period in August and September of each year. The Ribachy (Kola) explosions described in the current paper occur throughout the year and the infrasound observations therefore cover a far greater range of atmospheric variability. The ratio between the amplitudes of the infrasound and seismic signals recorded on the seismic sensors of ARCES is usually larger for the explosions discussed here than for the Finnish explosions and this is likely to result, at least in part, from the delay-fired source mechanism.

The observation of the infrasound signals at Lycksele demonstrates that the acoustic signals are observable at far greater distances than the corresponding seismic signals. The similarity of the patterns of infrasound observation for ARCES and Lycksele (see Fig. 9) suggests that the detection or non-detection at both sites is more a function of atmospheric factors than of, for example, local noise or site effects. ARCES and Lycksele, both to the west of the explosion site, detect infrasound for most of the events during summer, and very rarely for events during winter. This is consistent with the conclusions of LE PICHON et al. (2009) who point out that the detectability of infrasound globally is governed primarily by the stratospheric winds.

We emphasize the importance of exploiting the synergies between different monitoring techniques. In Fig. 5, we illustrate a convincing correlation detection on the ARCES array for which no signal is visible in the waveform data in any frequency band. If there is indeed a signal present, the SNR is less than unity. Careful examination of the seismic data at ARCES revealed a weak signal approximately 14 min after the event origin time indicated by the correlation procedure, which propagates over the array with air sound speed and with the correct backazimuth. This infrasound detection on the seismic sensors is sufficient to all but confirm the presence of a small explosion approximately co-located with the master event, despite the absence of seismic observations other than the trigger on the correlation coefficient stack.

Acknowledgments

Maps were created using GMT software (WESSEL and SMITH, 1995). This material is based upon work supported by the Department of Energy (National Nuclear Security Administration) under Award Number DE-FC52-05NA26604. This report was prepared as an account of work sponsored by an agency of the United States Government. Neither the United States Government nor any agency thereof, nor any of their employees, make any warranty, express or implied, or assumes any legal liability or responsibility for the accuracy, completeness, or usefulness of any information, apparatus, product, or process disclosed, or represents that its use would not infringe privately owned rights. Reference herein to any specific commercial product, process, or service by trade name, trademark, manufacturer, or otherwise does not necessarily constitute or imply its endorsement, recommendation, or favoring by the United States Government or any agency thereof. The views and opinions of authors expressed herein do not necessarily state or reflect those of the United States Government or any agency thereof. We are grateful to Professor Ludwik Liszka at the Swedish Institute of Space Physics, Umeå, Sweden, for providing access to the infrasound data from the IRF station network. We thank colleagues at the Kola Regional Seismological Center for providing the requested segments of infrasound data from the mircobarograph array at Apatity, Russia. We acknowledge Tormod Kværna of NORSAR and Gary Steele of TU-Delft for technical assistance and two anonymous referees for very constructive reviews which have significantly improved the paper. We acknowledge the IRIS Data Management Center for waveform data from the KEV station in Finland which is operated by the Institute of Seismology at the University of Helsinki. Copies of NORSAR technical reports can be obtained by contacting the authors or by sending an email to info@norsar.no

REFERENCES

ARROWSMITH, S. J., HEDLIN, M. A. H., ARROWSMITH, M.D., and STUMP, B. W. (2007), Identification of delay-fired mining explosions using seismic arrays: Application to the PDAR array in Wyoming, USA, Bull. Seismol. Soc. Am. 97, 989–1001

ARROWSMITH, S. J., WHITAKER, R., TAYLOR, S. R., BURLACU, R., STUMP, B., HEDLIN, M., RANDALL, G., HAYWARD, C., and REVELLE, D. (2008), Regional monitoring of infrasound events using multiple arrays: Application to Utah and Washington State, Geophys. J. Int. 175, 291–300

BONNER, J. L., PEARSON, D. C., and BLOMBERG, W. S. (2003), Azimuthal variation of short-period Rayleigh waves from cast blasts in northern Arizona, Bull. Seismol. Soc. Am. 93, 724–736

BROWN, D. J., KATZ, C. N., LE BRAS, R., FLANAGAN, M. P., WANG, J., and GAULT, A. K. (2002), Infrasonic signal detection and source location at the prototype International Data Centre, Pure Appl. Geophys. 159, 1081–1125

BUNGUM, H., HOKLAND, B. K., HUSEBYE, E. S., and RINGDAL, F. (1979), *An exceptional intraplate earthquake sequence in Meløy, northern Norway*, Nature *280*, 32–35

CHE, I.-Y., SHIN, J. S., and KANG, I. B. (2009), *Seismo-acoustic location method for small-magnitude surface explosions*, Earth, Planets, and Space *61*, e1–e4

EVERS, L. G. and HAAK, H. W. (2007), *Infrasonic forerunners: Exceptionally fast acoustic phases*, Geophys. Res. Lett. *34*, L10806, doi:10.1029/2007GL029353

FRANKEL, A., HOUGH, S., FRIBERG, P., and BUSBY, R. (1991), *Observations of Loma Prieta aftershocks from a dense array in Sunnyvale, California*, Bull. Seismol. Soc. Am. *81*, 1900–1922

GARCÉS, M. A., HANSEN, R. A., and LINDQUIST, K. G. (1998), *Travel times for infrasonic waves propagating in a stratified atmosphere*, Geophys. J. Int. *135*, 255–263

GELLER, R. J. and MUELLER, C. S. (1980), *Four similar earthquakes in central California*, Geophys. Res. Lett. *7*, 821–824

GIBBONS, S. J. (2006), *On the identification and documentation of timing errors: An example at the KBS station, Spitsbergen*, Seism. Res. Lett. *77*, 559–571

GIBBONS, S. J. and RINGDAL, F. (2006), *The detection of low magnitude seismic events using array-based waveform correlation*, Geophys. J. Int. *165*, 149–166

GIBBONS, S. J. and RINGDAL, F. (2007), *A Case Study of Seismic Event Identification: Explosions in NW Russia using the ARCES seismic array*, NORSAR Scientific Report: Semiannual Technical Summary No. 1-2007, NORSAR, Kjeller, Norway. 73–81

GIBBONS, S. J., KVÆRNA, T., and RINGDAL, F. (2005), *Monitoring of seismic events from a specific source region using a single regional array: A case study*, J. Seismol. *9*, 277–294

GIBBONS, S. J., BØTTGER SØRENSEN, M., HARRIS, D. B., and RINGDAL, F. (2007a), *The detection and location of low magnitude earthquakes in northern Norway using multi-channel waveform correlation at regional distances*, Phys. Earth Planet. Inter. *160*, 285–309

GIBBONS, S. J., RINGDAL, F., and KVÆRNA, T. (2007b), *Joint seismic-infrasonic processing of recordings from a repeating source of atmospheric explosions*, J. Acoust. Soc. Am. *122*, EL158–EL164

GIBBONS, S. J., KVÆRNA, T., and RINGDAL, F. (2009), *Considerations in phase estimation and event location using small-aperture regional seismic arrays*, Pure Appl. Geophys., doi:10.1007/s00024-009-0024-1

HARRIS, D. B. (1991), *A waveform correlation method for identifying quarry explosions*, Bull. Seism. Soc. Am. *81*, 2395–2418

HICKS, E. C., KVÆRNA, T., MYKKELTVEIT, S., SCHWEITZER, J., and RINGDAL, F. (2004), *Travel-times and attenuation relations for regional phases in the Barents Sea region*, Pure Appl. Geophys. *161*, 1–19

ISRAELSSON, H. (1990), *Correlation of waveforms from closely spaced regional events*, Bull. Seismol. Soc. Am. *80*, 2177–2193

KENNETT, B. L. N., *The Seismic Wavefield*. Volume II: Interpretation of Seismograms on Regional and Global Scales. (Cambridge University Press, Cambridge, 2002)

KENNETT, B. L. N., ENGDAHL, E. R., and BULAND, R. (1995), *Constraints on seismic velocities in the Earth from travel times*, Geophys, J. Int. *122*, 108–124

KIM, W.-Y., AHARONIAN, V., LERNER-LAM, A. L., and RICHARDS, P. G. (1997), *Discrimination of earthquakes and explosions in southern Russia using regional high-frequency three-component data from the IRIS/JSP Caucasus Network*, Bull. Seismol. Soc. Am. *87*, 569–588

KVÆRNA, T. and RINGDAL, F. (1986), *Stability of various f-k estimation techniques*, NORSAR Scientific Report: Semiannual Technical Summary No. 1-1986/1987, NORSAR, Kjeller, Norway, pp. 29–40

LE PICHON, A., VERGOZ, J., HENRY, P., and CERANNA, L. (2008), *Analyzing the detection capability of infrasound arrays in Central Europe*, J. Geophys. Res. *113*, D12115, doi:10.1029/2007JD009509

LE PICHON, A., VERGOZ, J., BLANC, E., GUILBERT, J., CERANNA, L., EVERS, L., and BRACHET, N. (2009), *Assessing the performance of the International Monitoring System's infrasound network: Geographical coverage and temporal variabilities*, J. Geophys. Res. *114*, D08112, doi:10.1029/2008JD010907

MACCARTHY, J., HARTSE, H., GREENE, M., and ROWE, C. (2008), *Using waveform cross-correlation and satellite imagery to identify repeating mine blasts in Eastern Kazakhstan*, Seism. Res. Lett. *79*, 393–399

MAERCKLIN, N., MYKKELTVEIT, S., SCHWEITZER, J., ROCK, D., and HARRIS, D. B. (2005), *Data from deployment of temporary seismic stations in northern Norway and Finland*, NORSAR Scientific Report: Semiannual Technical Summary No. 1-2005, NORSAR, Kjeller, Norway, pp. 77–85

MCKENNA, M. H., STUMP, B. W., HAYEK, S., MCKENNA, J. R., and STANTON, T. R. (2007), *Tele-infrasonic studies of hard-rock mining explosions*, J. Acoust. Soc. Am. *122*, 97–106

MCLAUGHLIN, K. L., BONNER, J. L., and BARKER, T. (2004), *Seismic source mechanisms for quarry blasts: Modelling observed Rayleigh and Love wave patterns from a Texas quarry*, Geophys. J. Int. *156*, 79–93

MYKKELTVEIT, S. and RINGDAL, F., Phase identification and event location at regional distances using small-aperture array data. In *Identification of Seismic Sources—Earthquake or Underground Explosions* (eds. E. S. Husebye and S. Mykkeltveit) (Reidel Publishing Company, 1981), pp. 467–481

RICHARDS, P. G., WALDHAUSER, F., SCHAFF, D., and KIM, W.-Y. (2006), *The applicability of modern methods of earthquake location*, Pure Appl. Geophys. *163*, 351–372

RINGDAL, F. and KVÆRNA, T. (1989), *A multi-channel processing approach to real time network detection, phase association, and threshold monitoring*, Bull. Seismol. Soc. Am. *79*, 1927–1940

RINGDAL, F., GIBBONS, S. J., KVÆRNA, T., ASMING, V., VINOGRADOV, Y., MYKKELTVEIT, S., and SCHWEITZER, J., Research in regional seismic monitoring. In *Proc. 27th Seismic Res. Rev., Rancho Mirage, California, September 20–22, 2005. Ground-Based Nuclear Explosion Monitoring Technologies*, LA-UR-05-6407 (2005), pp. 423–432

RINGDAL, F., GIBBONS, S. J., and HARRIS, D. B., Adaptive waveform correlation detectors for arrays: Algorithms for autonomous calibration. In *Proc. 30th Monitoring Research Review, Portsmouth, Virginia, September 23–25, 2008. Ground-Based Nuclear Explosion Monitoring Technologies*, LA-UR-08-05261 (2008), pp. 465–474

SCHAFF, D. P. (2008), *Semiempirical statistics of correlation-detector performance*, Bull. Seismol. Soc. Am. *98*, 1495–1507

SCHULTE-THEIS, H. and JOSWIG, M. (1993), *Clustering and location of mining induced seismicity in the Ruhr Basin by automated Master Event Comparison based on Dynamic Waveform Matching (DWM)*, Computers and Geosci. *19*, 233–241

SCHWEITZER, J. (2001), *HYPOSAT-An enhanced routine to locate seismic events*, Pure Appl. Geophys. *158*, 277–289

SHELLY, D. P., BEROZA, G. C., and IDE, S. (2007), *Non-volcanic tremor and low-frequency earthquake swarms*, Nature 446, 305–307

SKORVE, J., *The Kola satellite image atlas: Perspectives on arms control and environmental protection*, Security Policy Library (The Norwegian Atlantic Committee, 1991). ISBN 82-90161-37-9

STEVENS, J. L., GIBBONS, S., RIMER, N., XU, H., LINDHOLM, C., RINGDAL, F., KVAERNA, T., and MURPHY, J. R. (2006), *Analysis and simulation of chemical explosions in nonspherical cavities in granite*, J. Geophys. Res. *111*, B04306, doi:10.1029/2005JB003768

STORCHAK, D. A., SCHWEITZER, J., and BORMANN, P. (2003), *The IASPEI standard seismic phase list*, Seismolog. Res. Lett. *74*, 761–772

STUMP, B. W. and REINKE, R. E. (1988), *Experimental confirmation of superposition from small-scale explosions*, Bull. Seismol. Soc. Am. *78*, 1059–1073

SZUBERLA, C. A. L. and OLSON, J. V. (2004), *Uncertainties associated with parameter estimation in atmospheric infrasound arrays*, J. Acoust. Soc. Am. *115*, 253–258

THOMSON, D. J. (1982), *Spectrum estimation and harmonic analysis*, Proc. IEEE *70*, 1055–1096

WESSEL, P. and SMITH, W. H. F. (1995), *New version of the Generic Mapping Tools*, EOS Trans. Am. Geophys. Union *76*, 329

(Received November 20, 2008, revised August 7, 2009, accepted August 18, 2009, Published online December 15, 2009)

Pure Appl. Geophys. 167 (2010), 437–453
© 2010 US Government
DOI 10.1007/s00024-010-0080-6

| Pure and Applied Geophysics

The Temporal Morphology of Infrasound Propagation

Douglas P. Drob,[1] Milton Garcés,[2] Michael Hedlin,[3] and Nicolas Brachet[4]

Abstract—Expert knowledge suggests that the performance of automated infrasound event association and source location algorithms could be greatly improved by the ability to continually update station travel-time curves to properly account for the hourly, daily, and seasonal changes of the atmospheric state. With the goal of reducing false alarm rates and improving network detection capability we endeavor to develop, validate, and integrate this capability into infrasound processing operations at the International Data Centre of the Comprehensive Nuclear Test-Ban Treaty Organization. Numerous studies have demonstrated that incorporation of hybrid ground-to-space (G2S) enviromental specifications in numerical calculations of infrasound signal travel time and azimuth deviation yields significantly improved results over that of climatological atmospheric specifications, specifically for tropospheric and stratospheric modes. A robust infrastructure currently exists to generate hybrid G2S vector spherical harmonic coefficients, based on existing operational and emperical models on a real-time basis (every 3- to 6-hours) (DROB *et al.*, 2003). Thus the next requirement in this endeavor is to refine numerical procedures to calculate infrasound propagation characteristics for robust automatic infrasound arrival identification and network detection, location, and characterization algorithms. We present results from a new code that integrates the local (range-independent) τp ray equations to provide travel time, range, turning point, and azimuth deviation for any location on the globe given a G2S vector spherical harmonic coefficient set. The code employs an accurate numerical technique capable of handling square-root singularities. We investigate the seasonal variability of propagation characteristics over a five-year time series for two different stations within the International Monitoring System with the aim of understanding the capabilities of current working knowledge of the atmosphere and infrasound propagation models. The statistical behaviors or occurrence frequency of various propagation configurations are discussed. Representative examples of some of these propagation configuration states are also shown.

Key words: Infrasound, atmospheric variability, climatology, automated event detection, source location, CTBTO, IDC, IMS.

1. Background

The purpose of the automated infrasound processing developed at the IDC is to detect coherent signals measured on individual IMS sensors (Christie *et al.*, 2001), highlight the most significant detections as "phases" (as opposed to "noise"), and subsequently group these phases to form and locate hypocenters, so-called "events". The phases are determined using the progressive multi-channel correlation (PMCC) method (Cansi, 1995) which distinguishes the coherent signals produced by natural and man-made sources from incoherent ambient background noise which may also be of natural, cultural, or instrumental origin. A wide variety of sources are regularly recorded worldwide by the IMS network; ocean activity, mountain associated waves, volcanic eruptions, earthquakes, thunderstorms, meteors, avalanches, auroras, rocket launches and re-entries, aircraft, mine-blasts, accidental explosions, and industrial noise. It is important for the IDC to detect, locate, and categorize these sources to contrast with nuclear explosions; the task of the organization.

The detection, location, and characterization algorithms (henceforth DLC) described by Brown *et al.*, (2002a) may be used to locate the terminal burst point of exploding meteors, the origin time of volcanic eruptions, and the location of avalanches and rock slides, as well other null sources relevant to CTBTO operations (Le Pichon *et al.*, 2008b; Hedlin *et al.*, 2002). Although the various natural events represent false alarms for the CTBTO, they also provide valuable ground-truth information that can be

[1] Space Science Division, Naval Research Laboratory, Washington, DC, USA. E-mail: douglas.drob@nrl.navy.mil
[2] Infrasound Laboratory, University of Hawaii, Manoa, USA. E-mail: milton@isla.hawaii.edu
[3] Laboratory for Atmospheric Acoustics, University of California, San Diego, USA.
[4] International Data Centre, Provisional Technical Secretariat, CTBTO, Vienna, Austria.

used to constantly fine-tune and check the integrity of the system, insuring verifiability of the treaty. For example, terminal bursts of meteors with an average yield of one kiloton occur in the earth's atmosphere several times per year (NEMTCHINOV et al., 1997; BROWN et al., 2002b).

To facilitate rapid computation, today's automated DLC algorithms rely upon precompiled station travel-time information (BROWN et al., 2002a). The precompiled information, which is typically average propagation velocity and azimuth deviation, describes how an observed signal was affected by the background environment on its way from source to receiver. The application of this knowledge provides improved source location and signal association estimates.

Our objective is to reduce IDC false alarm rates and improve detection capability by reducing the sources of uncertainty in the existing model physics and DLC methodologies. Analyses of ground-truth events have shown that observationally constrained atmospheric specifications are superior to average climatology (LE PICHON et al., 2002, 2005; HERRIN et al., 2006). This is particularly true for tropospheric and stratospheric propagation for which there are several global operational numerical weather prediction systems such as the NOAA Global Forecast System (KALNAY et al., 1990) and ECMWF (COURTIER et al., 1998; BECHTOLD et al., 2008). Unfortunately, these systems are currently limited in their altitude extent due to the unavailability of routine operational satellite observations above approximately 80 km. The ground-to-space (G2S) environmental specification system was therefore developed to provide a compact numerical weather prediction post-processor and infrasound propagation calculation preprocessor (DROB et al., 2003) to account for all altitudes pertinent to infrasound propagation. The system serves as a placeholder until operational numerical weather prediction models and data sets include the lower thermosphere.

In addition to requiring up-to-date knowledge about the atmospheric state for improving DLC algorithms, an acoustic wave propagation model is required to compute how observed infrasound signals relate back to their source. In discrete inverse theory (MENKE, 1989) this is known as a forward model. There are a number of propagation modeling

techniques available such as ray tracing (GOSSARD and HOOKE, 1975), parabolic equations (LINGEVITCH et al., 2002), and normal modes (PIERCE, 1967). Unfortunately with detailed physics comes greater complexity. The approach must not be so primitive that any value added from near-real-time atmospheric specification has no meaningful influence; in turn, the technique must not be so complicated that implementation is impractical in automated DLC algorithms. The data and procedures must be readily available and simple enough to integrate into operational monitoring systems.

2. The τp Equations

At present the τp equations of GARCES et al., (1998) provide a good balance of simplicity and geophysical information content for automated DLC algorithms, particularly in conjunction with near-real-time atmospheric specifications. In general, the acoustic ray-tracing approach represents the propagation or translation and rotation of an acoustic wavefront through space and time. The τp equations are an expression of the Eikonal ray-tracing equations (LIGHTHILL, 1978; GOSSARD and HOOKE, 1975) in integral form with the approximations of range independence and no vertical wind. For this, each ray or wavefront element can be uniquely represented by an invariant ray parameter (p),

$$p = \frac{k_z}{c_o}\left(1 + \frac{k_z u_o}{c_o}\right)^{-1}, \qquad (1)$$

which depends on the static sound speed at the receiver (c_o), the vertical wave number $k_z = \sin(q)$ where q is elevation angle, and u_o the horizontal wind velocity along the direction of propagation at the receiver. This ray parameter is also the reciprocal of the intrinsic horizontal phase velocity of the wave $V_\theta = 1/p$, therefore;

$$V_\theta = \frac{c_o}{k_z}\left(1 - \frac{k_z}{c_o}\right). \qquad (2)$$

The equation for the along track range travelled between bounces, i.e., propagation from the bottom of the atmospheric duct to the top and back down again in a phase loop is

$$R(z,p) = 2 \int_{z_o}^{z(p)} \psi(z,p) \left[\frac{p}{(1-u(z)p)} + u(z)\zeta(z) \right] dz,$$

(3)

where z_o is the lower limit of integration (typically zero or the surface altitude) and $z(p)$ is the upper limit, which is the first root above z_o of the characteristic function.

$$\psi(z,p) = \left[\zeta(z) - \frac{p^2}{(1-u(z)p)^2} \right]^{-1/2}.$$

(4)

This root represents the turning point of the ray following from classical WKB ray theory which states that a ray will turn when its horizontal phase velocity (V_θ) matches that of the background effective sound speed, $c(z) + u(z)$ where these are the adiabatic sound velocity and horizontal wind speed along the direction of propagation, respectively. The infrasound propagation characteristics in Eqs. 3 and 4 are a function of the local vertical profiles of $\zeta(z) = 1/c^2(z)$ and $u(z)$ as well. The corresponding travel time (T) for a phase loop is similarly

$$T(z,p) = 2 \int_{z_o}^{z(p)} \psi(z,p)\zeta(z)dz.$$

(5)

The celerity (V), or average group velocity from the source to the receiver is simply $V = R/T$. Lastly, the apparent azimuth deviation is computed as $\Omega = \arctan^{-1}(Q/R)$ where the transverse offset (Q) for a phase loop is

$$Q(z,p) = \int_{z_o}^{z(p)} \psi(z,p)\zeta(z)v(z)dz,$$

(6)

where $v(z)$ is the horizontal wind component transverse to the direction of propagation. Throughout this paper, all of the results are computed in the frame of reference of the receiver as opposed to the source by simply reversing the sign of the wind fields. It can be shown that this is also equivalent to integrating the Eikonal ray equations in differential form with a negative time step.

A discussion of the methodology for the estimation of the eigenrays associating a received signal with a given source is beyond the scope of the present work. Relevant, however, is the fact that for a specific eigenray the measured azimuth deviation is an apparent effect similar to that of an airplane yawing in a crosswind in order to maintain a constant bearing. The acoustic wavefront must be skewed at some angle Ω with respect to the great circle path in order to offset the net lateral advection from transverse wind components; in other words so that the net transverse offset at the top and bottom of the phase loop are zero. Although the total path does not deviate from the true great circle path between source and receiver, in the presence of transverse winds the wavefront can depart from the great circle path anywhere else in the phase loop. Over one range of altitudes a crosswind may push the ray off the great circle path, while at other altitudes a crosswind in the opposite direction may push the ray back onto the great circle path. Furthermore, the ray may deviate to one side of the path on the up leg and the other side of the path on the down leg. This is also an important factor that needs to be considered when signals are observed from a source at altitude such as a bolide. Care must be taken in automated and interactive DLC algorithms when applying calculated values of Ω to correct the array observations for apparent azimuth deviation. If topography and atmospheric range variations are included then the actual path can be even more complicated, however the overall characteristics remain the same.

The crux of solving the τp integrals accurately is the treatment of the inverse square-root singularity in the function $\psi(z)$ at the upper limit of the domain. At any point in the domain (i.e. for any given p) the root is calculated by first bracketing it with a grid search and then applying Brent's method to approximate the root to a high degree of accuracy (PRESS, 1989). Knowing the location of the square-root singularity $z(p)$, Eqs. (3)–(6) can then be integrated with a Romberg method employing a modified midpoint rule that can handle inverse square-root singularities at the upper bound (PRESS, 1989). For all of the numerical results presented, continuous functional values for $u(z)$, $w(z)$, $c(z)$, and $\zeta(z)$ are calculated by cubic spline interpolation of gridded G2S values with $\Delta z = 125$ m. Note that the calculation of vertical derivatives is not required to integrate Eqs. (3)–(6).

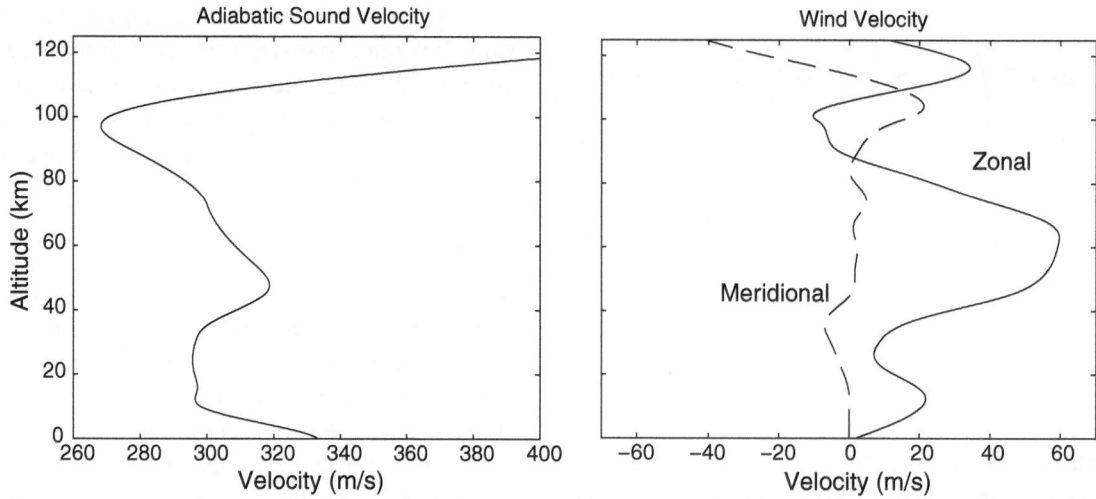

Figure 1
Climatological profiles; adiabatic sound velocity (*left*), zonal wind (*right solid*), and meridional wind (*right, dashed*) from the HWM93/MSISE-00 empirical models (HEDIN *et al.*, 1996; PICONE *et al.*, 2002) for January 1, 2005 at 00:00 UT

These integrals may also be evaluated for sources at altitude such as bolides by including a second term for the fractional part of the propagation phase loop where the limits of integration have been adjusted accordingly.

Two factors are relevant to current IDC DLC processing with respect to signals generated by sources well above the ground where the expected number of detectable phases can double at distant recording stations owing to the fact that a source will have both upward and downward directed acoustic components. First, current automated procedures focus on picking the onset time of the fastest infrasound arrival, and when and where closely spaced multiple arrivals exist, they tend to be averaged into a single characteristic by the parameters of the automated PMCC feature extraction algorithms. Secondly, progress has been made to demonstrate that it is possible to estimate hypocenter altitude at the level of interactive IDC analyst expert review, where manual identification of tropospheric, stratospheric, and thermospheric phases is possible; and in particular in conjunction with additional information from detections by the seismic, and auxiliary seismic components of the IMS network (EDWARDS and HILDEBRAND, 2004; ARROWSMITH *et al.*, 2007; LE PICHON *et al.*, 2008a). As experience progresses, it will eventually be possible to implement these considerations into automated IDC processing, however

as is the case in seismology, hypocenters depths are difficult to compute without a dense local network or specific depth phases. Automated hypocenter height estimates via infrasound will thus likely only be approximate until an IDC analyst can refine them.

Example climatological profiles from the HWM93/MSISE-00 empirical models (HEDIN *et al.*, 1996; PICONE *et al.*, 2002) for January 1, 2005 at 00:00 UT for a typical northern hemisphere mid-latitude station (I56US) are shown in Fig. 1. Illustrative τp calculations corresponding to these example profiles are shown in Fig. 2. These calculations are performed over all observable azimuths for elevation angles from 0° to 35°. The horizontal phase velocity (V_θ) of the parameter space over the domain is shown in panel A. Recall that this is only a function of the atmospheric conditions at the detector. Panel B shows a contour plot of the effective sound velocity as a function of altitude (z) and backazimuth. The wintertime stratospheric wind jet near 55 km can be observed, with winds toward the detector at −90°, and away from the detector at +90°. The turning points of all incoming rays at elevation angles of 5°, 10°, and 15° are also indicated. The corresponding turning heights of the rays, over the entire domain, is shown in panel C. Two predominant ducts are present at this time and location, the thermospheric duct for all backazimuths and the eastward stratospheric duct from the wintertime stratospheric zonal wind jet. The

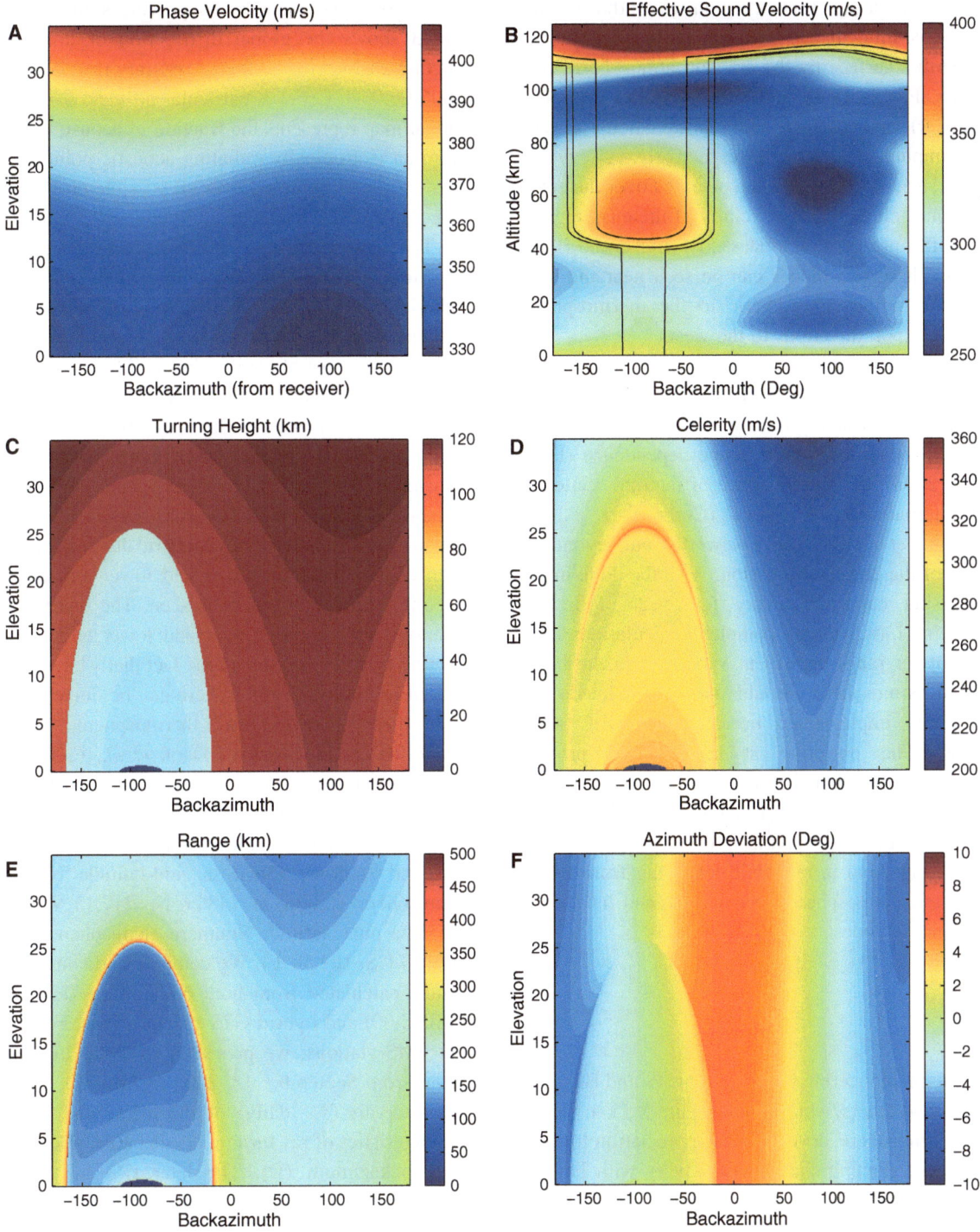

Figure 2

τp ducting characteristics for I56US (48.2640°N, 117.1257°W) on January 1, 2006 0:00 UTC as function of backazimuth at the detector. Panel A show the horizontal phase velocity for all elevations from 0° to 35°, Panel B shows the effective sound speed velocity in the direction of propagation, Panel C the ray turning height, panel D the celerity, and E the range to first bounce, and F the apparent azimuth deviation. The contour lines in panel B indicate the turning height of rays launched at elevation angles of 5°, 10°, and 15°, respectively dark blue area represents a region where the acoustic energy is immediately refracted downward toward the earth's surface by the ambient atmospheric conditions. Panels D, E, and F show the celerity (V), range (R), and apparent azimuth deviation (Ω) over the computational domain, respectively. A very stable and accurate result can be obtained over the entire domain, including in the vicinity of the cusp regions where fast propagation modes exist (EVERS and HAAK, 2007). These fast modes can be seen near the transition from the stratospheric to the thermospheric ducts for look directions to the west.

Several limitations resulting from the various approximations in the τp method such as the shortcomings of linear ray-tracing theory (MILLET et al., 2007), the lack of explicit range dependence (DROB et al., 2003), and the influence of internal scattering by internal gravity waves (CHUNCHUZOV, 2004; OSTASHEV et al., 2005) are noteworthy but beyond the scope of the present discussion. With the doubling of processing capacities every few years following Moore's Law, more complex calculations that account for range-dependent variations in the background atmosphere should be investigated and eventually implemented in automated DLC algorithms. The objective at hand is to provide geophysical insight to demonstrate that the utilization of climatological travel times in DLC algorithms is at best outdated in comparison to the possibility of calculating them in real time following from recent progress in atmospheric specification and infrasound propagation codes.

3. Results

Classic pioneering work by GEORGES and BEASLEY, (1977) and others, which relied on limited knowledge of the atmosphere, developed an appreciation for how infrasound ducting characteristics vary with latitude over the year. With more recent information from thirty-five years of satellite and ground-based atmospheric wind and temperature measurements, DROB et al., (2003) investigated how infrasound propagation characteristics varied over the globe at a given universal time; in particular, how acoustic energy is partitioned between the troposphere, stratospheric,

and thermosphere ducts. To provide some context, Fig. 3 shows the global distribution of infrasound ducting characteristics for an arbitrary time of 05/24/2006 00:00 UT. For a particular altitude level, each global map represents the fraction of acoustic energy from an isotropically radiating acoustic point source on the ground, summed over all possible propagation directions (see DROB et al., 2003).

The regions where tropospheric ducting occurs are shown in the top panel. Ducting along the tropospheric jet stream can also be seen in the Southern Hemisphere. Marine inversion layers also occurred off the west coast of California and Africa, as well as near the Korean Peninsula. Stratospheric ducting (middle panel) is seen in the Northern And Southern Hemisphere mid-latitude regions but is absent in the equatorial regions. The lower panel shows the remaining thermospheric ducting fractions. Interesting correlations in the thermospheric ducting fractions with continental landmasses and lower atmospheric ducting fractions can also be seen. The inverse correlations between the upper and lower atmospheric ducting fractions are due to the fact that what was not ducted in the lower atmosphere can be ducted in the upper mesosphere and lower thermosphere.

Following the work of DROB et al., (2003), we now present several case studies based on the calculation of a multiyear time series of infrasound propagation characteristics for two of the IMS infrasound stations; I56US a mid-latitude Northern Hemisphere station at (48.26°N, 117.13°W), and I55US a polar latitude Southern Hemisphere station at (77.74°S, 167.58°E). We compare and contrast the results calculated from both climatology (HWM93/MSISE-00) and hybrid G2S specifications. For these two IMS stations, we present a five-year long-time series from September 13, 2002 to April 30, 2007 at 6 h intervals (4× daily) of the infrasound ducting characteristics of ray turning heights $z(p)$, celerity (V) and backazimuth (Ω). The later two have direct application in infrasound DLC algorithms.

In the detection algorithms described by BROWN et al., (2002a) currently in use at the IDC, backazimuths receive a slightly greater statistical emphasis (1.0) as compared to travel times (0.8) in the calculation of a metric (Σ) for the trigging of an automatic event ($\Sigma > 3.55$) and Reviewed Event Bulletin

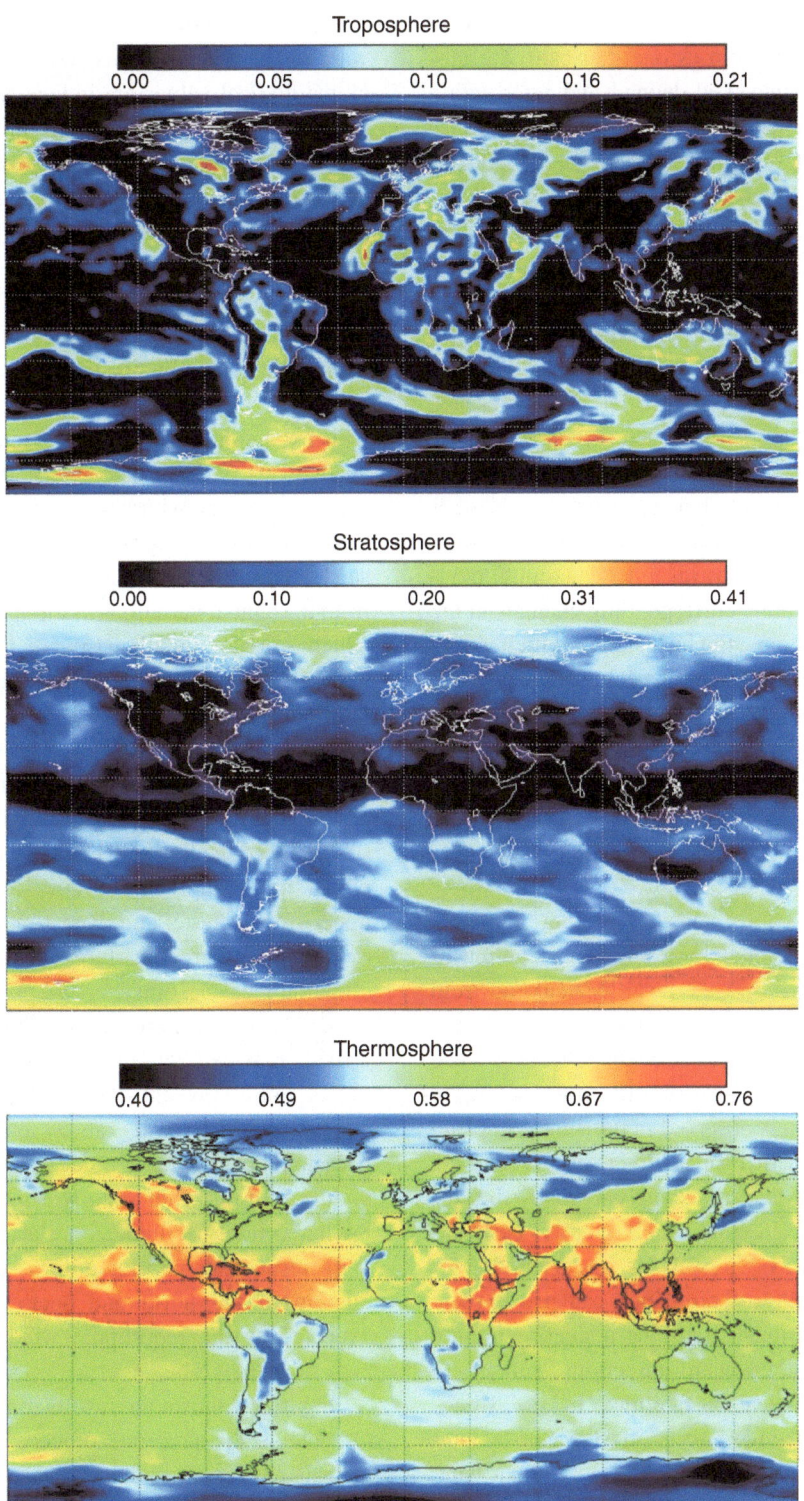

Figure 3
Tropospheric, stratospheric, and thermospheric infrasound ducting fractions for 05/24/2006 00:00 UT

(REB) ($\Sigma > 4.6$). This detection criteria effectively defines a significant event as one that can be established by at least two well-defined and intersecting back azimuths for which the associated travel times do not also violate causality (BROWN *et al.,* 2002a). More recently, a novel detection scheme was developed by ARROWSMITH *et al.,* (2008) that dynamically adjusts network detection thresholds in real time to account for the presence of correlated and varying background noise. Furthermore, ARROWSMITH *et al.,* (2008) demonstrated that the new algorithm has excellent performance characteristics in the presence of clutter, suggesting the approach provides a viable means to reduce the number of false alarms that need to be reviewed by a human analysis. Neither approach currently accounts for the hourly, daily, or seasonal changes of the travel time or azimuth deviation resulting from the corresponding changes in the atmospheric conditions; the inclusion of which would further allow for a more accurate calculation of the Σ metric thus improving the network sensitivity and reducing the number of false alarms.

4. I56US

Figure 4 shows the computed turning points of infrasound for arrivals at I56US for an elevation of 5°. The top panel shows the results calculated from the climatology and the lower panel the results calculated with the hybrid G2S specification. The alternating seasonal pattern, where eastward stratospheric propagation is observed in the wintertime and westward stratospheric propagation is observed in the summer time, is evident. Ducting caused by the tropospheric jet stream for predominantly eastward propagating arrivals, as well as occasionally for northward and southward directions, can also be seen. Furthermore, occasional stratospheric ducting in both the westward and eastward directions, related to global-scale dynamical instabilities in the stratosphere, can sometimes occur during the winter months. As would be expected but not shown here, the corresponding results for lower elevation angles exhibit more tropospheric and stratospheric ducting for lower incoming elevations and less for large elevation angles (more thermospheric ducting).

Figure 5a shows the azimuth deviation for westward arrivals from the five-year time series at I56US; climatological values are indicated in red and results from the hybrid specifications in blue. There is an average scatter in the hybrid specifications of about ±2°, on par with the climatological predictions, plus occasional excursions of up to ±4° during the winter months. The four interleaved bands in the climatologically predicted variations result from the different local times under the influence of the solar migrating tides as described by GARCES *et al.,* (1998). Figure 5b shows the azimuth deviation for southward arrivals with excursions up to 10° in January 2003, and on average up to 7.5° during wintertime. In addition, there is an asymmetry with respect to the summer months with deviations of up to −3°, which tend to be more stable. These wide ranging azimuth deviations result from the annual variations of the stratospheric wind jet which is predominantly eastward, lower, stronger, and variable in the wintertime, as compared to the summertime jet which is westward, higher, and stable.

Figure 6a shows a time series of celerity for I56US for eastward arrivals again at 5° elevation, calculated with hybrid G2S and empirical atmospheric specifications. A band of arrivals at 340 m/s, which are comprised of both lower tropospheric, upper tropospheric, and even stratospheric modes, is evident. Random departures of up to 30 m/s from climatological estimates and seasonal variations occur during wintertime for the other branch of arrivals between 250 and 320 m/s.

Figure 6b presents the comparison of celerity for all southward arrivals at 5° elevation. Note the occasional tropospheric modes (330 m/s) with a half-width of 20 m/s, including seasonal oscillations. The predicted tidal oscillations are also more significant. With respect to the climatology, lower atmospheric ducting to the north and south are generally not expected as the meridional wind fields average to zero over the globe.

Figure 6c shows the results for westward arrivals. Of note is the presence of occasional tropospheric arrivals (340 m/s) with clear seasonal variability. If not properly accounted for (i.e. given the appropriate statistical weighting) these could result in spurious associations and poor source localizations. The results also show that there is pronounced annual variability

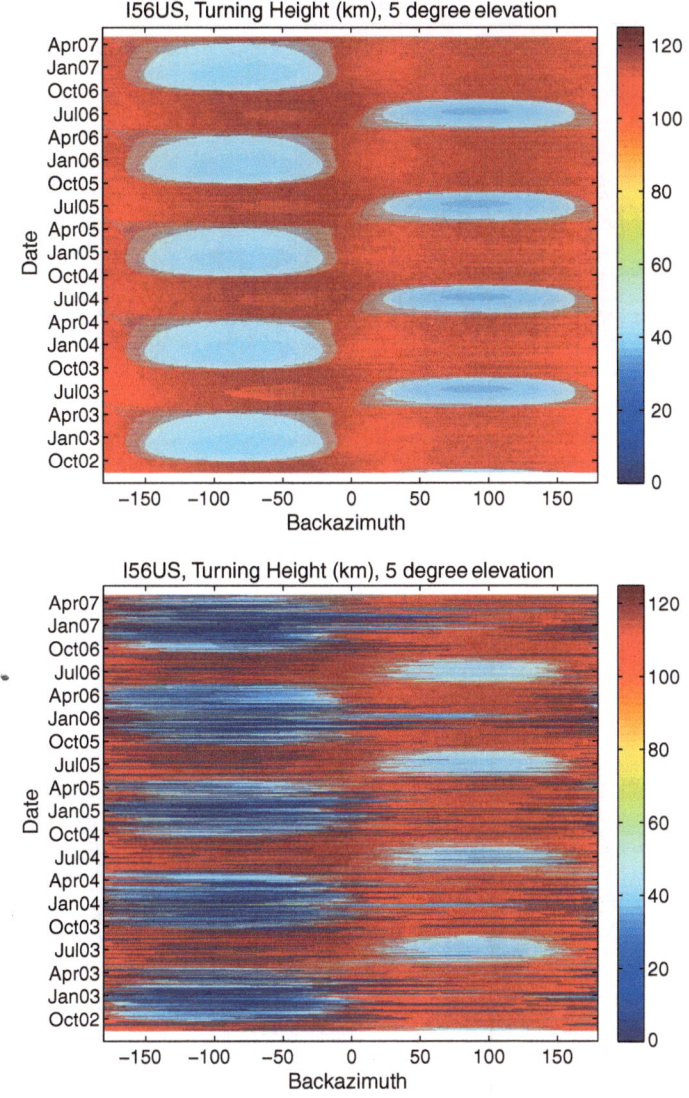

Figure 4
The turning height for all rays at I56US that enter the receiver at an elevation angle of 5°. The upper panel shows the results calculated from the HWM/MSIS climatology and the lower panel the results from the hybrid G2S atmospheric specifications

with stable stratospheric modes in the summer time, transitioning to thermospheric modes in the wintertime as was shown in Fig. 4. The existence of sporadic stratospheric modes occurring in both the eastward and westward directions in late winter are associated with the dynamical instability of the stratospheric wind jet driven by vertically propagating planetary waves. Disturbances associated with sudden stratospheric warmings (MANNEY *et al.,* 2008) can even result in prolonged intervals of westward winds in the stratosphere during the wintertime.

5. I55US

The second set of illustrative examples is for the polar Southern Hemisphere station I55US. The comparison of turning points calculated with climatology and the hybrid specifications for rays that will enter the detector at 5° elevation, as a function of backazimuth and time, are shown in Fig. 7. While the overall morphology of the climatological and hybrid specifications for the mid-latitude I56US station is generally similar in Fig. 4, this is not the case for

73

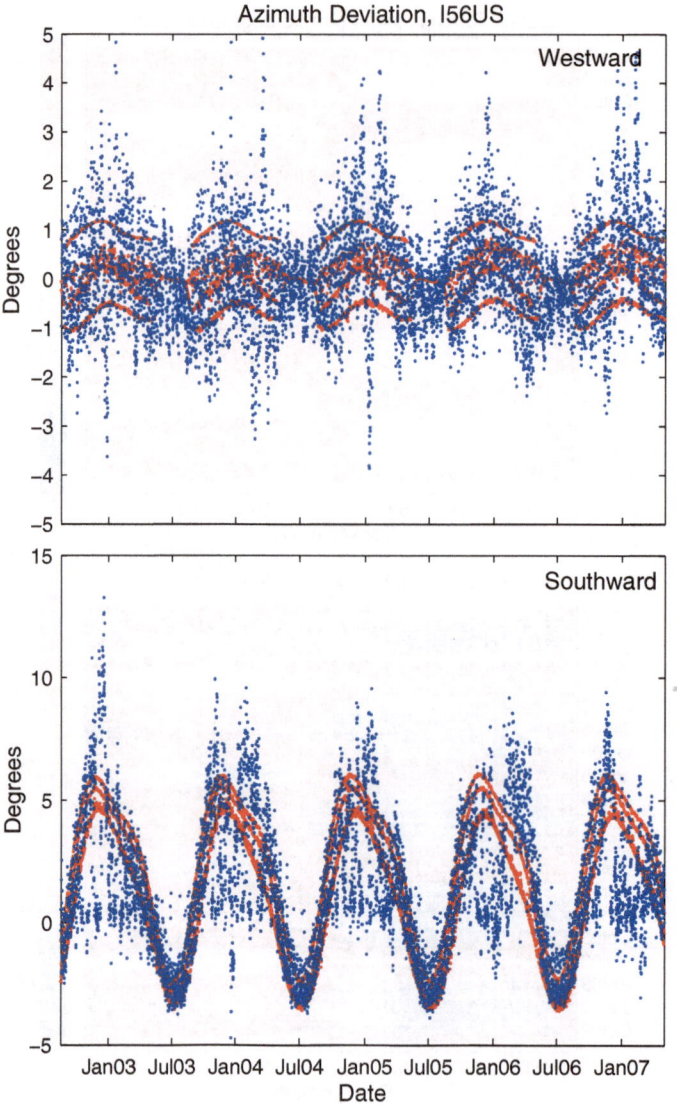

Figure 5
Time series of azimuth deviation of westward (*top*) and southward (*bottom*) arrival for I56US at an elevation angle of 5°. Predictions using climatology are in red. Predictions using the hybrid (G2S) model are in blue

I55US. Again, the day-to-day variability, whether for a tropospheric, stratospheric, or thermospheric duct, is more pronounced in the real atmosphere (G2S hybrid) than the calculations with the monthly average climatology would imply. From these examples it should be obvious that climatology does not accurately predict the occurrence of tropospheric and even stratospheric ducting in the region.

The time series of computed celerity for westward arrivals is shown in Fig. 8a. Annual variations of hybrid G2S characteristics follow the climatology of

the predominate stratospheric and thermospheric modes reasonably well, but not perfectly. Like for the Northern Hemisphere mid-latitudes, greater variability exists in the computed propagation characteristics during wintertime. A predominant but transient lower atmospheric mode, disappearing in summer, with an

Figure 6 ▶
Time series of the celerity for 5° elevation arrivals at I56US for eastward, southward, and westward directions, respectively. The color coding is as described in the caption for Fig. 5

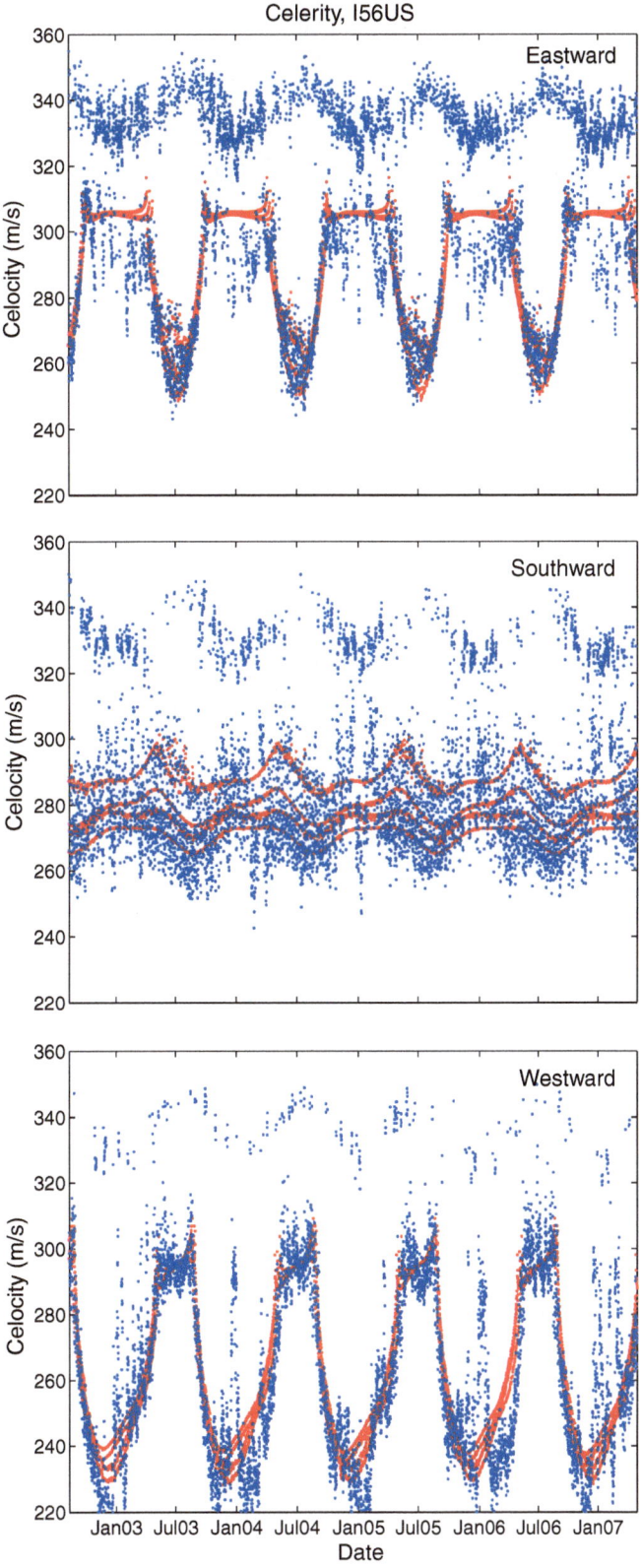

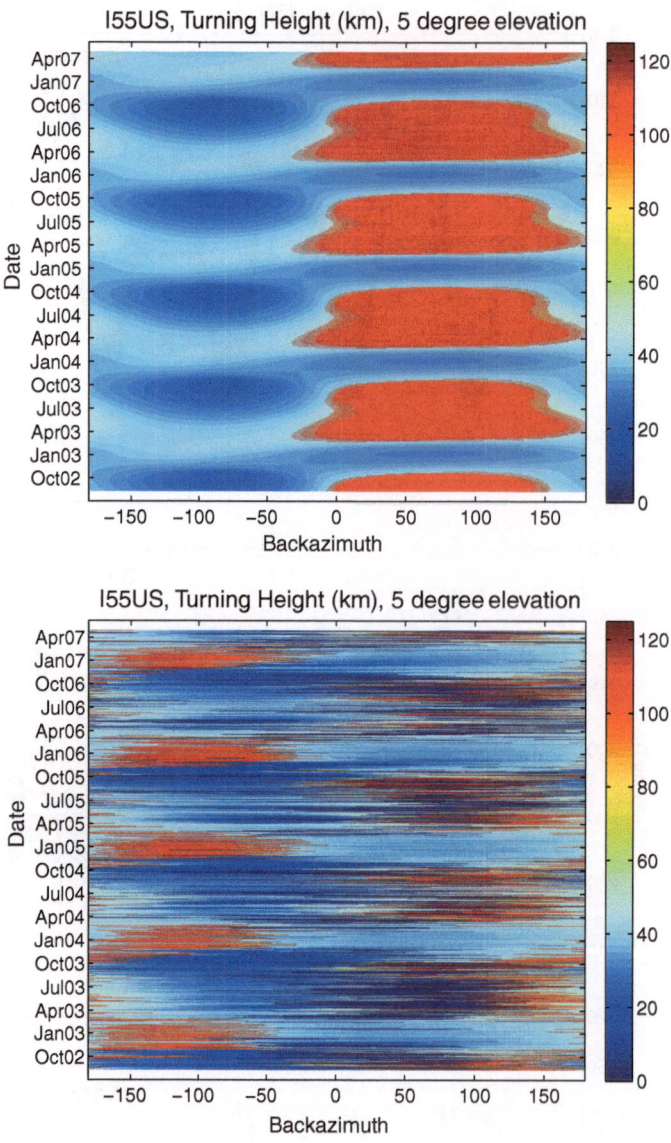

Figure 7

The turning height for all rays at I55US that enter the receiver at an elevation angle of 5°. The upper panel shows the results calculated from the HWM/MSIS climatology and the lower panel the results from the hybrid G2S atmospheric specifications

average celerity of about 310 m/s is also present. The 340 m/s celerities observed at I56US (Fig. 6) are generally not be observed at I55US as the polar troposphere is colder and the station is too far poleward to be influenced by the tropospheric jet stream. Figure 8b shows the results for southward arrivals, which vary from 260 to 325 m/s, again exhibiting significant departures from climatological predictions.

Lastly, we consider the azimuth deviations for IMS station I55US. The time series of southward arrivals is shown in Fig. 9a, for which there are asymmetric seasonal variations with occasional sporadic excursions of over 10°, and up to 7.5° on average. Significant local-time (tidal) variations of the thermospheric modes are again present in the climatology. The results shown in Fig. 9b for the westward

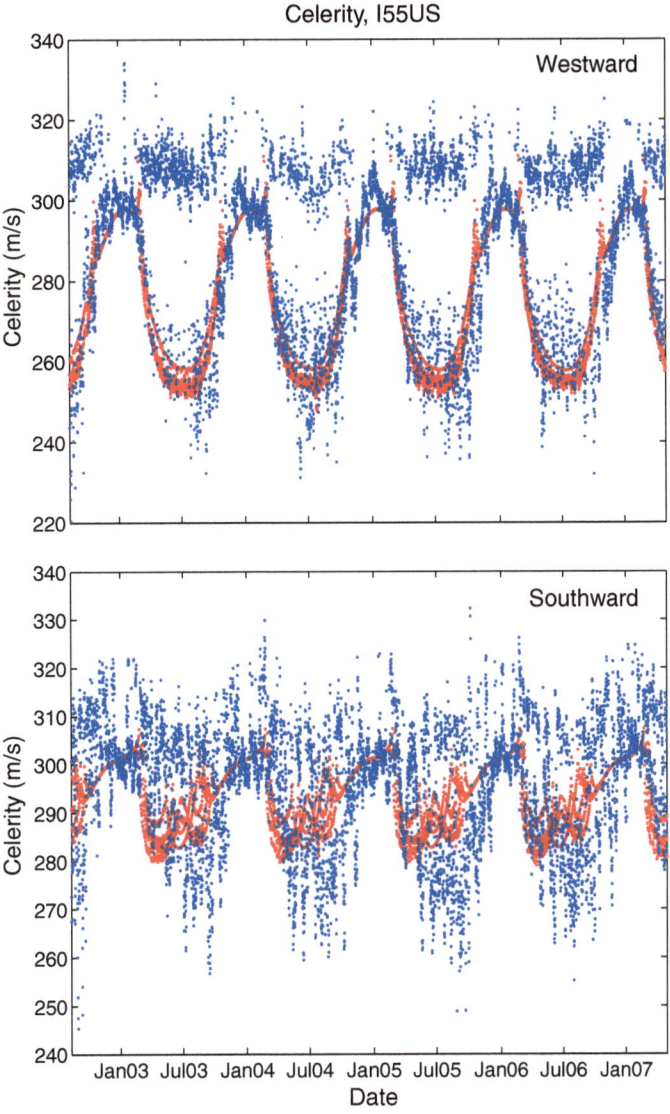

Figure 8
Time series of the celerity for 5° elevation arrivals at I55US for westward and southward, respectively. The color coding is as described in the caption for Fig. 5

arrivals at I55US depart widely from the average climatology going from +4° to −4° over a month.

6. Discussion/Conclusion

As described in BROWN *et al.,* (2002a) one could imagine tables of statistical propagation characteristics comprised of several dominant modes that could be implemented in operational DLC algorithms; a constant phase at 310–340 m/s and an annual varying one with stratospheric and thermospheric phases. In future IDC software updates, these could and should also be a function of day of the year, look direction, and station. Histogram analysis could be used to establish preferred propagation modes with uncertainties and assigned probabilities based on half-widths; however, direct utilization of the procedures we have outlined here on a daily basis is just as easy to implement. At present the IDC uses travel-time tables which are independent of season, though do depend on the elevation of arrival.

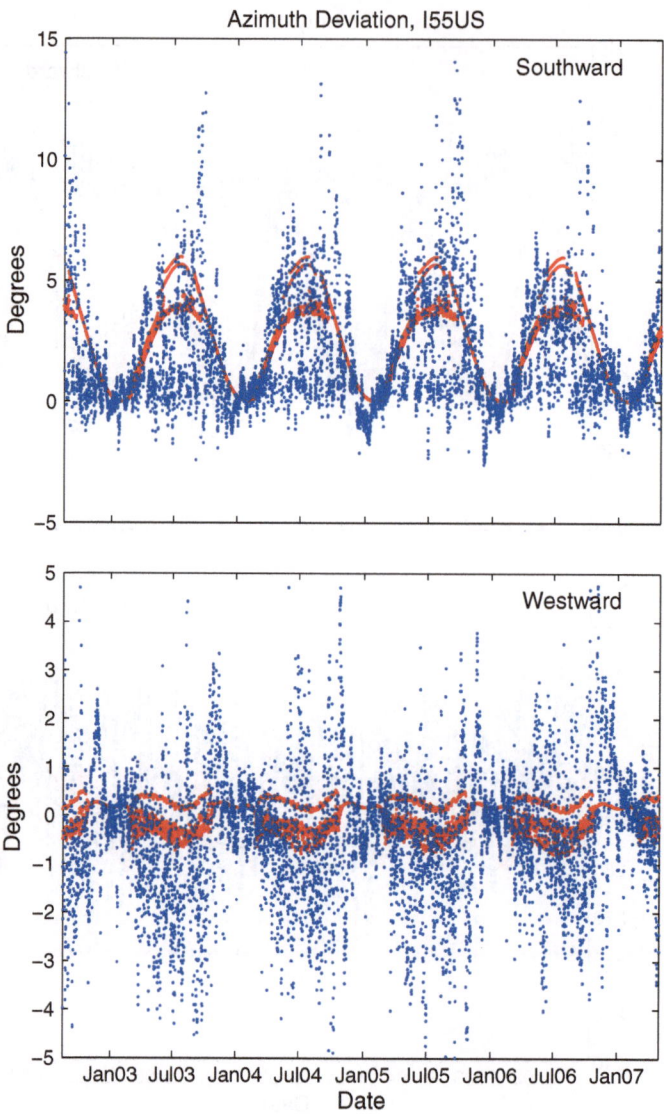

Figure 9
Time series of southward and westward arrivals for I55US at an elevation angle of 5°, respectively. The color coding is as described in the caption for Fig. 5

Furthermore, from the consideration of the variability of backazimuth and celerity presented, it is clear that the current seasonal averaged travel-time tables provide a poor representation of the day-to-day and month-to-month variations, and thus limit the full potential of the CTBTO automated infrasound DLC algorithms. The inherent variability is simply lost in the histogram analysis. It should be noted that in concert with the other monitoring technologies of the CTBTO, the current infrasound algorithms are passable, but improvable, as has been demonstrated by numerous researchers and results. Additional work, following examples such as ARROWSMITH *et al.,* (2008); LE PICHON *et al.,* (2008c) should be undertaken in order to ascertain the value added to the system in terms of false alarm rates and network detection thresholds with careful consideration of the computational complexity and burden introduced into the existing operational system.

With respect to caveats for the calculations presented here, for certain locations and times the dynamical variability of the upper mesospheric and lower thermospheric modes may be even greater in reality than calculated here after accounting for observed and documented non-migrating tides (FORBES et al., 2003), day-to-day and inter-seasonal tidal variability (FRITTS and ISLER, 1994; LIEBERMAN et al., 2007), and the existence of vertically propagating and stationary mesosphere lower thermosphere planetary waves (SMITH, 1996; McLANDRESS et al., 2006) above 55 km. These effects are not yet fully included in either the empirical climatology or hybrid G2S atmospheric specification. The HWM93 model is also known to underestimate the magnitude of the migrating tides for certain seasons and latitudes. A recent update of HWM (HWM07) by DROB et al., (2008) resolves most of the issues related to the amplitude and phases of the migrating solar tides in the mesosphere and lower thermosphere. Given a proper statistical treatment of the present uncertainties, limitations of the present atmospheric specifications above 55 km should not invalidate the applicability of the work described here to DLC algorithms. In particular, tropospheric and stratospheric modes, which have a much greater signal to noise ratio are more likely to be detected and not subject to these problems.

The τp and G2S software which produced the results presented here are specifically designed to be utilized at the IDC in batch mode via a shell script in order to compute the local propagation characteristics for all IMS infrasound stations in real time. These codes can also be invoked interactively by an IDC analyist. For a given station and time the τp calculations only require a few seconds to complete. The process is relatively fast compared to the periodic calculations of the global G2S coefficient set from available atmospheric data sources, requiring several minutes to downloading available global weather fields and perform the vector spherical harmonic transforms. With the addition or allocation of a single dedicated compute node that is accessible to the operational DLC system, the travel-time characteristics for all sixty IMS infrasound stations could be updated every 60 min. Compared to automated DLC calculations from climatological travel-time tables gains in system performance could thus be achieved.

Recent ground-truth events investigated by the infrasound research community have clearly demonstrated that accurate atmospheric specifications are required to properly explain observed infrasound signals. In tandem, the atmospheric science community is continually improving and currently producing reliable specifications of the atmospheric state that can be utilized to improve automated DLC algorithms. Given the advances and availability of low-cost computing resources, and the reliable openly available real-time atmospheric specifications such as provided by NOAA and ECMWF there are no good reasons, technical or otherwise, why near-real-time travel-time tables should not be used in infrasound propagation calculations. Of course when these systems are brought online it is important to continually monitor and evaluate the performance with the many naturally occurring geophysical, as well as coincidental man-made, ground-truth events. One possibility as already demonstrated by similar research is the example of the analysis of the multi-year time series of volcanic observations at I22FR (LE PICHON et al., 2005). Lastly for robust event location and screening of automated event bulletins by human analysts (BROWN et al., 2002a), we recommend that detailed propagation modeling techniques that account for range dependence (GOSSARD and HOOKE, 1975), topography (ARROWSMITH et al., 2007), and other effects such as interval gravity waves (OSTASHEV et al., 2005; CHUNCHUZOV, 2004) be utilized.

This paper has presented time series of infrasound propagation characteristics. A number of physical approximations where made to keep these calculations simple and robust. Through these calculations, we have argued that precomputed monthly average travel-time tables are poor for operational DLC algorithms. To this end, we have advanced progress toward the integration of real-time infrasound propagation travel-time tables into automated IMS processing.

Acknowledgments

The methodologies and tools presented here were developed in part in an effort supported by the Office of Naval Research to investigate whether signals from infrasound ground-truth events could be inverted to obtain information about the atmosphere, i.e., ground-to-space infrasound acoustic tomography, such as is currently in routine use in oceanography and seismology.

REFERENCES

ARROWSMITH, S.J., DROB, D.P., HEDLIN, M.A.H., and EDWARDS, W. (2007), *A joint seismic and acoustic study of the Washington State bolide: observations and modeling*, J. Geophys. Res.-Atmos *112*.

ARROWSMITH, S.J., WHITAKER, R., TAYLOR, S.R., BURLACU, R., STUMP, B.W., HEDLIN, M.A.H., RANDALL, G., HAYWARD, C., and REVELLE, D.O. (2008), *Regional monitoring of infrasound events using multiple arrays: application to Utah and Washington State*, Geophys. J. Int. *175*, 291–300.

BECHTOLD, P., KOHLER, M., JUNG, T., DOBLAS-REYES, F., LEUTBE-CHER, M., RODWELL, M.J., VITART, F., and BALSAMO, G. (2008), *Advances in simulating atmospheric variability with the ECMWF model: from synoptic to decadal time-scales*, Quart. J. Roy. Meteorol. Soc. *134*, 1337–1351.

BROWN, D.J., KATZ, C.N., LE BRAS, R., FLANAGAN, M.P., WANG, J., and GAULT, A.K. (2002a), *Infrasonic signal detection and source location at the Prototype International Data Centre*, Pure Appl. Geophys. *159*, 1081–1125.

BROWN, P., SPALDING, R.E., REVELLE, D.O., TAGLIAFERRI, E., and WORDEN, S.P. (2002b), *The flux of small near-Earth objects colliding with the Earth*, Nature *420*, 294–296.

CANSI, Y. (1995), *An Automatic Seismic Event Processing for Detection and Location—the PMCC Method*, Geophys. Res. Lett *22*, 1021–1024.

CHRISTIE, D.R., VELOSO, J.A.V., CAMPUS, P., BELL, M., HOFFMANN, T., LANGLOIS, A., MARTYSEVICH, P., DEMIROVIC, E., and CARVALHO, J. (2001), *Detection of atmospheric nuclear explosions: the infrasound component of the International Monitoring System*, Kerntechnik *66*, 96–101.

CHUNCHUZOV, I.P. (2004), *Influence of internal gravity waves on sound propagation in the lower atmosphere*, Meteorol. Atmos. Phys. *85*, 61–76.

COURTIER, P., ANDERSSON, E., HECKLEY, W., PAILLEUX, J., VASILJEVIC, D., HAMRUD, M., HOLLINGSWORTH, A., RABIER, E., and FISHER, M. (1998), *The ECMWF implementation of three-dimensional variational assimilation (3D-Var). I: Formulation*, Quart. J. Roy. Meteorol. Soc. *124*, 1783–1807.

DROB, D., EMMERT, J.T., CROWLEY, G., PICONE, J.M., SHEPHERD, G.G., SKINNER, W., HAYS, P., NICIEJEWSKI, R.J., LARSEN, M., SHE, C.Y., MERIWETHER, J.W., HERNANDEZ, G., JARVIS, M.J., D. P. SIPLER, TEPLEY, C.A., O'BRIEN, M.S., BOWMAN, J.R., WU, Q., MURAYAMA, Y., KAWAMURA, S., REID, I.M., and VINCENT, R.A. (2008), *An Empirical Model of the Earth's Horizontal Wind Fields: HWM07*, J. Geophys. Res.-Space Phys., in press.

DROB, D.P., PICONE, J.M., and GARCES, M. (2003), *Global morphology of infrasound propagation*, J. Geophys. Res.-Atmos. *108*.

EDWARDS, W.N. and HILDEBRAND, A.R. (2004), *SUPRACENTER: Locating fireball terminal bursts in the atmosphere using seismic arrivals*, Meteor. & Planet. Sci. *39*, 1449–1460.

EVERS, L.G. and HAAK, H.W. (2007), *Infrasonic forerunners: Exceptionally fast acoustic phases*, Geophys. Res. Lett. *34*.

FORBES, J.M., ZHANG, X.L., TALAAT, E.R., and WARD, W. (2003), *Nonmigrating diurnal tides in the thermosphere*, J. Geophys. Res.-Space Phys. *108*.

FRITTS, D.C. and ISLER, J.R. (1994), *Mean motions and tidal and 2-day structure and variability in the mesosphere and lower thermosphere over Hawaii*, J. Atmos. Sci. *51*, 2145–2164.

GARCES, M.A., HANSEN, R.A., and LINDQUIST, K.G. (1998), *Traveltimes for infrasonic waves propagating in a stratified atmosphere*, Geophys. J. Internat. *135*, 255–263.

GEORGES, T.M. and BEASLEY, W.H. (1977), *Refraction of infrasound by upper-atmospheric winds*, J. Acoust. Soc. Am. 61, 28-34.

GOSSARD, E.E. and HOOKE, W.H. *Waves in the Atmosphere: Atmospheric Infrasound and Gravity Waves: Their Generation and Propagation* (Elsevier Scientific Pub. Co., Amsterdam; New York 1975).

HEDIN, A.E., FLEMING, E.L., MANSON, A.H., SCHMIDLIN, F.J., AVERY, S.K., CLARK, R.R., FRANKE, S.J., FRASER, G.J., TSUDA, T., VIAL, F., and VINCENT, R.A. (1996), *Empirical wind model for the upper, middle and lower atmosphere*, J. Atmos. Terre. Phys. *58*, 1421–1447.

HEDLIN, M.A.H., GARCES, M., BASS, H.E., HAYWARD, HERRIN, G., OLSON, G., and WILSON, C. (2002), *Listening to the secret sounds of the earth's atmosphere*, Eos Trans. AGU *83*, 564–565.

HERRIN, E.T., KIM, T.S., and STUMP, B.W. (2006), *Evidence for an infrasound waveguide*, Geophys. Res. Lett. *33*.

KALNAY, E., KANAMITSU, M., and BAKER, W.E. (1990), *Global Numerical Weather Prediction at the National-Meteorological-Center*, Bull. Am. Meteor. Soc. *71*, 1410–1428.

LE PICHON, A., ANTIER, K., CANSI, Y., HERNANDEZ, B., MINAYA, E., BURGOA, B., DROB, D., EVERS, L.G., and VAUBAILLON, J. (2008a), *Evidence for a meteoritic origin of the September 15, 2007, Carancas crater*, Meteor. Planet. Sci. *43*, 1797–1809.

LE PICHON, A., BLANC, E., DROB, D., LAMBOTTE, S., DESSA, J.X., LARDY, M., BANI, P., and VERGNIOLLE, S. (2005), *Infrasound monitoring of volcanoes to probe high-altitude winds*, J. Geophy. Res.-Atmos. *110*.

LE PICHON, A., GARCES, M., BLANC, E., BARTHELEMY, M., and DROB, D.P. (2002), *Acoustic propagation and atmosphere characteristics derived from infrasonic waves generated by the Concorde*, J. Acoust. Soc. Am. *111*, 629–641.

LE PICHON, A., VERGOZ, J., HERRY, P., and CERANNA, L. (2008b), *Analyzing the detection capability of infrasound arrays in Central Europe*, J. Geophys. Res.-Atmos. *113*, 9.

LE PICHON, A., VERGOZ, J., HERRY, P., and CERANNA, L. (2008c), *Analyzing the detection capability of infrasound arrays in Central Europe*, J. Geophys. Res.-Atmos. *113*.

LIEBERMAN, R.S., RIGGIN, D.M., ORTLAND, D.A., NESBITT, S.W., and VINCENT, R.A. (2007), *Variability of mesospheric diurnal tides and tropospheric diurnal heating during 1997-1998*, J. Geophys. Res.-Atmos. *112*, 17.

LIGHTHILL, M.J. *Waves in Fluids,* (Cambridge University Press, Cambridge [Eng.]; New York 1978).

LINGEVITCH, J.F., COLLINS, M.D., DACOL, D.K., DROB, D.P., ROGERS, J.C.W., and SIEGMANN, W.L. (2002), *A wide angle and high Mach number parabolic equation*, J. Acoust. Soc. Am. 111, 729–734.

MANNEY, G.L., KRUGER, K., PAWSON, S., MINSCHWANER, K., SCHWARTZ, M.J., DAFFER, W.H., LIVESEY, N.J., MLYNCZAK, M.G., REMSBERG, E.E., RUSSELL, J.M., and WATERS, J.W. (2008), *The evolution of the stratopause during the 2006 major warming: Satellite data and assimilated meteorological analyses*, J. Geophys. Res.-Atmos. *113*.

McLandress, C., Ward, W.E., Fomichev, V.I., Semeniuk, K., Beagley, S.R., McFarlane, N.A., and Shepherd, T.G. (2006), *Large-scale dynamics of the mesosphere and lower thermosphere: An analysis using the extended Canadian Middle Atmosphere Model*, J. Geophys. Res.-Atmos. *111*.

Menke, W. *Geophysical Data Analysis: Discrete Inverse Theory*, Rev. edn. (Academic Press, San Diego 1989).

Millet, C., Robinet, J.C., and Roblin, C. (2007), *On using computational aeroacoustics for long-range propagation of infrasounds in realistic atmospheres*, Geophys. Res. Lett. *34*.

Nemtchinov, I.V., Svetsov, V.V., Kosarev, I.B., Golub, A.P., Popova, O.P., Shuvalov, V.V., Spalding, R.E., Jacobs, C., and Tagliaferri, E. (1997), *Assessment of kinetic energy of meteoroids detected by satellite-based light sensors*, Icarus *130*, 259–274.

Ostashev, V.E., Chunchuzov, I.P., and Wilson, D.K. (2005), *Sound propagation through and scattering by internal gravity waves in a stably stratified atmosphere*, J. Acoust. Soc. Am. *118*, 3420–3429.

Picone, J.M., Hedin, A.E., Drob, D.P., and Aikin, A.C. (2002), *NRLMSISE-00 empirical model of the atmosphere: Statistical comparisons and scientific issues*, J. Geophys. Res.-Space Phys. *107*.

Pierce, A.D. (1967), *Guided infrasonic modes in a temperature- and wind-stratified atmosphere*, J. Acoust. Soc. Am. *41*, 597.

Press, W.H. *Numerical Recipes: The Art of Scientific Computing*, (Cambridge University Press, Cambridge; New York 1989).

Smith, A.K. (1996), *Longitudinal variations in mesospheric winds: Evidence for gravity wave filtering by planetary waves*, J. Atmos. Sci. *53*, 1156–1173.

(Received November 21, 2008, revised January 12, 2009, accepted April 21, 2009, Published online March 10, 2010)

Reprinted from the journal

Pure Appl. Geophys. 167 (2010), 455–461
© 2009 Birkhäuser Verlag, Basel/Switzerland
DOI 10.1007/s00024-009-0026-z

▌**Pure and Applied Geophysics**

Underground Nuclear Explosions and Release of Radioactive Noble Gases

Yuri V. Dubasov[1]

Abstract—Over a period in 1961–1990 496 underground nuclear tests and explosions of different purpose and in different rocks were conducted in the Soviet Union at Semipalatinsk and anovaya Zemlya Test Sites. A total of 340 underground nuclear tests were conducted at the Semipalatinsk Test Site. One hundred seventy-nine explosions (52.6%) among them were classified as these of complete containment, 145 explosions (42.6%) as explosions with weak release of radioactive noble gases (RNG), 12 explosions (3.5%) as explosions with nonstandard radiation situation, and four excavation explosions with ground ejection (1.1%). Thirty-nine nuclear tests had been conducted at the Novaya Zemlya Test Site; six of them – in shafts. In 14 tests (36%) there were no RNG release. Twenty-three tests have been accompanied by RNG release into the atmosphere without sedimental contamination. Nonstandard radiation situation occurred in two tests. In incomplete containment explosions both early-time RNG release (up to ~ 1 h) and late-time release from 1 to 28 h after the explosion were observed. Sometimes gas release took place for several days, and it occurred either through tunnel portal or epicentral zone, depending on atmospheric air temperature.

Key words: Underground nuclear explosion, noble gas radionuclides release, Semipalatinsk Test Site, Novaya Zemlya Test Site.

Abbreviations

RNG	Radioactive noble gas
UNE	Underground nuclear explosion
CCE	Complete contained explosion
ICE (RNG)	Incomplete contained explosion
ICE (NRS)	Incomplete contained explosion
NRS	Nonstandard radiation situation
$T_{1/2}$	The radionuclide half-life (sec, min, hour, etc.)
W	Yield (explosion energy release) in kilotons (kt) of trinitrotoluene equivalent

[1] V.G. Khlopin Radium Institute, 28, 2nd Murinskiy Ave., 194021 St. Petersburg, Russia. E-mail: yuri@dyuv.spb.su

1. Introduction

One of the main methods for detection of clandestine nuclear explosions is a radionuclide method based on the Xe radionuclides monitoring in atmospheric air.

Four hundred ninety-six underground nuclear tests and explosions of different purpose and in different rocks were conducted in the Soviet Union over a period in 1961–1990. Using available experimental data we shall consider the main possible variants of radioactive noble gas (RNG) release during and after underground nuclear explosions (UNE) in different rocks.

In our domestic practice (Mikhailov *et al.*, 1994) complete contained nuclear explosions (CCE) can be characterized according to radiation circumstances at the explosion site as follows:

CCE—(complete contained explosion)—An underground nuclear explosion of complete internal action accompanied by the formation of an underground cavity with a respective compaction, fragmentation, and cracking of rock around it, but the rock block prevents the egress of gaseous products.

ICE (RNG)—(incomplete contained explosion)—An explosion of complete internal action accompanied by joining of fissured and spalled zones on the Earth surface in the explosion epicentral area and by a ventilation, as a rule, insignificant seepage to the atmosphere of short-lived (RNG): ^{85m}Kr ($T_{1/2} = 4.5$ h), ^{87}Kr ($T_{1/2} = 76.3$ min), ^{88}Kr ($T_{1/2} = 2.84$ h), ^{131m}Xe ($T_{1/2} = 11.9$ day), ^{133}Xe ($T_{1/2} = 5.2$ day), ^{133m}Xe ($T_{1/2} = 2.2$ day), ^{135}Xe ($T_{1/2} = 9.09$ h), ^{135m}Xe ($T_{1/2} = 15.3$ min), ^{138}Xe ($T_{1/2} = 14.17$ min).

ICE (NRS)—(incomplete contained explosion)—An explosion of complete internal action with a nonstandard radiation situation (NRS) accompanied by an early and forced dynamic release to the atmosphere of explosion products in the gas and vapor phase caused by an occasional disturbance of the normal course of the test or by its consequences unforeseen in the project which could or did lead to exposure of people over established norms or to material damage (GORIN et al., 1993).

340 underground nuclear tests were conducted at the Semipalatinsk Test Site: 209 in tunnels and 131 in shafts (vertical holes). One hundred seventy-nine (52.6%) explosions among them are classified as CCE, 145 explosions (42.6%) as ICE (RNG) with weak seepage of RNG, and 12 explosions (3.5%) as explosions with nonstandard radiation situation–ICE (NRS). Four excavation explosions with ground ejection and full release of RNG were conducted in shafts #1004 (15 January, 1965), #1003 (14 October, 1965), #T-1 (21 October, 1968) and #T-2 (multiple explosion—three charges were detonated in three holes—12 November, 1968) for peaceful purposes (scientific and industrial applications in the national economy, not military) and for development of peaceful nuclear explosion technology, similar to the USA program "Plowsher" (NORDYKE, 1996; TELLER et al. 1968).

Thirty-nine nuclear tests were conducted at the Novaya Zemlya Test Site, six of them – in holes. In 14 tests (36%) there was no RNG release. Twenty-three tests have been accompanied by RNG release into the atmosphere without residual (sedimental) contamination of the area. In two tests accidental radiation condition occurred (MIKHAILOV, 1997; MIKHAILOV et al., 1994). (Table 1)

According to the subject of our study we shall be interested here in incomplete contained explosions accompanied by insignificant seepage of noble gases. Analyzing the data of GORIN et al., (1993), it should be noted that among 127 contained explosions in shafts at the Semipalatinsk Test Site, 44% of them are classified as CCE, 56% as ICE (RNG).

2. Methods of Analysis

Determination of the time and area of radioactive explosion products release in epicentral zone, as well as of radioactive substances spreading in the atmosphere was carried out using aircraft for radiation reconnaissance; in the nearest zone helicopter-dosimeters of Mi-8T type were used, and in the far-field zone flying laboratories installed on research planes AN-24RR and AN-30RR were used.

The helicopter was, as a rule, equipped both, with the Roentgen meter "Vozdukh-2M" ("GO-21" analogue) having gamma radiation dose rate measurements ranging from 0.05 mGy/h to-5 Gy/h (5 mR/h to 500 R/h), and portable dosimeter-radiometers KDG, SRP-68, KRBG-1.

Aero-gamma spectrometric complex AGSK, which has good technical characteristics was recently used for radiation surveying. Part of such complex was usually located at the aircrafts, and another part was installed on the field. Automated complex AGSK made it possible to carry out large measurement volume in real time, namely:

Table 1

Distribution of completely contained explosions

Test type	Number of explosions	CCE	ICE (RNG)	Excavation explosions
Conducted at the Semipalatinsk Test Site in radiation situation (GORIN et al., 1993; MIKHAILOV, 1996)				
Explosion in shaft	131	60 (45.8%)	67 (51.1%)	4 (3%)
Explosion in tunnel	209	126 (60%)	83 (40%)	–

Test type	Number of explosions	CCE	ICE (RNG)	ICE(NRS)-accidental
Conducted at the Novaya Zemlya Test Site in radiation situation (MIKHAILOV, 1997; MIKHAILOV et al., 1994)				
Explosion in shaft	6	4 (66.6%)	2 (33.4%)	–
Explosion in tunnel	33	10 (30.3%)	21 (63.6%)	2 (6.1%)

– to determine a radionuclide composition and the location of gas-aerosols release remotely based on a qualitative gamma-spectrometry method;
– to carry out express-measurements and estimation of the area contamination density by gamma-emission radionuclides;
– to register the dynamics of radioactive products release;
– to determine the size and contour of radioactive clouds in the atmosphere.

AGSK complex was also deployed at the aircraft laboratory aimed at monitoring radioactive air masses spreading from the tests area.

Radioactive noble gases (RNG) measurement was carried out in the areas of RNG release by taking air samples into a measuring chamber of "RAG" ("RANAG", later), 10 L by volume. Gases were also taken into pilot -balloon shells with further measuring conducted in test site laboratories by gamma spectrometers (in the 1960s of the last century scintillation spectrometers with NaI (Tl) detectors, and in the 1970s and later, Ge (Li), detector spectrometers).

At nuclear explosions of ICE type both early (up to ∼1 h) and late RNG release from 1 to 28 h after explosion was observed. Sometimes gas release lasted for several days, and depending on atmospheric air temperature it occurred through either the tunnel portal or epicentral zone. In tunnel tests radioactive gas release occurred through cracks in the epicentral zone, occasionally through the test tunnel portal; in shaft tests - through the head of the test shaft, cracks in the epicentral zone, intercable space, the space outside the metallic casing, sampling pipeline.

3. Results

3.1. Radiation Condition at the Semipalatinsk Test Site

Test in tunnel V-1. It was the first underground nuclear test conducted on 11 October, 1961, i.e., before signing the Moscow Treaty on Limited Test Ban Treaty in 1963. The explosion yield was ∼1 kt. Due to the late release of RNG (3–4 h after the explosion) (LOGACHEV, 1997) the explosion was of

low intensity through the epicentral zone and stemming tunnel complex; this test was classified as an ICE. However, there was no significant residual contamination of the rock tunnel portal and epicentral zone.

Test in tunnel I9 (25 March, 1967). Explosion debris was not detected in the atmosphere, RNG seepage was late and of low intensity. The test is classified as ICE (RNG) with low-intensity RNG seepage in the epicentral zone, outdoor air temperature T_{out} being higher than that inside the tunnel T_{in}. The level of exposure rate was not higher than 4 mGy/h (400 mR/h) in the epicentral zone in the area of rock fracture control 5 h after the explosion.

Test in tunnel I8 (04 August, 1967). Early RNG release occurred with the gas flow through the tunnel portal ∼7.5 h after the explosion; it resulted in a short-time increase of radiation exposure rate up to 15 mGy/h (1.5 R/h) near the tunnel portal.

Test in tunnel I04 (28 July, 1978). This was a simultaneous explosion, the maximum number of charges being 5. The test was classified as ICE with a late release and low intensity. Exposure rate at the tunnel portal was less than 7 mGy/h (0.7 R/h). The residual contamination of the portal rock face and tunnel by decay products of ^{90}Kr (^{90}Sr) and ^{137}Xe (^{137}Cs) occurred due to the late seepage and is observed currently.

Test in tunnel 184 (14 August, 1981). This explosion was classified as ICE (RNG). Within 10 min after the explosion early RNG release occurred in the epicentral zone; the gamma dose rate was not higher than 0.12 Gy/h (12 R/h). The character of seepage was conditioned by non-optimal depth of charge emplacement and the difference in outdoor and inner air temperature (12°C). Such process was forecasted for the summer period.

Test in shaft N⁰ 1366 (12 February, 1989). This test is classified as ICE (RNG). Gaseous radioactive products seepage began about 2 h after detonation along sampling mains (system), and within 1.5 h gases began to flow out from the cracks in the epicentral zone (Fig. 1). Radioactive gases jet was spreading in the air 50–200 m from the surface. The next day a jet was detected at a distance of 230 km from the explosion location. On the 4th day after the explosion a dose rate at the Komsomolskiy

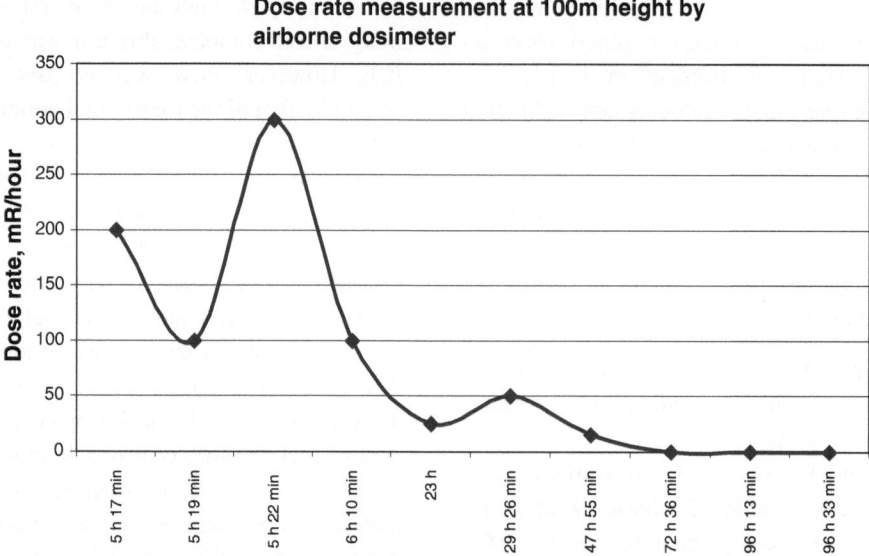

Figure 1
RNG Release from the epicentral area of the explosion in Shaft -1366 (STS, February, 1989)

settlement at a distance of 100 km from shaft 1366 did not differ from that of a background one. Analysis of gases taken from cracks in the epicentral zone revealed the presence of only ^{133}Xe and ^{135}Xe in air samples. During all seepage periods RNG volumetric activity in villages was lower than 3.7 Bq/L (10^{-10}Ci/L). According to the data of the air-borne survey (they are presented in Table 2), radioactive gas release in the epicentral zone was observed for 4 days; these data are characterized as temporal variations of the release rate. (MIKHAILOV, 1997, p. 193).

Some results of the air-borne radiation survey after four explosions are shown in Table 3; specific distances including, where an airborne Roentgen-meter did not record the exceeding of radiation background, that is, it put an end to noble gases monitoring (dose rate was lower than 0.01 μGy/h).

The analysis of gas samples showed the active component of spread to include in most cases the sum of Kr and Xe radionuclides: ^{85}Kr, ^{85m}Kr, ^{88}Kr, ^{133}Xe, ^{133m}Xe, ^{135}Xe.

Table 2

Gamma-radiation dose rate at a height of 100 m from an explosion epicenter in shaft 1366 (MIKHAILOV, 1997)

Time after explosion	Dose rate (mGy/h)
50 min	Background
5 h 10 min	0.0009
5 h 17 min	2.0
5 h 19 min	1.0
5 h 22 min	3.0
6 h 10 min	1.0
23 h	0.25
29 h 26 min	0.5
47 h 55 min	0.15
72 h 36 min	0.005
96 h 13 min	0.0005
96 h 33 min	0.002

3.2. *Radiation Condition at the Novaya Zemlya Test Site*

In the Novaya Zemlya Test Site tests an RNG seepage beyond the boundaries of the Test Site was observed in 11 tests with the total activity of RNG released being higher than $3.7 \cdot 10^{14}$ Bq (10^4 Ci). Maximum total activity did not exceed $3.7 \cdot 10^{17}$

Table 3

Some characteristics of radioactive gases after underground nuclear explosions at the Semipalatinsk Test Site

Test, date	NG stream length at the level of μGy/h, km		Airstream parameters			
	First day after explosion	Second day after explosion	Measurement moment for maximum dose rate after explosion (h)	Maximum dose rate at distance from epicenter (km)	Maximum width (km)	Maximum dose rate (mGy/h) (at a height of, m)
Tests in tunnel						
605 R (20 Sept. 1979)	60	40	5.75	10	12	0.012 (300)
113 R (14 Dec. 1978)	100	120	4.58	0.0	–	40 (600)
Tests in shaft						
1355 (13 Dec. 1987)	40	50	1.0	2	2	0.003 (50)
1341 (15 June, 1985)	70		2	5	4	2.5 (300)

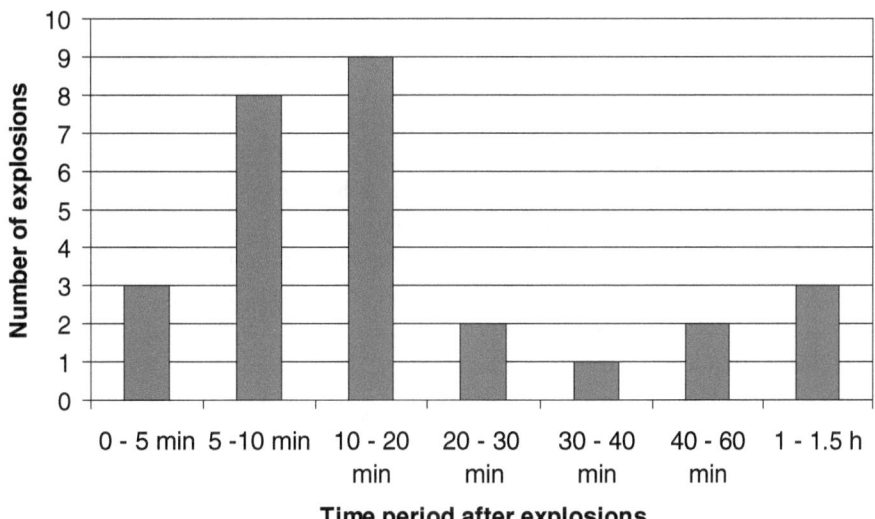

Figure 2
Beginning time of RNG release after UNE's at Novaya Zemlya Test Site

Bq (1 test). Total activity being lower than $3.7 \cdot 10^{13}$ Bq (10^3Ci), RNG were not registered beyond the Test Site boundary. In nearly half of the tests noble gases release began with in a time interval of 5–20 min. (Fig. 2) (MIKHAILOV *et al.*, 1994, 2004).

3.3. Radiation Condition at some Sites for Peaceful Nuclear Explosions

Under the Soviet Program for Peaceful Uses of Nuclear Explosions 124 explosions were carried out outside military test sites. Sometimes the release of radioactive noble gases took place after peaceful nuclear explosions. 39 of them with a yield ranging from 2 to 23 kt in vertical holes (shafts) at a depth of 400–1,000 m had been conducted for the core seismic sounding program (MIKHAILOV *et al.*, 1994). In 22 explosions there was no radioactive substance release at the day surface. Complete containment of explosions had been observed here.

The release of radioactive noble gases along detonation cables took place occasionally. Gamma radiation dose rate to closely monitor the cable was 1–5 μGy/h. Usually the RNG release began within 1–3 h after detonation and lasted several days, if appropriate measures of sealing (cutting of the monitor cable, its dropping into the free part of the shaft with subsequent placing of the cement plug)

Table 4

Characteristics of gas seepage after some peaceful underground nuclear explosions at "Galit" Site (MIKHAILOV et al., 1994; ADAMSKIY et al., 1998; KRIVOKHATSKY et al., 1994; VOLOSHIN et al., 2001)

Parameter	Explosion in shaft		
	A-I	A-II-1	A-VIII
Date	22, April, 1966	01 July, 1968	17 January, 1979
Depth (m)	161	600	995
Yield (kt)	1.1	27	65
Beginning of seepage (min)	12	30	45
Duration of seepage	20 days	9 days	26 h
Release RNG activity (Bq)	7.0×10^{15}	2.0×10^{17}	1.5×10^{17}
Release fraction from total RNG (%)	$3.5 \times 10^{-4}\%$	$3.9 \times 10^{-4}\%$	$1.2 \times 10^{-4}\%$
Volumetric activity of ^{133}Xe, Bq/L, at $T = 0$ min	–	3.7×10^{9}	3.7×10^{10}

were not taken (DUBASOV, 1998). Since this release was insignificant and was not hazardous either for personnel or the population, these explosions can be considered as CCE. RNG composition depended on the start time of their seepage.

In a number of cases a low-pressure seepage of radioactive gases occurred through: the space outside the metallic casing of the charge emplacement shaft on site "Globus-3" (10 July, 1971, Komi Province, 64°18′N, 55°17′E), site "Batolit-1" (01 November, 1980, Krasnoyarsk Province, near Baykit Village, 61°54′N, 96°40′E), and the depressurized shaft on site "Takhta-Kugulta" (26 September, 1969, Stavropol' Province, 45°51′N, 41°57′E), shaft 2T on site "Lira" (20 July, 1983, Kazakhstan, 51°18′N, 53°26′E, near Aksay town).

Table 4 presents the data of observations and measurements after explosions in the salt dome at the "Galit" Site (West Kazakhstan, Azgir village, 47°49′N, 47°35′E) where the RNG release occurred. Radioactive noble gases (RNG) measurement was carried out at sites near shafts or sampling holes by taking air samples into measuring chamber "RANAG". Gases were also taken into pilot-balloon shells with their further measuring in test site laboratories by gamma spectrometers (in the 1960s of the last century scintillation spectrometers with NaI (Tl) detectors, and in the 1970s and later, Ge (Li), detector spectrometers). The volume of flowing gases was measured during all periods of their egress and thus the total activity of each radionuclide was determined.

Radiation states in the epicentral zone of peaceful nuclear explosion in Quel'porr mountain massif, object "Dnieper-2" (27 August 1984, Murmansk Province, 67°42′N, 33°42′E), was a background one for 8 h after the explosion. The impulse character of low-pressure release of RNG was detected later and continued for 10 days. Radioactive gases seepage occurred along the cracks at the mountain surface (DUBASOV, 1998; VOLOSHIN *et al.*, 2001).

4. Conclusion

Experimental data presented show a significant number of explosions with radioactive noble gas release and seepage of this gas with different intensity and duration. Channels of gas release also have different sizes and characteristics.

Radioactive gases release control during underground nuclear tests was aimed at a staff and population radiation safety observance beyond the boundaries of military test sites and peaceful test sites. Therefore a sensitivity of ^{133}Xe concentration determination by techniques and equipment existing at that time was ~ 400 Bq/m^3, and it is significantly lower than that of the equipment of the new generation. Therefore it is difficult to say now, how long the RNG seepage period could go onto reach a concentration of approximately 1 mBq/m^3.

Information regarding RNG release and seepage in underground nuclear explosions in the USSR is given here. RNG release and seepage also took place in nuclear tests of the USA (HAWKINS, 2007).

REFERENCES

ADAMSKIY, V.B., *et al.* (1998), *Peaceful nuclear explosions in salt dome formation Bolshoy Azgir*, Institute of Nuclear Physics NNC RK, VNIPI Promtekhnologii, Almaty (in Russian).

GORIN, V.V., KRASILOV, G.A., *et al.* (1993), *Semipalatinsk Test Site: Chronology of underground nuclear explosions and their primary radiation effects (1961–1989)*, Information Bull. Centre for Public Information on Nuclear Energy. *9*, Special Issue, pp. 21–32 (in Russian).

HAWKINS, W. (2007) *Containment of underground nuclear tests.* Presentation for Noble Gas Workshop, 5–9 November, Las Vegas, NV, USA.

KRIVOKHATSKY, A.S., DUBASOV, Yu.V., DUBROVIN, V.S., SAVONENKOV, V.G., *et al.* (1994), *Analysis and Stage Reconstruction of Radiation Situation near the Object "Galit" to Estimate Possible Irradiation Of Population of near-by Situated Viladges during the Whole Period of the Object "Galit" Existence*, Report of RPA "V. G. Khlopin Radium Institute", St.-Petersburg (in Russian).

LOGACHEV, V.A., VOLOSHIN, N.P., DUBASOV, Yu. V., *et al.* (eds.), (1997), *Nuclear Tests in the USSR, Semipalatinsk Test Site.* Moscow, p. 273 (in Russian).

MIKHAILOV, V.N. (ed.), (1996), *Nuclear weapon tests and peaceful nuclear explosions in the USSR in 1949–1990*, Russian Federal Nuclear Center, VNIIEF, Sarov (in Russian).

MIKHAILOV, V.N. (ed.), (1997), *Nuclear Explosions in the USSR*, V2, *Nuclear Test Technology. Influence on the environment*, RFNC VNIIEF, Begell-Atom, LLC (in Russian).

MIKHAILOV, V.N., DUBASOV, Yu.V., and MATUSHCHENKO, A.M. (eds.), (1999), *Nuclear Explosions in the USSR. North Test Site.* Reference Material, 2nd edition, RPA "The Radium Institute", St. Petersburg (in Russian).

MIKHAILOV, V.N., DUBASOV, Yu.V., and MATUSHCHENKO, A.M. (eds.), (2004), *Nuclear Explosions in the USSR. North Test Site.* Reference Material. Version 3. Reproduced by the IAEA, Division of Radiation, Transportation and Waste Safety, Vienna, Austria.

MIKHAILOV, V.M., KEDROVSKY, O.L., and KRIVOKHATSKY, A.S. (eds.), (1994), *Nuclear Explosions in the USSR.* 4th issue, Peaceful Use of Nuclear Explosions. VNIPI Promtekhnologii, RPA "V.G.Khlopin Radium Institute", Moscow (in Russian).

NORDYKE M. D. (1996), *The Soviet Program for Peaceful Uses of Nuclear Explosions*, Livermor National Laboratory, Livermore, CA, USA.

TELLER, E., TALLEY, W.K., Higgins, G.H., and Johnson, G.W. (1968), The Constructive Uses of Nuclear Explosives (McGraw-Hill Book Co).

VOLOSHIN, N.P., DUBASOV, Yu.V., and LOGACHEV, V.A. (eds.), (2001), *Nuclear Explosions of the USSR. Peaceful Nuclear Explosions, General and Total Radiation Safety during the Explosions*, IsdAt, Moscow (in Russian).

DUBASOV, Yu. V. (1998), *Material Changes of Rocks in Underground Nuclear Explosion and Radioactive Contamination of Rock Massif.* Final Technical Report. Project ISTC 520-97, Moscow-St.-Petersburg.

(Received October 29, 2008, revised May 8, 2009, accepted July 20, 2009, Published online December 15, 2009)

Pure Appl. Geophys. 167 (2010), 463–470
© 2010 Birkhäuser Verlag, Basel/Switzerland
DOI 10.1007/s00024-009-0037-9

❙Pure and Applied Geophysics

Genesis and Equilibrium of Natural Lithospheric Radioxenon and its Influence on Subsurface Noble Gas Samples for CTBT On-site Inspections

SIMON HEBEL[1]

Abstract—During on-site inspections to verify the comprehensive nuclear-test-ban treaty (CTBT), soil gas samples may be taken and analysed for their content of the xenon isotopes ^{131m}Xe, ^{133}Xe, ^{133m}Xe and ^{135}Xe in order to identify a suspected underground nuclear test. These samples might contain natural radioxenon which is present as a trace gas in the ground. This work analyses the different production mechanisms of natural lithospheric radioxenon to assess theoretically the background concentration under different sampling conditions. The results imply that the equilibrium concentrations of the examined xenon isotopes can be measured in certain rock types using actual CTBTO on-site inspection equipment. Radioxenon production is dominated by spontaneous fission of ^{238}U, resulting in a reactor-like xenon isotopic signature rather than an explosion-like signature.

Key words: CTBT, on-site inspection, xenon, soil gas sampling, xenon background.

1. Introduction

When an underground nuclear test explosion is suspected based on seismological or radiological evidence, the comprehensive nuclear-test-ban treaty (CTBT) allows for an on-site inspection of the area identified by the CTBT Organisation's International Monitoring System (IMS). After an underground nuclear test, radioactive xenon isotopes are released that can travel to the surface immediately or within days or weeks, depending on the local geological properties and weather conditions, along air vents, such as cracks. Being noble gases, they are not prone to chemical reactions and their diffusion is not strongly hindered by layers of ground water.

To verify the nuclear nature of the detected event, soil gas samples are planned to be taken during on-site inspections by drilling a borehole of about 1.5–3 m depth, then introducing a tube to pump out the gas while sealing off several square metres of the ground above with foil to avoid accidental sampling of atmospheric air (CTBTO-PrepCom, 2007). A soil gas volume of about 1 m^3, depending on the sampling system utilised, is retrieved and analysed off-site for traces of the xenon isotopes ^{131m}Xe, ^{133}Xe, ^{133m}Xe and ^{135}Xe. The CTBTO Preparatory Commission is currently evaluating two xenon sampling and measurement systems for on-site inspection purposes: ARIX-3F and SAUNA.

While most known sources of radioxenon are anthropogenic, there are a number of natural production processes. These sources are generally deemed insignificant, but to the author's knowledge there are no detailed analyses of the natural subsoil radioxenon concentration to date. To assess the radioxenon background, processes capable of producing radioxenon have been identified and quantified to enable calculation of the background concentration in different types of rock. The main goal of this work was not to simulate detailed scenarios, but to assess whether the xenon background can be pronounced enough to affect the actual soil gas measurements.

2. Xenon Sources in Nature

While anthropogenic radioxenon is formed primarily by neutron induced fission of ^{235}U or ^{239}Pu in a bomb or reactor, a multitude of processes are theoretically capable of forming natural radioxenon,

---
[1] Carl Friedrich von Weizsäcker Centre for Science and Peace Research, University of Hamburg, Beim Schlump 83, 20255 Hamburg, Germany. E-mail: SHebel@physnet.uni-hamburg.de

from fission to neutron and cosmic particle induced nuclear reactions. These reactions have been identified and evaluated to determine which are capable of producing detectable amounts of radioxenon in natural underground environments.

Most of these reactions depend on neutron radiation, which originates in cosmic particle showers as well as (α, n) reactions by the alpha particles emitted during uranium decay. The cosmic neutron flux is dominant in shallow depths where soil gas samples for on-site inspections are usually taken. The production of (α, n) neutrons varies with parameters of rock structure such as granularity, elemental composition and uranium distribution (BALLENTINE and BURNARD, 2002).

2.1. Fission Reactions

Uranium and thorium are common in most geological configurations. Acidic rocks tend to have a higher uranium content than basic rocks. However, most geochemical environments allow wide ranges of uranium concentration. Tables exist providing the average content of uranium isotopes in different types of rock. Such tables have been utilised in this work, notably the data compiled by JUST (1989).

Uranium and thorium fission can produce the CTBT-relevant xenon isotopes directly or via intermediary elements of the same mass which eventually form xenon in a series of beta decays. While neutron-induced fission occurs in all uranium and thorium isotopes depending on the incident neutron energy, ^{238}U also undergoes spontaneous fission.

2.2. Neutron Reactions with Minerals

Apart from fission, a number of nuclear reactions yielding radioxenon exist. These are primarily neutron reactions, which have been identified by querying the ENDF database (CHADWICK et al., 2006) for reactions yielding isotopes of the masses 131, 133 and 135. Most of these are potential xenon precursors if the target nuclide of the respective reaction is naturally occurring, meaning it is practically stable.

These criteria have identified numerous reactions capable of producing xenon precursors. The according reaction rates can be calculated using the equation

$$r_i = \rho_{at} \int \sigma_i(E) \cdot \Phi_n(E) dE \qquad (1)$$

with the atomic density ρ_{at} of every target nuclide, the neutron cross section $\sigma_i(E)$ of every reaction i, and the neutron flux spectrum $\Phi_n(E)$. For different exemplary rock types taken from FABRYKA-MARTIN (1988), these reaction rates are compared with the xenon formation rates of induced uranium and thorium fission.

All calculations have been performed using a neutron flux spectrum shaped like the one in Fig. 1, but normalised to match the integrated neutron flux given in FABRYKA-MARTIN (1988) for each rock type at a depth of about 1 m.

The results in Table 1 show that most reactions are insignificant compared with the induced fission of uranium and thorium in common rock types. In ultramafic rocks, which have unusually low uranium (1 ppb) and thorium (4 ppb) concentrations, tellurium and caesium are also important sources of neutron induced radioxenon, although the impact of tellurium is reduced by the fact that the branching ratio from its successor ^{131}I to ^{131m}Xe is only 1.4%, resulting in ^{130}Te generally being less effective at producing ^{131m}Xe than induced fission.

According to this calculation, pure ^{130}Te would exceed the induced fission ^{131m}Xe production in granite by about three orders of magnitude, which is below the amount produced by spontaneous fission in granite assuming realistically low neutron fluxes in the order of $10^{-3} \frac{n}{s \cdot cm^2}$ (see Sect. 5). A detection of radioxenon caused by an elevated tellurium concentration is thus highly unlikely, especially taking into account the extreme rarity of tellurium. Pure caesium would exceed the induced fission radioxenon production in granite by almost three orders of magnitude, pure barium by one order of magnitude, making these elements practically incapable of producing a measurable radioxenon concentration even though they are more common than tellurium.

2.3. Muon Particle Reactions

Muons are tertiary products of cosmic high-energy proton radiation. They emit less bremsstrahlung than electrons due to their high mass and thus

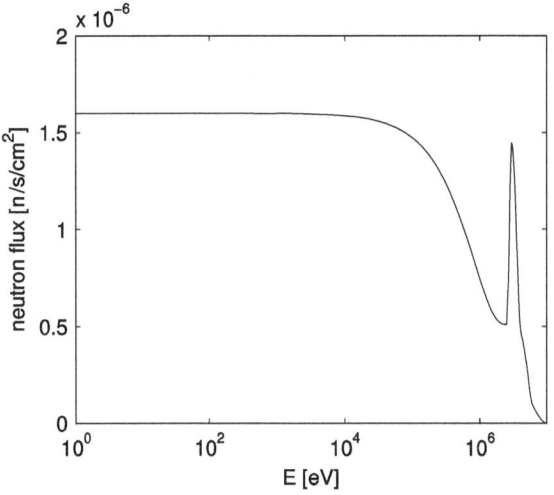

are able to penetrate hundreds of metres of ground. They can be captured by protons, producing neutrons and neutrinos:

$$p + \mu^- \rightarrow n + \nu_\mu.$$

A muon captured by a [133]Cs nucleus yields [133]Xe. This production path was evaluated using muon flux data from CHARALAMBUS (1971), resulting in a xenon production of less than 1% of that resulting from uranium, even when considering minerals with high caesium content.

3. Modelling Xenon Concentration

3.1. Model

The production rate of each nuclide i is the sum of contributions from the spontaneous fission of [238]U, P_{sp}, and the neutron-induced fission P_{ind} of [235]U, [238]U and [232]Th:

Figure 1
The above neutron flux spectrum has been postulated for use in this work based on the following considerations: first, the majority of neutrons are thermalised before being absorbed. Second, measurements by CHAZAL et al. (1998) suggest a peak at 2.4 MeV due to [238]U fission neutrons

Table 1

Comparison of the rates of reactions forming precursors of radioxenon

Reaction	Ultramafic	Basalt	Granite	Clay	Sandst.	Carbonate
[235]U(n, f)[131m]Xe	0.002	0.002	0.001	0.002	0.002	0.008
[238]U(n, f)[131m]Xe	0.039	0.039	0.028	0.046	0.042	0.203
[232]Th(n, f)[131m]Xe	1	1	1	1	1	1
[130]Te(n, γ)[131]Te	0.916	0.000	0.000	0.000	0.000	0.001
[133]Cs$(n, {}^3He)$[131]I	0.000	0.000	0.000	0.000	0.000	0.000
[135]Cs$(n, n\alpha)$[131]I	0.000	0.000	0.000	0.000	0.000	0.000
[235]U (n, f)[133]Xe	0.172	0.172	0.172	0.172	0.172	0.172
[238]U(n, f)[133]Xe	1	1	1	1	1	1
[232]Th(n, f)[133]Xe	0.003	0.003	0.004	0.002	0.002	0.000
[133]Cs(n, p)[133]Xe	0.725	0.007	0.010	0.027	0.008	0.002
[134]Ba$(n, 2p)$[133]Xe	0.000	0.000	0.000	0.000	0.000	0.000
[135]Ba$(n, {}^3He)$[133]Xe	0.000	0.000	0.000	0.000	0.000	0.000
[135]Cs$(n, {}^3He)$[133]I	0.000	0.000	0.000	0.000	0.000	0.000
[135]Cs(n, nd)[133]Xe	0.000	0.000	0.000	0.000	0.000	0.000
[135]Cs(n, t)[133]Xe	0.000	0.000	0.000	0.000	0.000	0.000
[136]Ba(n, α)[133]Xe	0.006	0.005	0.004	0.004	0.002	0.000
[137]Ba$(n, n\alpha)$[133]Xe	0.000	0.000	0.000	0.000	0.000	0.000
[235]U(n, f)[135]Xe	0.005	0.005	0.003	0.005	0.005	0.005
[238]U (n, f)[135]Xe	1	1	0.706	1	1	1
[232]Th(n, f)[135]Xe	1.000	1.000	1	0.859	0.944	0.193
[135]Cs(n, p)[135]Xe	0.000	0.000	0.000	0.000	0.000	0.000
[136]Ba$(n, 2p)$[135]Xe	0.000	0.000	0.000	0.000	0.000	0.000
[137]Ba$(n, {}^3He)$[135]Xe	0.000	0.000	0.000	0.000	0.000	0.000
[138]Ba(n, α)[135]Xe	0.029	0.024	0.014	0.018	0.008	0.000

For ease of comparison, the rates have been normalised to the highest fission reaction rate separately for each mass. The rock compositions are taken from FABRYKA-MARTIN (1988)

$$P_{sp,i} = (\rho_{at} \cdot V \cdot \lambda \cdot b^{sf} \cdot y^{sf})_{U238,i} \qquad (2)$$

$$P_{ind,i} = \sum_{\substack{target= \\ ^{235}U, ^{238}U, ^{232}Th}} V \cdot \rho_{target}$$
$$\int_E \sigma_i(E) \cdot \Phi_n(E) \cdot y^{ind}_{E,target,i} dE \qquad (3)$$

with the atomic densities ρ_{at}, the volume V, the decay constant $\lambda_{U238,i}$ and the ratio of decay by fission $b^{sf}_{U238,i}$ for ^{238}U. The spontaneous fission yields $y^{sf}_{U238,i}$ and the incident neutron energy dependent induced fission yields $y^{ind}_{E_n,i}$ are obtained as cumulative yields from ENGLAND and RIDER (1994). The fission cross section $\sigma_{(n,f)}(E_n)$ also depends on the energy E_n of the incoming neutrons and is obtained from the ENDF-VII.0 data base (CHADWICK et al., 2006). To account for the resonances of the cross sections and the neutron energy spectrum, the induced fission production rates $P_{ind,i}$ are calculated by integrating over the whole energy spectrum of the neutron flux $\Phi_n(E)$. All neutron flux spectra are constructed as described in Sect. 2 and Fig. 1.

The resulting concentration C^{Xe}_{vol} of xenon isotopes in the rock volume is multiplied by the xenon emanation coefficient E_{Xe} (see Sect. 2), which is the fraction of xenon escaping from the rock into the soil gas, and divided by the porosity to obtain the potential soil gas xenon concentration of

$$C^{Xe}_{gas} = E_{Xe} \cdot C^{Xe}_{vol} / \phi \qquad (4)$$

where the porosity ϕ is defined as the ratio of void space per volume of material.

Radioactive xenon isotopes, most prominently ^{135}Xe, have relatively high neutron cross sections, therefore the elimination of xenon by neutron radiation was included in the model. However, this has not significantly altered the xenon concentration for the relatively low neutron fluxes occurring in natural scenarios, as most fission is spontaneous.

3.2. Xenon Emanation Process

The emanation of radiogenic noble gases from rocks is complex and not well understood. While a vast amount of literature is available on the topic of radon emanation, xenon emanation is as yet not well researched. It can be argued, however, that the emanation process of xenon is similar to that of radon in many respects.

Most rocks containing uranium are of granular structure. Upon its creation by fission, a xenon precursor nucleus receives a recoil energy of about 74 MeV, being deposited either in its original grain, an adjacent grain, or the pore space between those grains containing the soil gas. A nucleus deposited in the pore space becomes part of the soil gas and can be acquired and measured by the sampling device. However, due to the low density and stopping power compared with the surrounding rock, only a small fraction of the xenon precursors are deposited in the pore space directly.

The emanation coefficient of a nuclide is defined as the fraction entering the pore gas. Radon nuclei are created by uranium decay with a recoil energy of about 100 keV. Still, experiments show radon emanation coefficients of up to $E_{Rn} = 0.5$, even for dense rocks, such as granites, a fact which can not be explained by simple diffusion of the noble gas through the rock (JUST, 1990).

This phenomenon is generally attributed to the effect of damage tracks along the path of the recoiled nucleus, along which the diffusibility is greatly increased (BALLENTINE and BURNARD, 2002), allowing a noble gas nucleus to escape if the damage track crosses the boundary and pore space between two adjacent grains. This effect should be similarly pronounced for xenon and radon, as a xenon precursor nucleus' range is more than half of an alpha particle's range, due to the high fission recoil energy, as shown in HEBEL (2008). In addition, the diffusion coefficient of xenon will generally be higher than that of radon due to its smaller mass. It is therefore possible to assume that the xenon emanation coefficient for a rock will be higher than the respective radon emanation coefficient. Consequently, a range of $E_{Rn} < E_{Xe} < 1$ will be considered in the following.

The emanation is further facilitated by the effects of humidity. Water-filled pores have a higher stopping power, increasing emanation coefficients by a factor of up to 4 (SEMKOW, 1990). Humidity also catalyses the concentration of uranium near the grain surface, placing a high percentage of uranium within recoil range of the pore space (MARTEL et al., 1990), enabling it to contribute to emanation. While

Table 2

Average xenon activities in mBq/m³ expected for some typical rock types

Rock type	ϕ (%)	C_U (ppm)	C_{gas}^{Xe31m}	C_{gas}^{Xe133}	C_{gas}^{Xe133m}	C_{gas}^{Xe135}
Granite	1.00	4.5	1.7E−1	152.2	4.3E+0	139.3
Granodiorite	1.00	3.9	1.8E−1	159.9	4.5E+0	146.3
Gneiss	1.00	4.0	6.7E−2	59.6	1.7E+0	54.5
Syenite	1.00	0.8	8.8E−3	7.8	2.2E−1	7.1
Pegmatite	1.00	3.2	1.6E−2	14.2	4.0E−1	13.0
Gabbro	0.15	0.8	2.4E−2	21.8	6.1E−1	19.9
Basalt	0.55	0.8	4.6E−3	4.1	1.2E−1	3.8
Quartzite	0.30	3.0	6.0E−2	53.8	1.5E+0	49.2
Sandstone	15.0	4.8	1.6E−3	1.5	4.1E−2	1.3
Marl	7.50	2.0	6.1E−4	0.5	1.5E−2	0.5
Tuff	27.0	29.7	1.4E−3	1.3	3.5E−2	1.2
Limestone	12.5	2.0	2.5E−4	0.2	6.2E−3	0.2
Soil, granite	25.0	4.5	5.9E−3	5.3	1.5E−1	4.8
Soil, lime	25.0	2.0	1.8E−3	1.6	4.4E−2	1.5
Soil, clay	25.0	4.8	4.4E−3	3.9	1.1E−1	3.6
Soil, sand	25.0	4.8	1.2E−3	1.0	2.9E−2	1.0
Soil, volcanic	25.0	29.7	4.4E−2	39.4	1.1E+0	36.0
Monazite sand	25.0	7.0E+3	8.86	7.8E+3	2.2E+2	7.2E+3
Uranium ore	10.0	1.2E+5	397	3.5E+5	9.9E+3	3.2E+5
MDC			3.4	3.0	2.0	9.0

The average porosities ϕ taken from PRESS and SIEVER ((2003) are given for each rock type, as well as the average uranium concentrations C_U in wt.-ppm from JUST (1989). The MDCs have been calculated for sampling volumes of 1 m³ (SAUNA system) by CTBTO-PREPCOM (2007)

estimating the influence of water content, one should be aware that underground nuclear tests are commonly conducted in arid environments for containment reasons (U.S. OTA, 1989).

3.3. Xenon Concentrations in Different Rock Types

Using the model described in Sect. 1, the expected concentrations for a representative selection of rock types have been calculated and listed in Table 2. They are based on average uranium concentrations for the respective rock types taken from JUST (1989), as well as the emanation coefficients shown in Table 3. However, it should be noted that rock properties are highly variable and the presented values are averages.

The results listed in Table 2 were compared with typical minimum detectable concentrations (MDCs) of state of the art sampling and measurement equipment, and suggest that natural lithospheric ^{133}Xe and [135]Xe can have a detectable concentration especially in rock types with a low porosity, in which the emitted radioxenon is concentrated into the small pore space. [133m]Xe could be detectable in granites

and granodiorites under favourable conditions, whereas the detection of [131m]Xe cannot be expected in common rock and soil types.

Detectability is not only a function of soil gas concentration, but also of permeability; the measure of how well fluids and gases are transmitted through the medium. The influence of permeability is complex. In rocks with a low permeability, the bulk of soil gas may be trapped in isolated pore space while only the fraction within the macroscopic conduits can be extracted. Due to this effect, the concentrations for solid rocks given in Table 2 are presumably higher than the respective measurement results would be. Gases in highly permeable soils and sands can leak into the atmosphere due to diffusion and pressure variations, also lowering the equilibrium concentration of radioxenon.

Rock formations of extremely high uranium content can contain concentrations which are well detectable, for example uranium ores. Uranium deposits are geologically diverse and may contain uranium concentrations from 0.01% for low grade ores up to 85% for pure uraninite (UO_2). As an example, the radioxenon concentration for a high

Table 3

Radon emanation coefficients for different types of rock, sediment and soil, compiled by JUST (1990)

	E_{Rn} (%)
Rock type	
Granite	<32.7
Granodiorite	<40.0
Gneis	<14.1
Syenite	9.3
Pegmatite	4.3
Gabbro	3.6
Basalt	2.5
Sediment type	
Quartzite	5.3
Sandstone	5.2
Marl	2.6
Tuff	1.7
Limestone	1.6
Soil type	
Granite	46.0
Lime	33.0
Clay	30.0
Sand	9.0
Vulcanic	49.0

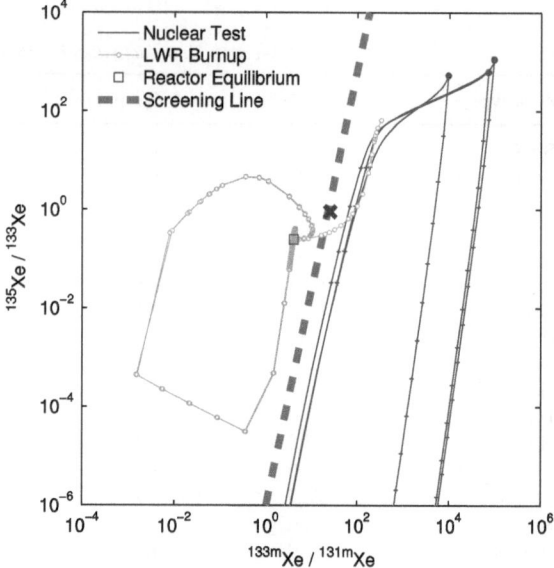

Figure 2

Isotope activity ratio (marked by X) resulting from spontaneous fission of ^{238}U

grade ore from Cigar Lake has been calculated based on the ore properties given by LIU and NERETNIEKS (1996). It is composed of uraninite and coffinite (USiO$_4$) in a clay matrix and has a high concentration of all relevant isotopes. Another example is monazite sand, which contains up to 0.7% uranium and 15% thorium and considerable radioxenon concentrations, nonetheless its high porosity could prevent a measurable concentration of all four isotopes from building. The monazite beaches in existence are considered a rare radioactive anomaly (EISENBUD and GESELL, 1997).

3.4. Concentration Ratios

KALINOWSKI and PISTNER (2006) have described how to utilise the activity ratios of the four xenon isotopes for differentiating between xenon samples originating in reactors and nuclear test explosions by regarding their position in a 4-isotope activity ratio plot. Spontaneous fission produces the constant equilibrium ratio plotted in Fig. 2. This ratio is on the screening line, and might be categorised as a test explosion signal in case of a large error margin. The

screening line runs parallel to decay, hence the isotope ratio will stay on the screening line even after the xenon is released from its source rock.

3.5. Influence of Neutron Flux Variations

While the spontaneous fission activity of uranium is only dependent on the prevalent ^{238}U concentration, the induced fission activity varies with the neutron flux. As shown in Fig. 3, a neutron flux of over $\Phi_n \approx 10^4 \frac{n}{\text{s} \cdot \text{cm}^2}$ is required for induced fission xenon to outweigh spontaneous fission xenon. Neutron fluxes of this magnitude are not found in nature, not even in high radiation background areas. Depending on uranium content and depth, natural lithospheric neutron fluxes will usually range from 10^{-6} to $10^{-3} \frac{n}{\text{s} \cdot \text{cm}^2}$. It can be concluded that the neutron flux is irrelevant to the xenon concentration. Xenon production from tellurium, barium and caesium, which was estimated in Sect. 2 to contribute less than the spontaneous fission of ^{238}U, is similarly irrelevant.

Consequently, the influence of the neutron flux on the xenon isotope ratios is equally small. For a high

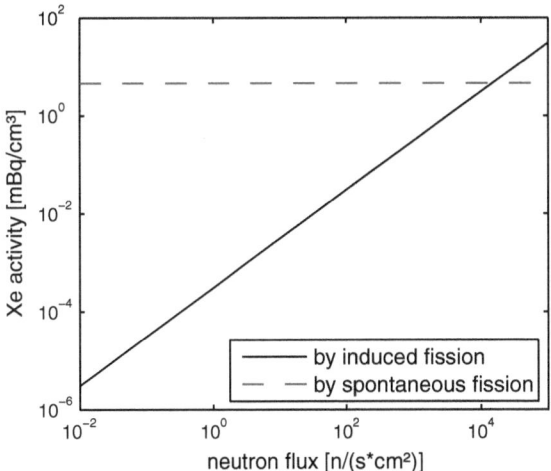

Figure 3

Comparison of ^{133}Xe concentrations generated from spontaneous and induced fission as a function of the absolute neutron flux, calculated for granite, before emanation

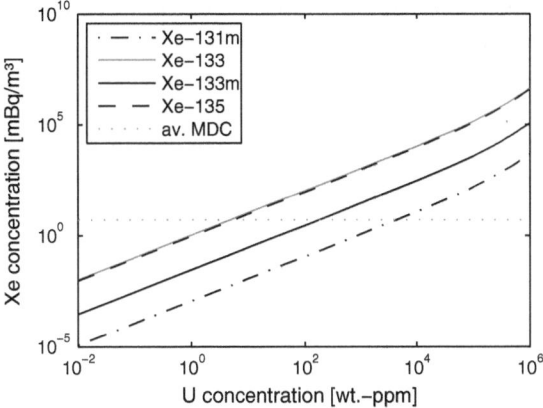

Figure 5

The activity concentration of xenon isotopes in granite before emanation at a depth of 1 m, dependent on the uranium concentration in wt.-ppm ($\frac{\mu g}{kg}$). In typical granite, the U concentration ranges from 0.1–30 ppm. An average MDC from the SAUNA measurement system, $5 \frac{mBq}{m^3}$, is given for comparison

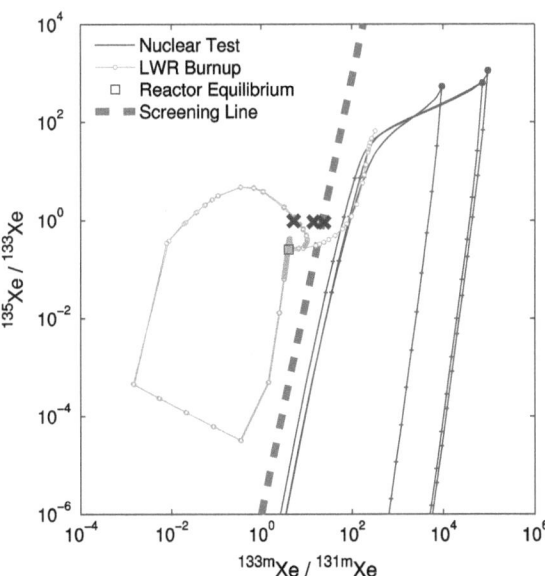

Figure 4

Isotope activity ratios in typical granite for neutron fluxes of $10^2 \frac{n}{s \cdot cm^2}$ (right), $10^{3.5} \frac{n}{s \cdot cm^2}$ (centre) and $10^5 \frac{n}{s \cdot cm^2}$ (left). As the flux increases, induced fission becomes the dominant xenon source, yielding xenon concentrations within the reactor domain

4. Conclusion

The calculations suggest that the dominant source of lithospheric radioxenon is the spontaneous fission of ^{238}U, due to the low neutron fluxes common in natural lithospheric environments. The resulting xenon background is proportional to the uranium content of the rock (see Fig. 5) and largely independent of depth and cosmic radiation. While this enables an estimation of the radioxenon equilibrium based on the uranium content, the theoretical xenon concentrations presented in this work are likely exaggerated, as they ignore the effects of permeability on the measurement results.

However, especially in minerals with a high uranium content such as uranium ore or monazite, a detectable radioxenon background is likely. High uranium environments can easily be identified using common radiation detection equipment. A small rock or soil sample from the site of soil gas measurement is sufficient to determine the uranium content using, for example, mass spectrometry methods.

The isotope ratio of the radioxenon background is constant as seen in Fig. 2 and resembles neither common reactor nor explosion emissions.

In this work, only rough estimations were made regarding the emanation and transport of radioxenon. As the emanation coefficient is directly proportional

neutron flux, the isotope ratios in the 4-isotope diagram are positioned within the reactor domain. If natural fluxes are present, the isotope ratio from ^{238}U is dominant (see Fig. 4).

to the concentration, further research of the emanation process and the influence of permeability is required to estimate the xenon background in soil gas samples more precisely. The measurement of actual subsoil radioxenon concentrations would be worthwhile to compare and verify the concentration levels derived in this article. It would also be useful to research the influence of advective gas transport due to atmospheric pressure variations on the equilibrium concentration.

Acknowledgments

This article is based on a diploma thesis by HEBEL (2008) at the Carl Friedrich von Weizsäcker-Centre for Science and Peace Research (ZNF), which is funded by the German Foundation for Peace Research (Deutsche Stiftung Friedensforschung, DSF).

REFERENCES

BALLENTINE, C.J., and BURNARD, P.G. (2002), *Production, release and transport of noble gases in the continental crust*. In *Noble Gases in Geochemistry and Cosmochemistry*, eds. DONALD P, Chris JB, and RAINER W, The Mineralogical Society of America, vol. 47 of *Reviews in Mineralogy and Geochemistry*, chap. 12, pp. 481–538. ISBN 0-939950-59-6.

CHADWICK, M.B., OBLOŽINSKÝ, P., HERMAN, M., *et al.* (2006), *ENDF/B-VII.0: next generation evaluated nuclear data library for nuclear science and technology*, Nuclear Data Sheets, *107*(12), 2931–3118.

CHARALAMBUS, S. (1971), *Nuclear transmutation by negative stopped muons and the activity induced by the cosmic ray muons*, Nuclear Phys. A, *166*(2), 145–161

CHAZAL, V., RISSOT, R., CAVAIGNAC, J.F., CHAMBON, B., DE JESUS, M., DRAIN, D., GIRAUD-HERAUD, Y., PASTOR, C., STUTZ, A., and VAGNERON, L. (1998) *Neutron background measurements in the underground laboratory of Modane*, Astroparticle Physics, *9*, 163–172.

CTBTO-PREPCOM (2007), *Development, Demonstration, Testing and Evaluation of On-Site Inspection Equipment for Xenon Sampling, Separation and Measurement*. Tech. Rep. CTBT/PTS/TR/2007-1, CTBTO Preparatory Commission.

EISENBUD, M., and GESELL, T. (1997), *Environmental Radioactivity—From Natural, Industrial, and Military Sources* (Academic Press, 1997) 4th edn. ISBN 0-12-235154-1.

ENGLAND, T.R., and RIDER, B.F. (1994), *Evaluation and Compilation of Fission Product Yields*. Tech. Rep. ENDF-349, Los Alamos National Laboratory, http://t2.lanl.gov/publications/yields. LA-UR-94-3106.

FABRYKA-MARTIN, JT. (1988), *Production of radionuclides in the earth and their hydrogeologic significance, with emphasis on chlorine-36 and iodine-129*. Ph.D. thesis, University of Arizona.

HEBEL, S. (2008), *Lithospheric radioxenon and its influence on the interpretation of on-site inspection measurements verifying the Comprehensive Nuclear-Test-Ban Treaty*. Master's thesis, University of Hamburg. Conducted at the Carl Friedrich von Weizsäcker-Centre for Science and Peace Research.

JUST, G. (1989), *Umweltschutz, Umweltgestaltung*—Heft 1: Physik und Chemie atmosphärischer Immissionen. Beiträge des Arbeitskreises "Ökologie und Umweltgestaltung" der Karl-Marx-Universität Leipzig.

JUST, G. (1990), *Fallstudie Erzgebirge: Natürlicher Strahlungsuntergrund, Radonbelastung, Austrag von Radionukliden aus dem Altbergbau*, Zusammenfassung der Ergebnisse, Schlussfolgerungen, Hinweise.

KALINOWSKI, MB., and CHRISTOPH P. (2006), *Isotopic signature of atmospheric xenon released from light water reactors*, J. Environ. Radioactivity, *88*, 215–235.

LIU, J., and NERETNIEKS, I. (1996), *A model for radiation energy deposition in natural uranium-bearing systems and its consequences to water radiolysis*, J. Nuclear Mater. *231*, 103–112.

MARTEL, D.J., O'NIONS, R.K., HILTON, D.R., and OXBURGH, E.R. (1990), *The role of element distribution in production and release of radiogenic helium: The Carnmenellis Granite, southwest England*, Chem. Geology *88*, 207–221.

PRESS, F. and SIEVER, R. (2003), *Allgemeine Geologie—Einführung in das System Erde* (Elsevier GmbH, Spektrum Akademischer Verlag), 3rd edn. ISBN-13: 987-3-9274-0304-0.

SEMKOW, T.M. (1990), *Recoil-emanation theory applied to radon release from mineral grains*, Geochimica et Cosmochimica Acta *54*, 425–440.

U.S. CONGRESS, OFFICE OF TECHNOLOGY ASSESSMENT (1989), *The containment of underground nuclear explosions*, Tech. Rep. OTA-ISC-414, Washington, DC: U.S. Government Printing Office.

(Received November 4, 2008, revised July 24, 2009, accepted July 27, 2009, Published online January 19, 2010)

Pure Appl. Geophys. 167 (2010), 471–486
© 2010 Birkhäuser Verlag, Basel/Switzerland
DOI 10.1007/s00024-009-0027-y

Ten Years of Development of Equipment for Measurement of Atmospheric Radioactive Xenon for the Verification of the CTBT

MATTHIAS AUER,[1] TIMO KUMBERG,[1] HARTMUT SARTORIUS,[1] BERND WERNSPERGER,[2] and CLEMENS SCHLOSSER[1]

Abstract—Atmospheric measurement of radioactive xenon isotopes (radioxenon) plays a key role in remote monitoring of nuclear explosions, since it has a high capability to capture radioactive debris for a wide range of explosion scenarios. It is therefore a powerful tool in providing evidence for nuclear testing, and is one of the key components of the verification regime of the Comprehensive Nuclear-Test-Ban Treaty (CTBT). The reliability of this method is largely based on a well-developed measurement technology. In the 1990s, with the prospect of the build-up of a monitoring network for the CTBT, new development of radioxenon equipment started. This article summarizes the physical and technical principles upon which the radioxenon technology is based and the advances the technology has undergone during the last 10 years. In contrast to previously used equipment, which was manually operated, the new generation of radioxenon monitoring equipment is designed for automated and continuous operation in remote field locations. Also the analytical capabilities of the equipment were strongly enhanced. Minimum detectable concentrations of the recently developed systems are well below 1 mBq/m^3 for the key nuclide ^{133}Xe for sampling periods between 8 and 24 h. All the systems described here are also able to separately measure with low detection limits the radioxenon isotopes ^{131m}Xe, ^{133m}Xe and ^{135}Xe, which are also relevant for the detection of nuclear tests. The equipment has been extensively tested during recent years by operation in a laboratory environment and in field locations, by performing comparison measurements with laboratory type equipment and by parallel operation. These tests demonstrate that the equipment has reached a sufficiently high technical standard for deployment in the global CTBT verification regime.

Key words: Comprehensive Nuclear-Test-Ban Treaty (CTBT), International Monitoring System (IMS), environmental monitoring, noble gas, radioxenon.

1. Radioxenon Measurements for Monitoring of Nuclear Tests

After many years of negotiation, the vote for the Comprehensive Nuclear-Test-Ban Treaty (CTBT) by the General Assembly of the United Nations on 10 September, 1996 marked a major milestone in nuclear nonproliferation (RAMAKER *et al.*, 2003; HANSEN, 2006). This treaty obliges its member states "not to carry out any nuclear weapon test explosion or any other nuclear explosion, and to prohibit and prevent any such nuclear explosion at any place under its jurisdiction of control" [Text of the CTBT, Article 1]. The implementation of this treaty is largely based on a verification system for monitoring nuclear tests underground, underwater and in the atmosphere in order to minimize the chance of clandestine testing. A crucial component of this verification regime is the International Monitoring System (IMS) which consists of 321 facilities worldwide and, once completed, will provide an unprecedented global coverage of facilities to monitor nuclear tests. The IMS is based on the combination and the synergy of facilities for seismic (BARRIENTOS and HASLINGER, 2001), hydroacoustic (LAWRENCE *et al.*, 2001), infrasound (CHRISTIE *et al.*, 2001) and radionuclide monitoring (MATTHEWS and SCHULZE, 2001). In order to fulfil the ambitious task of the IMS in an optimized manner, state-of-the-art equipment is deployed at IMS stations. Therefore, with the onset of the CTBT, development of new equipment was triggered in all four monitoring technologies. The radionuclide component of the IMS consists of eighty facilities for monitoring aerosol-bound radioactive fission products (denoted as particulates), of which forty facilities are planned to be

———————
[1] Federal Office for Radiation Protection, Rosastr. 9, 79098 Freiburg, Germany. E-mail: cschlosser@bfs.de
[2] Provisional Technical Secretariat of the Comprehensive Nuclear-Test-Ban Treaty Organization, 1400 Vienna, Austria.

equipped for monitoring radioactive xenon isotopes (denoted as radioxenon or noble gas) (see Fig. 1) and of 16 radionuclide laboratories, which provide further analysis of samples from the stations (KARHU and CLAWSON, 2001). At the first annual conference of state parties of the CTBTO, which will take place after entry into force of the CTBT (Text of the CTBT, Protocol, Part I, C, p. 130), the state parties will decide whether noble gas systems will be deployed throughout the IMS radionuclide network. This article focuses on the technical developments and advances in monitoring of radioactive xenon isotopes, which have been achieved during the last approximately 10 years and presents results that recommend a strong xenon role for monitoring the atmosphere.

Measurement of the radioactive xenon isotopes ^{131m}Xe ($T_{1/2} = 11.84$ days), ^{133}Xe ($T_{1/2} = 5.24$ days), ^{133m}Xe ($T_{1/2} = 2.19$ days) and ^{135}Xe ($T_{1/2} = 9.14$ h) (hereafter termed radioxenon) in the atmosphere is a crucial component of the IMS in particular for the detection of clandestine and underground tests (CARRIGAN et al., 1996; PERKINS and CASEY, 1996; DE GEER, 1996a): Release of fission products other than noble gases from an underground test is unlikely and restricted to accidentally venting tests or operational releases (BJURMAN et al., 1990), while for noble gases,

which are chemically inert, there is a considerably higher chance to be released through cracks and fissures in the ground. Xenon isotopes have the highest fission yield among noble gases with cumulative fission yields of 6–7% for ^{133}Xe, for a ^{235}U or ^{239}Pu fission device, respectively (ENGLAND and RIDER, 1993). This makes them the first choice as indicators for underground nuclear explosions. In addition, the only sink for radioxenon in the atmosphere is its radioactive decay, hence, the atmospheric residence time is equal to the mean radioactive lifetime, thus allowing for long-range transport in the atmosphere. Also, the typical atmospheric background level is low, e.g., in the order of 1 mBq/m^3 in Central Europe, Japan and US (DE GEER, 1996b, WEISS et al., 1997; BOWYER et al., 1997, 2002; IGARASHI et al., 2000), due to lack of significant natural sources and due to the relatively short half-lives of the radioxenon isotopes. In contrast to radioxenon, the high atmospheric background of ^{85}Kr, currently approximately 1.5 Bq/m^3 in the Northern Hemisphere (HIROTA et al., 2004) with a high temporal and geographical variability (WEISS et al., 1992; WINGER et al., 2005), makes the atmospheric concentration of this isotope relatively insensitive to emissions of nuclear tests. In addition, the fission yield of this isotope (cumulative

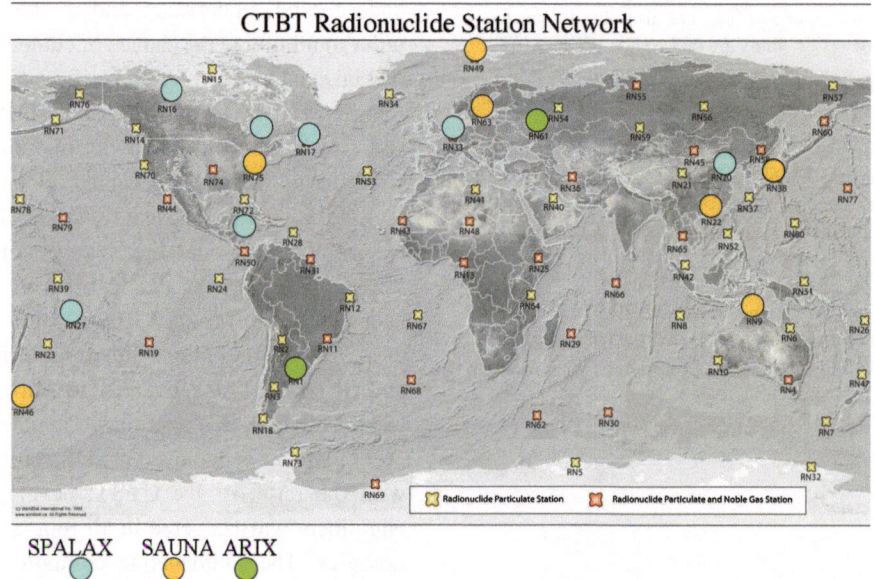

Figure 1
Radionuclide station network of the International Monitoring System. Blue, yellow and green dots indicate the system types at noble stations installed by September 2008 (*Source*: CTBTO IMS-Engineering and Development Section)

yield 0.2–0.4% (ENGLAND and RIDER, 1993) is approximately a factor 20 lower than the yield of, e.g., [133]Xe. Note that the fission yield for isobars of mass 85 is relatively high (approximately 1.5%), the low cumulative yield of [85]Kr is highly, because its precursor [85m]Kr beta-decays with a branching ratio of 79% to [85]Rb. The only other noble gas isotope, which aside from the xenon isotopes may be a sensitive indicator for an underground nuclear explosion is [37]Ar ($T_{1/2}$ = 35 days) (LOOSLI, 1992). This isotope has a negligible natural and a very low man-made background on the order of 1 mBq/m^3 in the atmosphere (LOOSLI, 1992) and approximately up to two orders of magnitude higher in soil gas (Purtschert, pers. comm.). [37]Ar is produced in underground nuclear explosions by neutron activation of [40]Ca contained in the rock surrounding the explosion. Measurement of [37]Ar is however technically challenging. Firstly, because the energy of the Auger electrons emitted from the electron capture decay of [37]Ar is low (2.82 keV) and secondly, the high Ar content in the air results in a large sample size. Currently there is only one laboratory worldwide where [37]Ar is measured routinely in environmental samples, which is the low-level underground laboratory of the University of Bern (LOOSLI et al., 1986). With the present available technology, [37]Ar can however, not be measured continuously in field-operated stations, although it is a very valuable and feasible technique for on-site inspections, where selective samples are taken close to the site of a presumed underground nuclear explosion.

What makes the measurement of radioxenon, and radionuclide measurements in general, an important and even indispensable component of the IMS is its unique capability to identify such a signal as originating from a nuclear explosion by measurement of the fission products. Using information on atmospheric transport processes, the signal measured by a radionuclide station can be attributed to a potential source region (WOTAWA et al., 2003). The localisation is also supported by taking into account information from seismic, hydro-acoustic or infrasound measurements. Measurement of radioxenon has a long history in the monitoring of nuclear activities. Already during the Second World War, the United States conducted over-flights in Germany in 1944 to

search for nuclear reactors (ZIEGLER and JACOBSON, 1995). Since then, measurements of radioxenon have been widely used to monitor the release of fission products either from nuclear explosions or from civil nuclear activities (e.g., SCHÖLCH et al., 1966; IGARASHI et al., 2000). A recent example for the importance of radioxenon may be the test conducted by the Democratic Peoples Republic of North Korea on 9 October, 2006. To date the most convincing evidence for the nuclear character of the explosion is the detection of radioxenon by different groups (SAEY et al., 2007; RINGBOM et al., 2009; BECKER et al., 2010).

2. General Requirements for IMS Noble Gas Equipment

Previous radioxenon measurement technology was based on manual analysis systems which were operated in a typical laboratory environment (EHHALT et al., 1963; SCHÖLCH et al., 1966; LUDWICK, 1966; STOCKBURGER et al., 1977; BERNSTRÖM and DE GEER, 1983; KUNTZ, 1989). With the recent development of the IMS, new requirements for monitoring equipment have been set. The noble gas component of the IMS is based on the principle of a globally uniform monitoring coverage at a high time resolution (the sampling time is equal or less than 24 h) and a high sensitivity of the measurement systems (detection sensitivity for ^{133}Xe less or equal to 1 mBq/m^3). The high time resolution is of particular importance first, in order to constrain the time window for the atmospheric transport calculations for source location and second, in order to reduce the chance of interference with background, e.g., from civil sources. Stations are also often at remote sites with very limited infrastructure and access. This requires a high technical standard and reliability of the equipment in order to meet the stringent requirements for IMS stations. In addition, manual operation of radioxenon equipment requires a high level of technical expertise, which is not necessarily available at IMS sites and which would involve extensive and costly training of operators. Automated systems are therefore a more suitable choice for IMS sites than manual systems. The previously used and laboratory-based equipment was not designed for such needs. This equipment usually required a high degree of manual operation by a system expert, which allowed the operator to interactively perform the sample analysis in order to adjust the analysis to the specific needs of the sample and optimize the results. This is not possible for the IMS, where a large amount of data has to be produced (40 noble gas stations with currently up to two samples per day). The data must be of uniform quality and quickly available at the International Data Center (IDC). The time allowed for reporting from the station to the IDC is 48 h after start of sampling.

In addition to these new operational challenges, the task to monitor nuclear tests, distinct from other nuclear activities, also sets very specific requirements for the analytical capabilities of the equipment. In particular for remote monitoring, high sensitivity of the equipment is required. Assuming venting of only independently produced ^{133}Xe, a 1 kT underground nuclear explosion may release in the order of 10^{14} Bq ^{133}Xe into the atmosphere. The order of distance from any possible source location on Earth to a radionuclide station is given by the typical distance between IMS radionuclide stations, which is approximately 1,000 km. Within this distance, the debris released from a nuclear explosion is diluted by a factor 10^{-14} to 10^{-18} (A. Becker, pers. comm.). Therefore, with a detection limit of 1 mBq/m^3 for ^{133}Xe, a release from a 1 kT nuclear explosion is likely to be detected by the IMS noble network (CD/NTB/WP.224, 1995).

The strength of a radioxenon signal alone is for many sampling locations an insufficient criterion, since civil sources may cause ambient concentrations of a few mBq/m^3 with short spikes of 10^2 to 10^3 mBq/m^3, in particular in areas with a high density of nuclear facilities, such as Northern Europe, Asia or Northern America (e.g., AUER et al., 2004; STOCKI et al., 2005). The distinction can only be made by the isotopic composition, which is usually different for nuclear explosions and other nuclear activities (KALINOWSKI et al., 2010) and by taking into account the characteristic concentrations and their typical variation at a specific site. For a long time, it was assumed that nuclear explosions are uniquely characterized by high ratios of ^{135}Xe/^{133}Xe or ^{133m}Xe/^{133}Xe. However, during a test of noble gas equipment in Freiburg in 2000, for the first time

relatively high ^{135}Xe/^{133}Xe ratios of nearly 10 were measured in ambient air (AUER et al., 2004). The estimated upper ratios of ^{133m}Xe/^{133}Xe were consistent with releases from a nuclear power reactor during the start-up phase (BOWYER et al., 2002) (during the start-up phase the radioisotope ratios are different from the equilibrium ratios). This indicated that single isotopic ratios are not sufficient for discrimination of nuclear tests from civil sources and also that the station specific history of concentrations and atmospheric backtracking must be taken into account. In order to classify a measurement as related to a nuclear explosion, detection of more than two isotopes provides the most reliable evidence (KALINOWSKI et al., 2010). This also sets very specific and stringent requirements to the detection capabilities of the equipment.

During the design phase of the IMS, a set of minimum requirements for IMS radioxenon systems has been defined by the policy-making organ (Preparatory Commission) of the CTBTO (see Table 1). These requirements partly focus on operational requirements of the IMS network, but also specify the detection requirements for IMS radionuclide stations,

Table 1

Minimum requirements for IMS radioxenon systems

Characteristics	Minimum requirements
Air flow	0.4 m^3/h
Total volume of sample	10 m^3
Collection time	$\leq$24 h
Measurement time	$\leq$24 h
Time before reporting	$\leq$48 h
Reporting frequency	Daily
Isotopes measured	^{131m}Xe, ^{133m}Xe, ^{133}Xe, ^{135}Xe
Measurement mode	Beta–gamma coincidence or high-resolution gamma spectrometry
Minimum detectable concentration[a]	1 mBq/m^3 for ^{133}Xe
State of health	Status data transmitted to International Data Centre
Communication	Two-way
Data availability	At least 95%
Down-time	No more than 7 consecutive days, No more than 15 days annually

[a] Minimum detectable concentrations for the other isotopes are not defined here since they critically depend on the detection system used

Table 2

Radioxenon systems developed for the International Monitoring System

System	Developer
Automatic Radioanalyzer for Isotopic Xenon (ARIX)	Khlopin Radium Institute (KRI), Russian Federation
Automated Radioxenon Sampler-Analyzer (ARSA)	Pacific National Northwest Laboratories (PNNL), USA
Swedish Automatic Unit for Noble Gas Acquisition (SAUNA)	Totalförsvarets Forskningsinstitut (FOI), Sweden
Système de Prélèvement Automatique en Ligne avec l'Analyse du Xénon (SPALAX)	Departement Analyse, Surveillance, Environnement du Commissariat à l'énergie atomique (CEA/DASE), France

which basically condense to the requirement of the aim of a 90% detection probability of a nuclear explosion of 1 kT TNT equivalent within 14 days after the explosion (SCHULZE *et al.*, 2000). The specific needs of IMS radioxenon stations set out in the minimum requirements triggered development of new types of equipment since the mid-1990s. Within only a few years, four new radioxenon systems have been developed, which all were explicitly designed to fulfil the requirement of IMS stations (Table 2). In addition to the development of automated and stationary equipment, mobile sampling and measurement units have been developed. This type of equipment was specifically developed for on-site inspections and is based on similar principles of the IMS-type equipment (CTBT/PTS/TR/2007-1). Furthermore, the development of IMS equipment also triggered development of new laboratory equipment.

3. Technical Principles of Noble Gas Measurement

3.1. Sampling and Processing of Radioxenon

The sampling principle of all IMS noble gas systems is extraction of xenon from atmospheric air by gas separation, purification and concentration on various adsorption and gas separation media. The sample consists mainly of stable xenon which is an atmospheric trace constituent with a concentration of 0.087 ppm by volume (ppmv). The commonly used main mechanism to extract xenon from air is adsorption on activated charcoal: xenon is more strongly absorbed on activated charcoal than the major atmospheric gas constituents. By passing air through a column filled with activated charcoal and by adjusting temperature and flow rate, xenon can be completely

separated from the air. Absorption is stronger at low temperatures, which is why usually laboratory systems but also the ARSA and ARIX systems perform the adsorption at low temperatures, typically below $-100°C$. Cooling is however not a necessary condition for xenon absorption. The relatively large amount of charcoal in the SAUNA sampling system allows sampling also with moderate cooling to $-5°C$, mainly for moisture removal, using thermoelectric elements (RINGBOM *et al.*, 2003). In the SPALAX system, the air is pre-enriched in Xe with a semi-permeable membrane to 1 ppmv relative to the atmospheric level of 0.087 ppmv (FONTAINE *et al.*, 2004). Due to this pre-enrichment, relatively large amounts of xenon (corresponding to 80 m^3 of air for a 24-h sampling period) can be sampled without cooling of the charcoal. In addition, in both systems sampling is switched between two parallel charcoal columns (6-h cycles for SAUNA and 2-h cycles for SPALAX) in order to compensate the less effective adsorption at high temperatures and to avoid a breakthrough of xenon. These subsamples are combined, in order to obtain composite samples with sampling times of 12 h (SAUNA) and 24 h (SPALAX).

After sampling, in order to remove the adsorbed xenon from the activated charcoal trap, the traps are heated typically to about 250–300°C and flushed with an inert purging gas like He or N_2. The purging gas acts as carrier of the collected xenon gas for further processing and measurement. For continuous sampling, two sampling lines are operated in parallel. After this first concentration step, the sample needs further treatment in order to remove trace impurities like Rn, H_2O and CO_2 and to increase the xenon concentration in the sample. This is done by different system specific arrangements of further gas traps.

3.2. Measurement of Radioxenon Activity Concentrations

Activity measurement is either done by high-resolution gamma-spectrometry or beta–gamma coincidence spectrometry. Currently, only the SPALAX system uses a high-resolution (HPGe) gamma spectrometry system (FONTAINE et al., 2004), the other systems all use the beta–gamma coincidence method.

In high resolution gamma spectrometry systems, the isotopes are quantified by their major gamma lines (Table 3). Due to the high selectivity of the sampling process only radioxenon and small remnants of atmospheric radon enter the detection system (more details on radon contamination below). Therefore, the resulting spectrum has relatively little complexity and peak location and identification is simplified by the limited number of possible peaks. However, the relatively low intensities of the gamma emissions for the decays of ^{131m}Xe and ^{133m}Xe of 1.95% and 10.0%, respectively, make the detection of these nuclides less sensitive than the detection of ^{135}Xe and ^{133}Xe. Analysis of X-rays helps to increase the sensitivity also for the meta-stable isotopes, however since ^{131m}Xe and ^{133m}Xe X-rays have the same energy, X-ray analysis alone gives only an integral measurement of ^{131m}Xe and ^{133m}Xe. Therefore, distinction between the two nuclides has to be made based on the gamma activity or detection limits based on gamma activity measurement. Here in particular the higher intensity of ^{133m}Xe gamma emissions can be used for the quantification of ^{131m}Xe, using the information on the absence or presence of ^{133m}Xe. The average minimum detectable activities (MDA), for example reached with the detector of the SPALAX system installed at the IMS radionuclide station at Schauinsland for a 22-h 40-min. measurement, are 600 mBq for ^{131m}Xe, 10 mBq for ^{133}Xe, 132 mBq for ^{133m}Xe and 47 mBq for ^{135}Xe if only gamma peaks are used for analysis. If only X-rays are present in the spectrum, probabilities for the occurrence of ^{131m}Xe and ^{133m}Xe can be given, based on a Bayesian statistical analysis (ZÄHRINGER and KIRCHNER, 2008).

For beta–gamma coincidence systems, the coincident decays listed in Table 3 are utilized for measurement. The noble gas systems currently deployed at IMS stations use a combination of a plastic scintillator tube as beta detector, which is also used as sample container, which is surrounded by a NaI(Tl) gamma detector. The coincidence measurement leads to a significant reduction of background; e.g., background count rates for an 18-h measurement are typically as low as 0.001 counts per second in the energy interval 70 to 90 keV, depending on the ambient background of the detector location.

Two different types of beta–gamma techniques have been used in IMS systems thus far. The "beta–gamma-energy correlation" method measures the energies of gamma rays and X-rays as well as the energies of beta-particles and conversion electrons. This is of particular relevance for the distinction of ^{131m}Xe and ^{133m}Xe X-rays, since by energy analysis, the conversion electrons of the ^{131m}Xe and ^{133m}Xe decay can be measured, which have relatively high intensities of 56% and 58%, respectively (REEDER and BOWYER, 1998). Thus the detection limit for these isotopes is strongly improved. This method first used in the ARSA system is also deployed in the SAUNA system (RINGBOM et al., 2003) and will be used in the newest generation of the ARIX system. The typical MDA reached with this method for an 18-h measurement is below 10 mBq for all four xenon isotopes. This neglects, however, the considerable memory effect of these systems (see below), which can increase the MDA significantly after a high activity measurement.

In the so-called "beta–gated gamma systems," as used in the earlier versions of the ARIX system, the detector only records gamma energy pulse height spectra but no beta energy spectra. The beta or conversion electrons pulses are only used to trigger recording of the gamma or X-rays. For this reason, ^{131m}Xe and ^{133m}Xe cannot be separated by energy analysis. Instead the different half-lives are used to analytically separate the two nuclides. The analytical power of this method is, however, strongly limited due to the fact that a long measurement time is required to separate ^{131m}Xe and ^{133m}Xe (half-lives 11.84 days and 2.19 days). For this reason, the beta–gated gamma method has recently been abandoned for the ARIX system, which in its newest version also uses a detector-based on the beta–gamma-energy correlation method.

Table 3

Characteristic energies for the decay of ^{131m}Xe, ^{133}Xe, ^{133m}Xe and ^{135}Xe [ENSDF]

	Decay energy (keV)	Branching ratio (%)
^{131m}Xe		
X-rays	29.46	15.4
	29.78	28.6
	33.60	10.2
	34.61	1.85
Gamma rays	163.93	1.95
Conversion electrons	129.4	61
Coincident decays	Sum of 29.46 to 34.61 keV X-rays and 129 keV e$^-$	56.1
^{133}Xe		
X-rays	30.62	14.1
	30.97	26.2
	35.00	9.4
	36.01	1.7
Gamma rays	80.99	37.0
Conversion electrons	45	55.1
Betas (maximum energy)	346	100
Coincident decays	31.63 keV X-ray + 45 keV e$^-$ + 346 keV beta	48.9
	80.98 keV gamma + 346 beta	37.2
^{133m}Xe		
X-rays	29.46	16.1
	29.78	29.8
	33.60	10.6
	34.61	1.9
Gamma rays	233.2	10.0
Conversion electrons	198.7	64
Coincident decays	Sum of 29.46–34.61 keV X-rays and 199 keV e$^-$	58.4
^{135}Xe		
X-rays	30.62	1.45
	30.97	2.69
	35.00	0.97
	36.01	0.185
Gamma rays	249.8	90
	608.2	2.90
Conversion Electrons	214	5.7
Betas (maximum energy)	910	100
Coincident Decays	31.63 keV X-ray + 214 keV e$^-$ + 910 keV beta	5.7
	249.8 gamma + 910 keV beta	90
^{214}Pb		
X-rays	74.82	4.8
	77.11	8.0
	86.83	1.0
	87.35	1.8
	89.78	0.67
	90.07	0.17
Gamma rays	241.95	7.43
	351.95	37.6
^{214}Bi		
X-rays	79.29	0.98

Also shown are gamma and X-ray energies from the ^{222}Rn daughter ^{214}Pb and ^{214}Bi, which may cause interference with radioxenon detection

Common to all beta–gamma systems is the relatively low resolution of the measurement, both in the beta/conversion electron detection (resolution requirements for IMS systems: better than 40 keV at 129 keV energy) and in the gamma detection (IMS resolution requirement: better than 15% FWHM at 80 keV) (CTBT-PTS-INF.921-Rev.3). These detectors can have a memory effect of around 5%, since the porosity of the plastic scintillator material allows xenon to diffuse through the walls of the detector (BEAN et al., 2007). In order to avoid false detection of radioxenon, in particular for a sample following a high activity measurement, the memory effect has to be accounted for by performing a background measurement prior to each sample measurement. The background counts from the background measurement must be subtracted from the counts determined in the sample measurement. This however, leads to a degradation of the minimum detectable activity in case of a measurement following a high activity sample. The high resolution gamma detector of the SPALAX systems does not have such a memory effect, since the sample container used for the activity measurement consists of aluminium.

For all types of detection systems, daughter activities of ^{222}Rn in the sample can cause interference with radioxenon signals. Ambient levels of ^{222}Rn in ground-level air are typically in the order of 10 Bq/m^3 (UNSCEAR, 2000). The half-life of ^{222}Rn is long enough (3.82 days) to survive the several-hour sample processing and thus it can enter the detector cell if not separated from the sample. The half-lives of the ^{222}Rn daughters ^{214}Bi ($T_{1/2} = 19.9$ min) and ^{214}Pb ($T_{1/2} = 26.8$ min), which have gamma and X-ray energies interfering with the radioxenon detection (Table 3), are short enough compared with the counting times of several hours to be in equilibrium with the radon, therefore any contamination with radon can cause a significant increase in gamma background. Due to its high atmospheric concentration, separation of ^{222}Rn has to be efficient, in all systems, the concentration in a sample is typically reduced by a factor of 10^5.

For determination of the activity concentration, the air volume of the sample has to be determined. This is done by quantification of the stable xenon volume in the activity counting cell. Since the concentration of stable xenon in the atmosphere is constant, the volume of stable xenon (V_{Xe}) can be converted into the corresponding air volume (V_{air}) by:

$$V_{air}[m^3] = V_{Xe}[cm^3]/0.087.$$

The stable xenon volume measurement is done either with a gas chromatograph or with a Thermal Conductivity Detector (TCD). The uncertainty of the stable gas measurement is typically 10–15%.

The activity concentration C in Bq/m^3 is calculated by:

$$C = \frac{\text{Sample activity [Bq]}}{\text{Xenon volume [cm}^3\text{]}} \times 0.087 \frac{[cm^3]}{[m^3]}.$$

4. Operational Experience during the Last 10 Years

4.1. The IMS Noble Gas Equipment Test

In parallel to the development of equipment, the Provisional Technical Secretariat (PTS) of the Preparatory Commission (PrepCom) for the CTBTO started the Noble Gas Equipment Test in 1999 (CTBT/PTS/INF.162), which was designed to ensure compliance of the newly developed equipment with the analytical and operational requirements of the IMS. In the course of the test, this programme was then extended to the International Noble Gas Experiment (INGE), which also includes aspects of data analysis, data interpretation, certification and operation of the radioxenon component of the IMS. The noble gas equipment test, should, aside from a technical assessment of the state of the art of noble gas technology, facilitate the transition of operation of the systems in a laboratory environment to automated routine operation in a remote environment under typical IMS-type operational conditions.

The Noble Gas Equipment Test has been divided into an equipment development phase (Phase I), an assessment of the analytical performance of the systems (Phase II) and a test of the operational (and also analytical) performance of the systems and of the network (Phase III), the latter phase being a still ongoing activity. The basic aims of the test show that:

• Noble gas monitoring is a reliable technique for monitoring nuclear tests (Phase II).
• The minimum requirements (Table 1) can be fulfilled (Phase II).
• The systems are suitable for operation under typical IMS conditions (Phase III).

The actual testing of the systems started with Phase II, during which the systems have been installed and operated side-by-side at the German Federal

Office for Radiation Protection (BfS) in Freiburg. Phase II started with the installation of an ARSA system in October 1999 and lasted until February 2001. The systems were tested in routine operation, measuring the ambient radioxenon concentrations, but also by performing spike experiments, i.e., injection of radioactive xenon samples containing mixtures of $^{131m}Xe/^{133}Xe$ and $^{135}Xe/^{133}Xe$ either into the air sampling systems or directly into the detectors of the systems. Occasionally, archived samples have also been re-analyzed at the noble gas laboratory of the BfS. The systems showed a good overall agreement within each other and with the measurements performed at the BfS laboratory. The experiment showed that all systems measured the activity of radioactive xenon isotopes with an accuracy approaching 10–20%. The minimum detectable activity concentrations for the isotope ^{133}Xe ranged between 0.1–1 mBq/m^3, showing that all four test systems fulfil the sensitivity requirements of the IMS. The measurement of archived samples with the BfS laboratory system also demonstrated the feasibility of independent re-analysis with a laboratory-based system for quality control. The test phase also provided a wealth of high resolution data in an area with a high density of nuclear facilities. New findings during this phase were the observation of xenon peak concentrations to over 100 mBq/m^3 with a duration of 12 h and less and the observation of relatively high $^{135}Xe/^{133}Xe$ ratios of up to 10. A detailed description and summary of the procedures and results of this phase of the test are given in AUER et al. (2004).

4.2. Field Testing and First Network Operation Experience

The transition of Phase II to Phase III of the test was accompanied by a transition of prototype systems provided by the system developers to commercial "off-the-shelf" type systems. Regarding the SAUNA and SPALAX systems, these new systems were provided by the commercial companies Gammadata Instrument AB (http://www.gammadata.se) for SAUNA and Environnement S.A. (http://www.environnement-sa.com/) for SPALAX, the new ARIX is still provided by its developer (Khlopin Radium Institute, http://www.khlopin.ru). The ARSA system is not provided as a commercial system, however major technological principles of this system, such as the beta–gamma coincidence detection method, have also been integrated in other systems.

Phase III of the Noble Gas Equipment Test is related to the testing of the operational performance of the systems installed at IMS radionuclide stations, i.e., examination of the verification of the operational parameters in the field, as well as the assessment of operational reliability of the systems. Besides this, the Phase III exercise comprises network-related aspects of the noble gas experiment such as establishment of a network-wide maintenance strategy, preparation of operational manuals, development of certification requirements, establishment of a QA/QC system, atmospheric transport modelling, as well as the development of a categorization scheme for noble gas monitoring. A further aspect of Phase III was the set-up of "mini-networks" in North America, Europe and Asia with four systems per mini-network in order to assess correlations between stations in areas with a relatively high density of nuclear facilities.

With regard to equipment testing, Phase III was divided into three sub-phases to reflect different stages of the field experience: Phase III/a consisted of the installation and tuning of the systems in the field. During this stage the system providers had unrestricted access to their systems for tuning and troubleshooting after installation. Next, Phase III/b was the first period where systems were operated by the station operator and the PTS. Per system type, a minimum time of 6 months for Phase III/b operation was considered necessary in order to collect sufficient data for the evaluation. After each system type has successfully completed Phase III/b, more systems have been installed and are running under Phase III/c with the scope of further testing and evaluation of equipment and the other Phase III/c tasks mentioned above.

The systems operating in Phase III (currently operated systems are shown in Fig. 1) show improved operational parameters compared to the Phase II prototype systems (Table 4).

Furthermore, IMS minimum requirements related to data transmission and operational performance are being assessed during Phase III: All data relevant for monitoring the operational status are transmitted and systems allow for two-way communication between

Table 4

Operational parameters of systems which were participating in Phase III testing

Characteristics	Minimum requirements	SPALAX	SAUNA	ARIX
Airflow	0.4 m^3/h	15–20 m^3/h	1.2 m^3/h	1.6 m^3/h
Total volume of sample	10 m^3	50–75 m^3	25–30 m^3	36 m^3
Collection time	≤24 h	≤24 h	12 h	12 h
Measurement time	≤24 h	≤24 h	11 h 10 min	18 h
Time before reporting	≤48 h	≤48 h	30 h	34 h
Reporting frequency	Daily	Daily	2 samples per day	2 samples per day
Isotopes measured	^{131m}Xe, ^{133}Xe, ^{133m}Xe, ^{135}Xe	^{131m}Xe, ^{133}Xe, ^{133m}Xe, ^{135}Xe	^{131m}Xe, ^{133}Xe, ^{133m}Xe, ^{135}Xe	^{131m}Xe, ^{133}Xe, ^{133m}Xe, ^{135}Xe
Measurement mode	β-γ-coincidence or high-res-γ- spectrometry	High-res-γ- spectrometry	β-γ-coincidence	β-γ-coincidence
MDC for ^{133}Xe	1 mBq/m^3	0.2–0.6 mBq/m^3	0.2–0.4 mBq/m^3	0.2–0.3 mBq/m^3

All values except MDCs are nominal operational values given by system suppliers and were verified during Phase III operation. MDC ranges were assessed during testing

station operators and the PTS. Data availability, down time and related reasons for operational failures are assessed. Data availability has been defined as the number of spectra that are received in a considered period and contain the essential information for spectral processing, compared to the total number of spectra expected for the considered period. Downtime causes are categorized according to the component where a failure occurred, as there are: (1) sampling and gas processing system, (2) nuclear measurement system, (3) gas quantification system, (4) power system, (5) computer and software, (6) human errors, (7) organisational problems and (8) other.

Assessment during the early years of Phase III operation has shown that the main causes of downtime are related to problems in the HPGe detector cooling of the nuclear measurement system, in the gas processing system of the equipment and in the software. Whereas failures in the detector resulted in down-time periods of several weeks to months, software problems and failures in the gas-processing system could be fixed within a couple of days by the station operator, provided that appropriate spare parts were present at the station. The PTS together with the system suppliers are establishing a network-wide maintenance strategy within the scope of Phase III/c to overcome such deficiencies. With a view to data availability during Phase III the systems have shown that they are able to operate according to IMS requirements over several months.

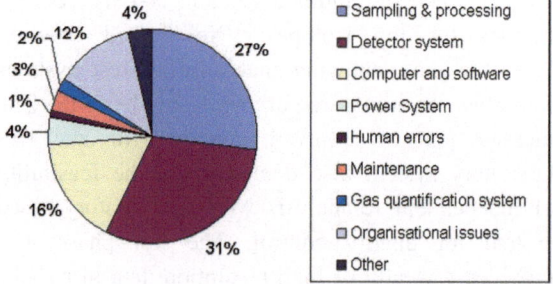

Figure 2
Network wide down-time causes, categorized according to the subsystems where they occurred

However, since the maintenance and data analysis system of the IMS noble gas network is not yet fully operational, quick and effective repair mechanisms are not fully in place and down-times can be several months. For the total IMS noble gas network, down time was 2,487.5 days out of total 9,605 days or 25.9% of operational days between July 2004–July 2008. An overview of typical error causes and their frequencies is shown in Fig. 2.

5. Operational Experience and Data from the Noble Gas Station at the IMS Radionuclide Site at Schauinsland

The IMS location RN33 at Schauinsland, Germany (47°54′N, 7°54′E) is one of the forty radionuclide stations which host, beside a particulate system, noble gas equipment. The site is located on a mountain ridge (approx. 1,200 m asl) in the Black Forest overlooking the Rhine Valley. It has a long history in monitoring of nuclear explosions. Initially used for measurement of cosmic radiation in the 1950s, already in 1953 debris from atmospheric nuclear tests was detected (SITTKUS, 1955) and afterwards a programme to continuously monitor radioactivity from natural sources and fission products from weapons test fall-out has been established which continues to date (STOCKBURGER and SITTKUS, 1966; BIERINGER and SCHLOSSER, 2004). Also at the site an automated IMS particulate station (RASA) is operated which was certified by the PTS in 2004 (ZÄHRINGER et al., 2008). Radioxenon measurements have been continuously performed by the BfS at the Schauinsland since 1980 using manually operated equipment. The sampling equipment consists of absorbers filled with activated charcoal and cooled with liquid nitrogen. Sampling time is 1 week, the analysis of the samples, which contain xenon from approximate 10 m^3 air is done at the central BfS laboratory in Freiburg. The minimum detectable ^{133}Xe concentrations for these samples are approximately 1 mBq/m^3. The median ^{133}Xe activity concentration in ground level air at this site over the last 28 years is 4.6 mBq/m^3, with peak concentrations reaching approximately 40 Bq/m^3 observed after the Chernobyl accident (Fig. 3).

In February 2004 a SPALAX system was installed at Schauinsland (Fig. 4). Since then, the system has been operated almost continuously (Fig. 5) with a major interruption of operation from 15.04.2006 until 24.02.2007 due to a breakdown of the electrical detector cooling system. Another long interruption occurred in spring 2008 due to a compressor failure and the delivery time for spare parts.

The ^{133}Xe concentrations measured from 2004 to 2008 are shown in Fig. 6. These data have been analyzed with the Aatami nuclear spectral analysis software (CTBTO, Preparatory Commission, 2007). Also shown in the figure are the weekly ^{133}Xe concentrations measured with the laboratory system of the BfS during the corresponding period. The ^{133}Xe activity concentrations reported by the BfS system are integral values of the activity of all radioxenon isotopes. Since the atmospheric radioxenon

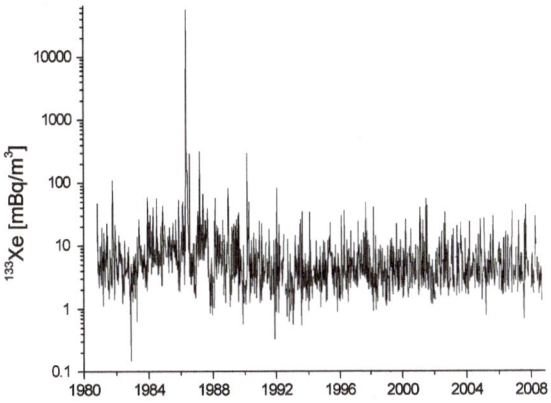

Figure 3
^{133}Xe concentrations measured at Schauinsland, Germany, since 1980. The large peak in 1986 is due to xenon released during the Chernobyl accident

Figure 4
SPALAX noble gas system installed at the Radionuclide Station RN33 at Schauinsland, Germany

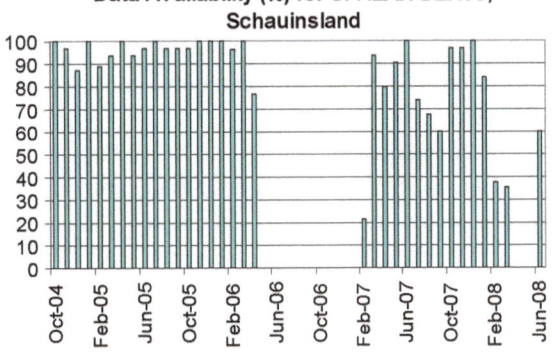

Figure 5
Data availability of SPALAX at the Radionuclide Station RN33 Schauinsland, Germany since start of operation

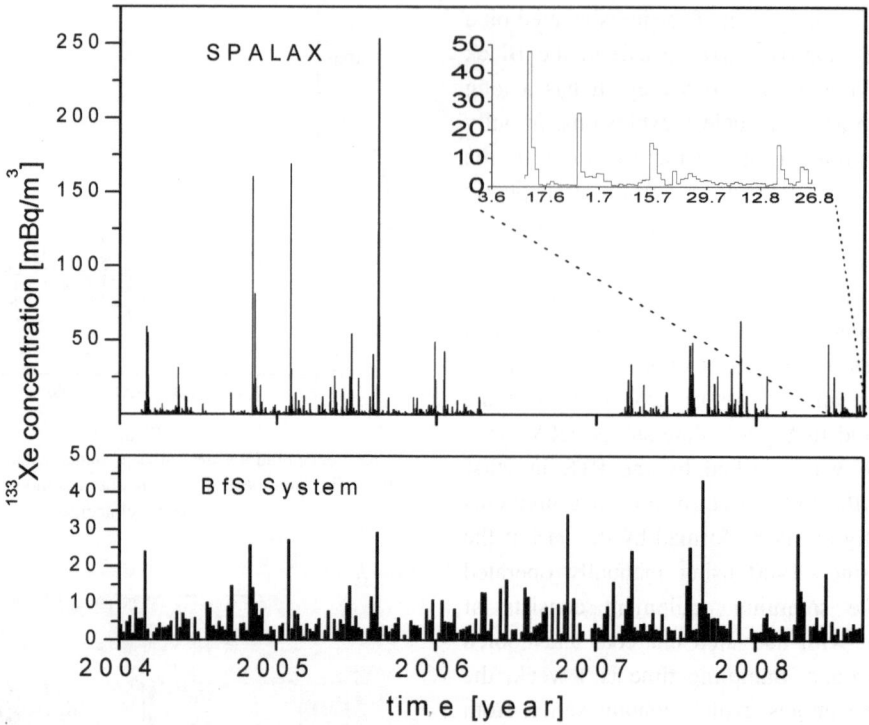

Figure 6

^{133}Xe activity concentration time series measured at Schauinsland with SPALAX and the BfS noble gas system. The *insert* shows an enlarged time interval for the period 03.06.2008 to 26.08.2008

concentrations are strongly dominated by ^{133}Xe, the error due to neglecting ^{135}Xe, ^{133m}Xe or ^{131m}Xe is small. This is supported by the SPALAX data which indicate that the sum of the activity of ^{131m}Xe, ^{133m}Xe and ^{135}Xe in all samples is less than 10% of the activity of ^{133}Xe, showing that taking the activities measured with the BfS system as ^{133}Xe activity is a sufficiently good approximation. As may be expected, the longer sampling duration of the BfS systems, combined with the short duration of ^{133}Xe peaks, results in lower peak concentrations of the BfS system relative to the concentrations measured by SPALAX.

For comparison of the SPALAX results with the BfS data, the SPALAX data have been combined to weekly values taking into account the radioactive decay: For each SPALAX sample, the activity concentration has been decay-corrected to the end of the BfS sampling cycle and the mean of the seven daily samples has then been used to calculate the mean activity during the corresponding 7-day period, assuming a constant concentration during that period.

For the comparison only data concentrations above 2 mBq/m^3 (i.e., approximate 2 times higher than the MDC of the BfS detection system) have been taken into account. The two data sets are clearly correlated (Fig. 7), although there is a bias of 0.73 ± 0.08 of the concentrations measured by the BfS system relative to the SPALAX system. The bias was estimated with a weighted least-squares fit method which takes into account uncertainties in both coordinates (REED, 1988). Note that the fit yields, within the uncertainties, the same results if the two highest data points are not allowed for.

Thirty-seven archived SPALAX samples have also been re-analyzed at the noble gas laboratory of the BfS (Fig. 8). A similar bias between SPALAX measurements and BfS system, which already has been observed in the comparison of independent samples measurements, is also observed in the re-analysis of the archive bottles. Here, the bias is 0.81 ± 0.01. The bias is most likely caused by systematic differences in the determination of the volume of stable xenon in the sample between

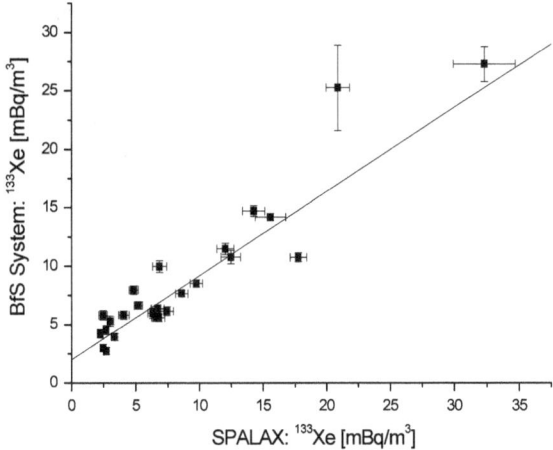

Figure 7

^{133}Xe weekly activity concentrations measured by the SPALAX system and by the BfS laboratory system. The line shows the uncertainty weighted least-squares fit of the data. The concentrations from SPALAX are calculated from daily values

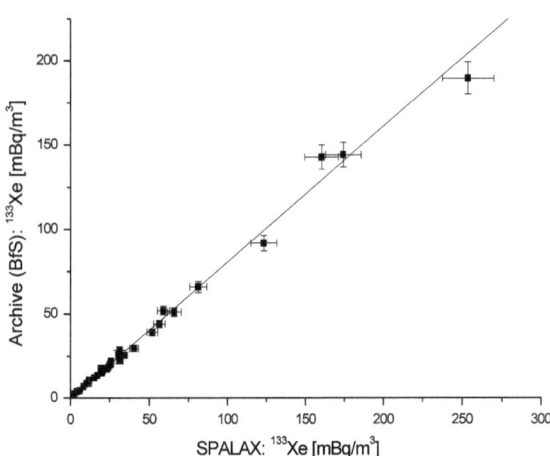

Figure 8

^{133}Xe concentrations measured by SPALAX and compared with re-measurements of the archived samples with the BfS laboratory system. The *line* shows the uncertainty weighted linear least-squares fit of the data

SPALAX and the BfS laboratory, since the re-analysis of the stable xenon concentration frequently yielded a higher stable xenon content than the analysis by SPALAX. The similar bias of the two comparisons shows on the other hand the good reproducibility of the atmospheric radioxenon measurements by independent samplers. Taking into account the bias between the two measurement methods, the data from the BfS long term time series can be used to characterize the radioxenon

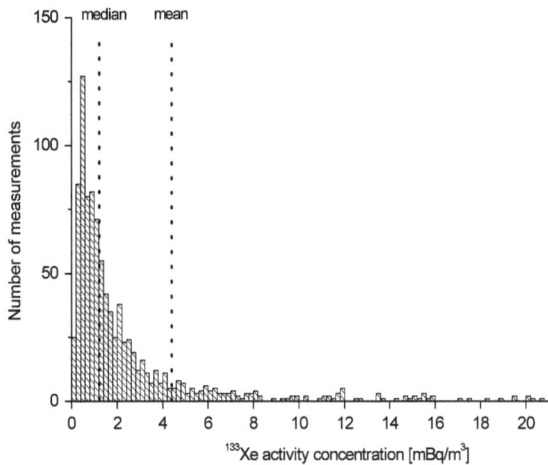

Figure 9

Frequency distribution for atmospheric ^{133}Xe concentrations measured with the SPALAX system at Schauinsland. Dotted lines indicate the median and the mean of the data set

background for evaluation of the data measured with the SPALAX system.

The concentrations measured by the SPALAX system range from below 0.1 mBq/m^3 in only three of the 995 samples of the data series up to (253 ± 16) mBq/m^3. It is worth noting that in only two of the samples, the activity was below the critical limit for detection, showing that the sensitivity of the SPALAX system is well suited for measurement of even the lowest ambient levels of radioxenon for a location such as Schauinsland. The range and distribution of concentrations are shown in Fig. 9. The typical background level for ^{133}Xe ranges between 0.2 and 1.3 mBq/m^3. The median is at 1.28 mBq/m^3, the mean at 4.4 mBq/m^3. The peaks of ^{133}Xe activity concentrations have a typical duration of one to three days as shown in the insert of Fig. 6.

In 10 of the total of 995 samples, the activity of all four isotopes was above the critical limit for detection. Those multiple isotope detections were associated with relatively high ^{133}Xe concentrations ranging between 6 mBq/m^3 and 175 mBq/m^3 and a median concentration of 41 mBq/m^3 with a mean uncertainty of 10%. The ^{133m}Xe/^{133}Xe ratios range between 0.015 and 0.13 (mean of the uncertainty 23%), the ^{135}Xe/^{133}Xe ratios range between 0.007 and 0.069 (mean of the uncertainty 38%) and the ^{133m}Xe/^{131m}Xe ratios range between 0.7 and 3.3 (mean of the uncertainty 37%). Those ratios are well within the expected values from reactor operations or with emissions from radiopharmaceutical production plants (KALINOWSKI *et al.*, 2010).

6. Conclusion and Outlook

The installation of the IMS for the CTBTO triggered new developments of equipment for measurement of radioxenon isotopes during the last decade. Compared to the equipment which was previously available and which was based on manually operated laboratory systems, significant progress has been made. The equipment which has been developed is automated and measures with a high sensitivity the four isotopes ^{131m}Xe, ^{133}Xe, ^{133m}Xe and ^{135}Xe. The analytical capabilities of these newly developed systems fulfil and even exceed the stringent detection requirements of the IMS. From this point of view, these systems have proven to be excellent tools for the verification system of the CTBT. A major requirement for monitoring the CTBT is also operational stability and reliability, in particular since the systems are frequently operated at remote sites with limited infrastructure. The operational characteristics of the systems under "field conditions" at various locations worldwide have been tested during the last 5 year and provided insight also into the technical long-term performance.

As of October 2008, 16 radioxenon systems are sending data to the PTS (7 SPALAX systems, 7 SAUNA systems and 2 ARIX), additional six stations are under installation, hence about 50% of the IMS noble gas network will be operational soon. Overall, the performance and reliability of the systems operation is high, however the evaluation of system down-times is difficult at present, since the structures and logistic measures for maintenance of these systems are not fully operational yet. Analysis in particular of reoccurring failures, may allow even further enhancement of the operational reliability.

The data produced by the already installed noble gas systems provide an unprecedented coverage of worldwide radioxenon activity concentrations. This allows a comprehensive assessment of global background data in regions with substantially different distribution and strength of sources. The density of the network also allows for investigation of the correlation of signals measured at neighbouring stations in order to improve the determination of the source location. These data are the basis for a future scheme for the evaluation and analysis of radioxenon concentrations with respect to their significance as an indicator for nuclear tests.

Acknowledgments

The progress described in the paper is largely the outcome of the effort of the members of the INGE (International Noble Gas Experiment) collaboration. The help of Mr. Schür for his assistance in system operation and of Ms. Schmid and Ms. Konrad for assistance in laboratory analysis of samples is also greatly appreciated.

REFERENCES

AUER, M., AXELSSON, A., BLANCHARD, X., BOWYER, T. W., BRACHET, G., BULOWSKI, I., DUBASOV, Y., ELMGREN, K., FONTAINE, J. P., HARMS, W., HAYES, J. C., HEIMBIGNER, T. R., McINTYER, J. I., PANISKO, M. E., POPOV, Y., RINGBOM, A., SARTORIUS, H., SCHMID, S., SCHULZE, J., SCHLOSSER, C., TAFFARY, T., WEISS, W., and WERNSPERGER, B. (2004), *Intercomparison experiments of systems for the measurement of xenon radionuclides in the atmosphere*, Appl. Rad. Isotopes 60, 863–877.

BARRIENTOS, S. and HASLINGER, F. (2001), *Seismological monitoring of the Comprehensive Nuclear-Test-Ban Treaty*, Kerntechnik 66 (3), 82–89.

BEAN, M., ST-AMANT, N., and UNGAR, R. K. (2007), *Understanding the Memory Effect in SAUNA's Plastic Scintillator and Its Impact on the Measurement*, Poster presented at the Informal Noble Gas Workshop, Las Vegas.

BECKER, A., WOTAWA, G., RINGBOM, A., and SAEY, P. J. R. (2010), *Backtracking of noble gas measurements taken in the aftermath of the announced October 2006 event in North Korea by mean of PTS methods in nuclear source region estimation and scenario reconstruction*, Topical Volume *Recent Advances in Nuclear Explosion Monitoring*, Pure Appl. Geophys. 167, 4/5 (2010, in press).

BERNSTRÖM, B. and DE GEER, L. E. (1983), *Mätning, Av Sma Mängder Xenon-133 I Luft*, FOA- Rapport C 20515-A1.

BIERINGER, J. and SCHLOSSER, C. (2004), *Monitoring ground-level air for trace analysis: methods and results*, Analyt. Bioanalyt. Chem. 379 (2), 234–241.

BJURMAN, B., DE GEER, L.-E., VINTERSVED, I., RUDJORD, A. L., UGLETVEIT, F., AALTONEN, H., SINKKO K., RANTAVAARA, A., NIELSEN, S. P., AARKROG, A., and KOLB, W. (1990), *The detection of radioactive material from a venting underground nuclear explosion*, J. Environ. Rad. 11, 1–14.

BOWYER, T. W., ABEL, K. H., HENSLEY, W. K., PANISKO, M. E., and PERKINS, R. W. (1997), *Ambient ^{133}Xe Levels in the Northeast US*, J. Environ. Rad. *37*, 143–153.

BOWYER, T. W., SCHLOSSER, C., ABEL, K. H., AUER, M., HAYES, J. C., HEIMBIGNER, T. R., MCINTYRE, J. I., PANISKO, M. E., REEDER, P. L., SARTORIUS, H., SCHULZE, J., and WEISS, W. (2002), *Detection and analysis of xenon isotopes for the comprehensive Nuclear-Test-Ban treaty international monitoring system*, J. Environ. Rad. *59*, 139–151.

CARRIGAN, C. R., HEINLE, R. A., HUDSON, G. B., NITAO, J. J., and ZUCCA, J. J. (1996), *Trace gas emissions on geological faults as indicators of underground nuclear testing*, Nature *382*, 528–531.

CHRISTIE, D. R., VIVAS VELOSO, J. A., CAMPUS, P., BELL, M., HOFFMANN, T., LANGLOIS, A., MARTYSEVICH, P., DEMIROVIC, E., and CARVALHO, J. (2001), *Detection of atmospheric nuclear explosions. The infrasound component of the International Monitoring System*, Kerntechnik *66* (3), 96–101.

CD/NTB/WP.224 (1995), *International Monitoring System Expert Group Report based on Technical Discussions held from 6 February to 3 March 1995*, Report of the Peer Review of the Conference on Disarmament.

CTBT, Text of the Comprehensive Nuclear Test Ban Treaty.

CTBTO, Preparatory Commission (2007), *Reference Manual of Radionuclide Analysis and Evaluation Software Aatami*, Version 4.10, Comprehensive Nuclear-Test-Ban Treaty Organization, Office of the Executive Secretary, Evaluation Section, Vienna, 2007.

CTBT/PTS/INF.162 (1999), Proposal for a programme to test and evaluate IMS noble gas equipment.

CTBT/PTS/INF.921-Rev.3 (2008), *Certification of noble gas equipment at IMS radionuclide stations*.

CTBT/PTS/TR/2007-1 (2007), *Development, Demonstration, Testing and Evaluation of On-Site Inspection Equipment for Xenon Sampling, Separation and Measurement*, Technical Report.

DE GEER, L. E. (1996a), *Sniffing out clandestine tests*, Nature *382*, 107.

DE GEER, L. E. (1996b), *Atmospheric radionuclide monitoring: A Swedish perspective. In Monitoring a Comprehensive-Test-Ban Treaty* (eds. Husebye, E.S. and Dainty, A.M.) (Kluwer Academic Publishers, Dordrecht 1996) pp. 157–177.

ENGLAND, T. R. and RIDER, B. F. (1993), Los Alamos National Laboratory, LA-UR-94-3106; ENDF-349.

ENSDF, Evaluated Nuclear Structure Data File, http://www.nndc.bnl.gov/ensdf/.

EHHALT, D., MÜNNICH, K. O., ROETHER, W., SCHÖLCH, J., and STICH, W. (1963), *Artificially produced radioactive noble gases in the atmosphere*, J. Geophys. Res. *68* (13), 3817–3821.

FONTAINE, J.-P. POINTURIER, F., BLANCHARD, X., and TAFFARY, T. (2004), *Atmospheric xenon radioactive isotope monitoring*, J. Environ. Rad. *72*, 129–135.

HANSEN, K. A., *The Comprehensive Nuclear Test Ban Treaty – An Insider's Perspective* (Stanford University Press, Stanford 2006).

HIROTA, M., NEMOTA, K., WADA, A., IGARASHI, Y., AOYAMA, M., MATSUEDA, H., HIROSE, K., SARTORIUS, H., SCHLOSSER, C., SCHMID, S., WEISS, W., and FUJII, K. (2004), *Spatial and temporal variations o atmospheric ^{85}Kr during 1995-2001 in Japan: Estimation of Atmospheric ^{85}Kr Inventory in the Northern Hemisphere*, J. Rad. Res. *45* (3), 405–413.

IGARASHI, Y., SARTORIUS, H., MIYAO, T., WEISS, W, FUSHIMI, K., AOYAMA, M., HIROSE, K., and INOUE, H. Y. (2000), *^{85}Kr and $^{133}Xe*

Monitoring at MRI, Tsukuba and its importance, J. Environ. Rad. *48*, 191–202.

KALINOWSKI, M. B., AXELSSON, A., BEAN, M., BLANCHARD, X., BOWYER, T. W., BRACHET, G., MCINTYRE, J. I., PETERS, J., PISTNER, C., RAITH, M., RINGBOM, S., SAEY, P., SCHLOSSER, C., STOCKI, T. J., TAFFARY, T., and UNGAR, R. K. (2010), *Discrimination of nuclear explosions against civilian sources based on atmospheric xenon isotopic activity ratios*, Topical Volume, *Recent Advances in Nuclear Explosion Monitoring*, Pure Appl. Geophys. *167*, 4/5 (2010, in press).

KARHU, P. and CLAWSON, R. (2001), *Radionuclide laboratories supporting the network of radionuclide stations in verification of the Comprehensive Nuclear-Test-Ban Treaty*, Kerntechnik *66* (3), 126–128.

KUNTZ, C. (1989), *Xe-133: Ambient air concentrations in upstate New York*, Atmos. Environ. *23* (8), 1827–1833.

LAWRENCE, M., GALINDO, M., GRENARD, P., and NEWTON, J. (2001), *Hydroacoustic monitoring system for the Comprehensive Nuclear-Test-Ban Treaty*, Kerntechnik *66* (3), 90–95.

LUDWICK, J. D. (1966), *Xenon 133 as an atmospheric tracer*, J. Geophys. Res. *71* (20), 4743–4748.

LOOSLI, H. H., MÖLL, M., OESCHGER, H., and SCHOTTERER, U. (1986), *Ten years low-level counting in the underground laboratory in Bern, Switzerland*, Nucl. Inst. Meth. *B17*, 402–405.

LOOSLI, H. H., *Applications of ^{37}Ar, ^{39}Ar and ^{85}Kr in hydrology, oceanography and atmospheric studies. In Isotopes of Noble Gases as Tracers in Environmental Studies*, Proc. Consultants Meeting, Vienna, 29 May–2 June, 1989 (International Atomic Energy Agency, Vienna 1992) pp. 74–84.

MATTHEWS, M. and SCHULZE, J. (2001), *The radionuclide monitoring system of the Comprehensive Nuclear-Test-Ban Treaty Organisation: From sample to product*, Kerntechnik, *66*(3), 102–112.

PERKINS, R. W. and CASEY, L. A. (1996), *Radioxenons: Their role in monitoring a comprehensive Nuclear-Test-Ban Treaty*, Technical Report, DOE/RL-96-51; PNNL-SA—27750.

RAMAKER, J., MACKBY, J., MARSHALL, P. D., GEIL R.; *The Final Test, A History of the Comprehensive Nuclear-Test-Ban Treaty Negotiations* (Provisional Technical Secretariat of the Preparatory Commission for the Comprehensive Nuclear-Test-Ban Treaty Organization, Vienna 2003).

REED, B. C. (1988), *Linear least-squares fits with errors in both coordinates*, Am. J. Phys. *57*(7), 642–646.

REEDER, P. L. and BOWYER, T. W. (1998), *Xe isotope detection and discrimination using beta spectroscopy with coincident gamma spectroscopy*, Nucl. Inst. Meth. A *408*, 582–590.

RINGBOM, A., LARSON, T., AXELSSON, A., ELMGREN, K., and JOHANSSON, C. (2003), *SAUNA—A system for automatic sampling, processing, and analysis of radioactive xenon*, Nucl. Inst. Meth. A *508*, 542–553.

RINGBOM, A., ELMGREN, K., LINDH, K., PETERSON, J., BOWYER, T. W., HAYES, J. C., MCINTYRE, J. I., PANISKO, M., and WILLIAMS, R. (2009), *Measurements of radioxenon in ground level air in South Korea following the claimed nuclear test in North Korea on October 9, 2006*, J. Radioanal. Nucl. Chem. *282*, 773–779, doi: 10.1007/s10967-009-0271-8.

SAEY, P., BEAN, M., BECKER, A., COYNE, J., d'AMOURS, R., DE GEER, L. E., HOGUE, R., STOCKI, T. J., UNGAR, K., and WOTAWA, G. (2007), *A long distance measurement of radioxenon in Yellowknife, Canada, in late October 2006*, Geophys. Res. Lett. *34*, L20802, doi:10.1029/2007GL030611.

SCHÖLCH, J., STICH, W., and MÜNNICH, K. O. (1966), *Measurement of radioactive xenon in the atmosphere*, Tellus *XVIII*(2), 298–300.

SCHULZE, J., AUER, M., and WERZI, R. (2000), *Low level radioactivity measurement in support of the CTBTO*, Appl. Rad. Isotopes *53*, 23–30.

SITTKUS, A. (1955), *Beobachtung an radioaktiven Schwaden von atomtechnischen Versuchen im Jahre 1953/1954*, Die Naturwissenschaften, Heft *17*, 478–482.

STOCKBURGER, H. and SITTKUS, A. (1966), *Unmittelbare Messungen der natürlichen und künstlichen Radioaktivität der atmosphärischen Luft*, Zeitschrift für Naturforschung, Band *21*, Heft 7, 1128–1132.

STOCKBURGER, H., SARTORIUS, H., and SITTKUS, A. (1977), *Messung der Krypton-85 und Xe-133 Aktivität in der atmosphärischen Luft*, Zeitschrift für Naturforschung *32*, 1239–1253.

STOCKI, T. J., BLANCHARD, X., D'AMOURS, R., UNGAR, R. K., FONTAINE, J. P., SOHIER, M., BEAN, M., TAFFARY, T., RACINE, J., TRACY, B. L., BRACHET, G., JEAN, M., and MEYERHOF, D. (2005), *Automated radioxenon monitoring for the Comprehensive Nuclear-Test-Ban Treaty in two distinctive locations: Ottawa and Tahiti*, J. Environ. Rad. *80*, 305–326.

UNSCEAR (2000), *Sources and Effects of Ionizing Radiation*, Volume I, Sources, 103, United Nations Publications, ISBN, 92-1-142238-8.

WEISS, W., SARTORIUS, H., and STOCKBURGER, H. (1992), *Global Distribution of Atmospheric ^{85}Kr. In Isotopes of Noble Gases as Tracers in Environmental Studies*, Proc. of a Consultants Meeting, Vienna, 29 May–2 June 1989 (International Atomic Energy Agency, Vienna 1992), pp. 29–62.

WEISS, W., SARTORIUS, H., and SCHLOSSER, C. (1997), *The background levels of radionuclides of the noble gas Xenon in the environment and the existing source detection capability*, Informal Radionuclide Workshop on Radionuclide IMS Network Specifications, US DOE Environmental Monitoring Laboratory, New York.

WINGER, K., FEICHTER, J., KALINOWSKI, M. B., SARTORIUS, H., and SCHLOSSER, C. (2005), *A new compilation of the atmospheric ^{85}Krypton inventories from 1945 to 2000 and its evaluation in a global transport model*, J. Environ. Rad. *80*, 183–215.

WOTAWA, G., DE GEER, L. E., DENIER, P., KALINOWSKI, M., TOIVONEN, H., D'AMOURS, R., DESIATO, F., ISSARTEL, J.-P., LANGER, M., SEIBERT, P., FRANK, A., SLOAN, C., and YAMAZAWA, H. (2003), *Atmospheric transport modelling in support of CTBT verification—Overview and basic concepts*, Atmos. Environ. *37*, 2529–2537.

ZÄHRINGER, M., BIERINGER, J., and SCHLOSSER, C. (2008), *Three years of operational experience from Schauinsland CTBT monitoring station*, J. Environ. Rad. *99*, 596–606.

ZÄHRINGER, M. and KIRCHNER, G. (2008), *Nuclide ratios and source identification from high-resolution gamma-ray spectra with Bayesian decision methods*, Nucl. Inst. Meth. Phys. Res. A *594*, 400–406, doi:10.1016/j.nima.2008.06.044.

ZIEGLER, C. A. and JACOBSON, D., *Spying Without Spies: Origins of America's Secret Nuclear Surveillance System* (Praeger Publishers, Westport, CT, USA 1995), ISBN-13: 978-0275950491.

(Received October 30, 2008, revised February 2, 2009, accepted February 16, 2009, Published online January 20, 2010)

Pure Appl. Geophys. 167 (2010), 487–498
© 2009 Birkhäuser Verlag, Basel/Switzerland
DOI 10.1007/s00024-009-0028-x

▌Pure and Applied Geophysics

Krypton and Xenon Radionuclides Monitoring in the Northwest Region of Russia

Yuri V. Dubasov[1] and Nikolay S. Okunev[1]

Abstract—Monitoring of Xe and Kr radionuclides was conducted from August 2006 to 30 July 2008 within the framework of ISTC Project #2133. Cherepovets City in Vologda Province and St. Petersburg were chosen as monitoring locations. Kr–Xe concentrate samples were obtained as a result of processing of several thousand m^3 of atmospheric air. New results of ^{85}Kr monitoring show, that for last 15 years, the ^{85}Kr volumetric activity in the atmospheric air of the northwest region of Russia has increased approximately 50% and has achieved a level of 1.5 Bq/m³. This value correlates well with similar data for Western Europe and Japan. The xenon fraction (80–160 cm³ under STP) is adsorbed on charcoal in the ampoule, which is measured in the well of HPGe gamma detector. Minimum detectable concentration (MDC) of ^{133}Xe for this technique is 0.008 mBq/m³, and it is the most sensitive method used today. The ^{133}Xe concentration in the atmospheric air of Cherepovets City varied in the monitoring period ranging from 0.09 to 2.5 mBq/m³. During the period of March 2007–30 July 2008, ^{133}Xe activity concentration in the atmospheric air of St. Petersburg changed from background values (0.2–0.3 mBq/m³) to 185 mBq/m³ and for approximately 20% of the samples ^{135}Xe was also measured with the ^{135}Xe/^{133}Xe activity ratio varied within the range of 0.03–3.5.

Key words: Xe radionuclides, ^{85}Kr monitoring, Xe radionuclides background in the air of St.-Petersburg, noble gas radionuclides release from nuclear power plants, high sensitivity methods of measuring Xe radionuclides.

1. ^{85}Kr monitoring in the former USSR and Russia

Nuclear power plants, nuclear reprocessing plants, nuclear explosions, and the radiopharmaceuticals production facilities are the primary sources of Xe and Kr radionuclides in atmospheric air. Increases in xenon radionuclide concentration in the atmosphere document the violation of gas purification procedures in nuclear power plants and at other nuclear facilities. In the USSR and Russia ^{85}Kr ($T_{1/2} = 10.752$ years)

monitoring was conducted during the period of 1977–1993 by the Experimental Meteorology Institute—RPA «Typhoon» (Makhon'ko, 1990, 1992).

Shortly after the disintegration of the USSR, ^{85}Kr monitoring programs ceased. As a matter of fact, ^{85}Kr monitoring has been conducted worldwide since the late 1950s. The atmospheric ^{85}Kr concentration has grown steadily during this period.

In the late 1980s to early 1990s of the last century, workers of the Khlopin Radium Institute conducted noble gas radionuclide monitoring under the guidance of Prof. Anatoliy S. Krivokhatsky, both within the USSR and in neutral waters of the Atlantic Ocean. A large series of research efforts on ^{85}Kr monitoring was carried out in the region of Chelyabinsk City, as well as in the English Channel (Pakhomov et al., 1990, 1991). Atmospheric ^{85}Kr volumetric activity was on the average about 0.9 Bq/m³ at that time.

RPA «Typhoon» sampled krypton fraction monthly at certain plants for further determination of atmospheric activity. The data on the values of average ^{85}Kr concentration for the last years of monitoring are given in Table 1.

2. Materials and Methods

Kr-85 monitoring was resumed at V.G. Khlopin Radium Institute in 2006 within the framework of International Science and Technology Center (ISTC) Project #2133 "Development of methodical bases and mobile equipment for monitoring of Xe and Kr radionuclides in the northwest region of Russia" far from nuclear reprocessing plants.

With this purpose, krypton–and–xenon fractions were sampled at one of the plants for atmospheric

[1] V.G. Khlopin Radium Institute, 28, 2nd Murinskiy ave., 194021 St. Petersburg, Russia. E-mail: yuri@dyuv.spb.su

Table 1

^{85}Kr *mean concentration in the atmosphere of Cherepovets, Novomoskovsk, and Mariupol, Bq/m^3 (MAKHON'KO, 1990; MAKHON'KO, 1992)*

Years	Cherepovets City, Vologda Province, Russia (59.12 N, 37.85 E)	Novomoskovsk City, Tula Province, Russia (54°2′N, 38°16′E)	Mariupol City, Ukraine (47°9′N, 37°36.5′E)
1989	0.96 ± 0.04	0.87 ± 0.02	–
1990	0.93 ± 0.02	0.98 ± 0.02	–
1991	0.95 ± 0.06	–	1.0 ± 0.04
1992	0.92 ± 0.04	–	–

oxygen extraction in Cherepovets City of the Vologda region and subjected to analysis at our laboratory. Sampling collection time was 50–72 h. Kr and Xe separation was carried out by means of low-temperature (cryogenic) adsorption on activated carbon (charcoal) SKT-3 and its further concentration during preparation for spectrometric analysis.

Gamma-spectrometers with Ge-detector were used for the measurements for ^{85}Kr. Since gamma quantum emission probability, $E_\gamma = 514$ keV for ^{85}Kr beta decay is only 0.435%, its measurement by gamma spectrometry method offers serious limitations, because of the nearby annihilation peak of 511 keV. That is why it seems to be reasonable to use a krypton sample obtained in the course of processing a large volume sample of atmospheric air in order to measure ^{85}Kr. The krypton fraction obtained from processing an air sample of ~1,500–2,000 m^3 in volume was subjected to adsorption on activated charcoal inside of an evacuated sampling container. The container with the krypton sample was placed on coaxial Ge-detector; duration of measuring was no less than 3 h, and every sample was measured 3–5 times. Two to four atmospheric samples of ^{85}Kr were obtained from Cherepovets atmospheric air and measured each month. Gamma spectroscopy with a well-HPGe detector was also used for the measurements.

Gamma-spectrometers were calibrated by a specially prepared standard sample using containers with a geometry similar to real samples. The specific activity of ^{85}Kr in a standard sample (Russian) was 25.6 ± 0.7 Bq/cm^3 at STP (273 K and 1,013 hPa). and the activity of the calibration sample was 13,300 ± 400 Bq. Intercomparison of ^{85}Kr atmospheric samples from Germany (BfS) and the northwest region of Russia, Cherepovets City was carried out in 2007–2008 and shows good

correspondence of ±2% between measurement equipment of BfS and the Radium Institute.

The increase in the concentrations of short-lived radionuclides of krypton (^{85m}Kr, ^{88}Kr) and xenon (^{131m}Xe, ^{133m}Xe, ^{133}Xe, ^{135}Xe,) at sampling sites can indicate the violation of standard operating conditions for NPP's gas purification system or a release of Xe radionuclides by Isotope Production Facilities, as well as nuclear explosions. Radionuclides of ^{133}Xe ($T_{1/2} = 5.245$ days), ^{135}Xe ($T_{1/2} = 9.1$ h), ^{133m}Xe ($T_{1/2} = 2.19$ days), and ^{131m}Xe ($_{1/2} = 11.9$ days) are main xenon radionuclides, generated as a result of nuclear fission reaction, and these four radioxenons are relevant ones for the verification of the Comprehensive Nuclear Test-Ban Treaty (CTBT). ^{133}Xe is the most abundant isotope among this family, as it has the highest yield under nuclear fission and relatively long half-life that allows it to be detected at considerable distances from its place of release.

In order to assure the sampling of the Xe radionuclide for monitoring, high sensitivity measurement techniques were specially developed. The xenon fraction recovered from the krypton–and–xenon mixture was fixed on SKT-3 charcoal in a small ampoule for measuring in the well of HPGe-detector. The spectrometer was calibrated using a standard with the same geometry and density as that of real samples and radionuclides that exceed the energy range of gamma quantums emitted by Xe radionuclides. The ^{133}Xe minimum activity, which can be detected by our well HPGe spectrometer, is 12 mBq for a measurement time of 24 h (at a confidence level of 95%).

The technique that we have developed allows for the measurement of radioxenon isotopes obtained from the processing of air sample volumes up to 2,000 m^3. Therefore this current technique provides a record-level of high sensitivity, and MDC of ^{133}Xe has reached an extremely low value of 0.008 mBq/m^3 (at a confidence level of 95%).

Approximately 40–80 m^3 of air was sampled twice a week in St. Petersburg and about 1,500–2,000 m^3 in Cherepovets 2–4 times a month, and the Xe and Kr fraction was separated from those samples. While monitoring in St. Petersburg we had to develop a new technique for Kr and Xe fraction separation from a sample of liquid technical oxygen, as such oxygen is usually very highly enriched with Kr and Xe as a consequence of the liquefaction of air. The MDC of ^{133}Xe for the aforementioned technique can reach the value of 0.2 mBq/m^3.

3. Results of ^{85}Kr Monitoring

We hoped that ^{85}Kr concentration stabilization or even its decrease in the atmosphere would occur after the cessation of nuclear tests and weapons-grade plutonium fabrication both in the USSR and the USA: however this did not happen. The first measurements, obtained in October–November 2006 indicated the ^{85}Kr volumetric activity at the Cherepovets monitoring point to be of 1.5 Bq/m^3, i.e. it had significantly increased when compared with sampling from the early 1990s. Published data (HIROTA et al., 2004; SMITH et al., 2005; WINGER et al., 2005) show that the period from 1995 to 2003 is characterized by a significant increase of ^{85}Kr concentration in the atmosphere of both Western Europe and Japan. The ^{85}Kr mean concentration increased to 1.5–1.6 Bq/m^3, which in certain countries can be related to irradiated and spent fuel processing.

The ^{85}Kr concentration activity in the air was calculated by taking into account the stable Kr concentration in atmospheric air of $1.14 \times 10^{-4}\%$ (volumetric). The volume of krypton in the measurement container was from 500 to 1,380 cm^3 (850 cm^3 on average) which corresponds to a sample of atmospheric air of 450–1,200 m^3 by volume. ^{85}Kr concentration activity during that period varied from 1.31 to 1.79 Bq/m^3, with a mean of 1.55 ± 0.13 Bq/m^3 (1 sigma, $n = 45$) and the measure of variability, v ($v = 1\sigma$/mean value $\times$ 100%) was of 8%, with the median magnitude being close to mean value. The diagram of ^{85}Kr concentration variation of atmospheric air at Cherepovets is

given in Fig. 1. As can be seen, during the period from September 2006 to July 2008, ^{85}Kr concentration activity did not increase; however, certain variations could be observed within the limits of experimental errors. Therefore, it can be supposed that ^{85}Kr concentration in the northwest region of the Russian Federation reached of 1.55 Bq/m^3 and has not seen any significant changes for about 3 years.

However, for the last 15 years ^{85}Kr concentration activity has grown almost 50% in the northwest region of Russia.

3.1. Results of Xe Radionuclides Monitoring

Shown in Fig. 2 is a diagram of ^{133}Xe concentration variation in the atmospheric air of Cherepovets since the initiation of monitoring.

The results of xenon radionuclide monitoring in the atmospheric air of Cherepovets City show that the ^{133}Xe concentrations vary from a minimum value of 0.09–2.5 mBq/m^3, with the average being 0.79 ± 0.48 mBq/m^3 (1 sigma, $n = 62$), and the median of 0.68 mBq/m^3, which is lower than average value. Thus the minimal level (may be background) for the region of Cherepovets City can be considered to be 0.09 mBq/m^3, however, it should be noted that our technique's MDC was 0.008 mBq/m^3.

St. Petersburg is located 80 km from the Leningrad NPP. In contrast to Cherepovets City, the ^{133}Xe concentration in St. Petersburg was found to be substantially higher and more varied during the aforementioned time period within a range from 0.2 to 185 mBq/m^3, and the ^{133}Xe mean concentration was 11.8 mBq/m^3 ($n = 126$), the median value being 1.9 mBq/m^3. In 20% of the samples, another shorter-life xenon isotope ^{135}Xe ($T_{1/2} = 9.14$ h) was observed, in addition to ^{133}Xe. The activity ratio of these isotopes ^{135}Xe/^{133}Xe, which identifies both locations of the release source and to some of its release parameters, varied from 0.03 to 3.5. The results of ^{133}Xe monitoring are shown in Fig. 3.

Table 2 includes only the analyses where the two Xe radionuclides were detected. It can be seen from the data of Table 2 that the concentration activity of ^{133}Xe varies from 1.9 to 185 mBq/m^3, with the median being 25.3 mBq/m^3, i.e., substantially higher than that for the entire period of monitoring; and the

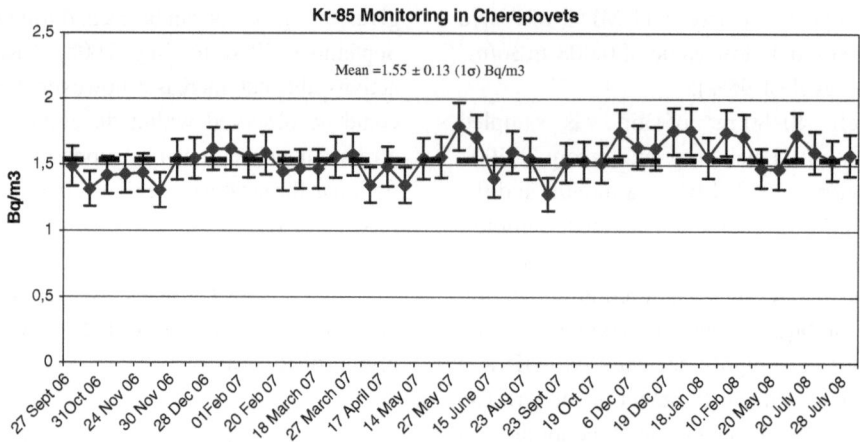

Figure 1
[85]Kr concentration activity in Cherepovets city atmospheric air; *dashed line*—mean concentration activity

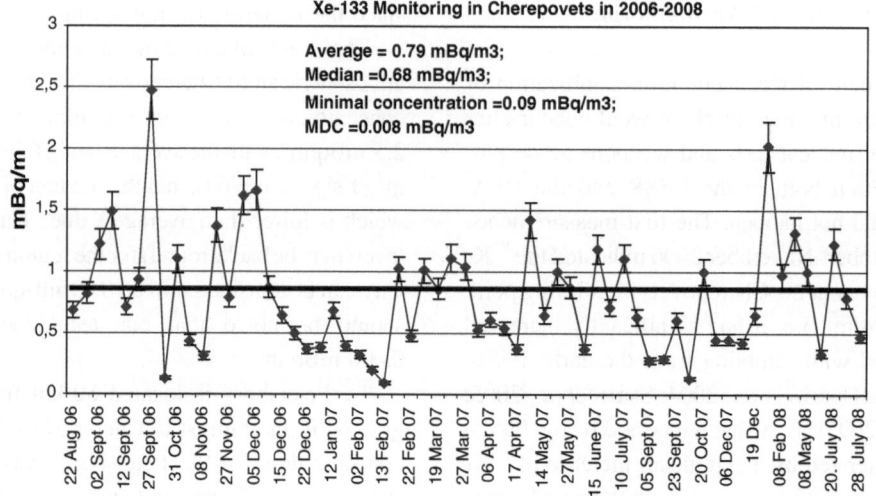

Figure 2
[133]Xe concentration activity in Cherepovets air during 2006–2008; *black bold line*—mean concentration activity

mean value is also higher and equals to 48.7 mBq/m³. However, for some samples the ^{133}Xe concentration activity was only 1.9 and 2.6 mBq/m³ (April 10 and May 12, 2008 correspondingly). The [135]Xe/^{133}Xe activity ratio varied from 0.033 to 3.5. A correlation between these two radio-xenon concentrations is not observed. The xenon radionuclides activity ratio for reactor equilibrium is reported to be about 0.3–0.24 (BOWYER *et al.*, 1998; FINKELSTEIN, 2001).

It can be seen from the data presented in Fig. 4 that in most cases the [135]Xe/^{133}Xe activity ratio was lower than 0.5, the median being 0.23, and only in 7 analyses of the 20 total, did the value of this ratio exceed 0.5. High values of the [135]Xe/^{133}Xe activity ratio can be affected

by release of either the starting of the reactor operating period with fresh-loaded fuel or from other sources.

Previous years (2000, 2003, and 2004) of sampling by the ARIX-01and ARIX-02 installations (Analyser of Radioactive Isotope of Xenon—automatic installations for analysis of Xe radionuclides—developed by the Khlopin Radium Institute, Russia) over a period of a month, were also times when both ^{133}Xe and [135]Xe radionuclides were detected in the atmospheric air. At St. Petersburg, increases of regional level of ^{133}Xe concentration activity (up to 100 mBq/m³) and more seldom of [135]Xe (up to 30 mBq/m³) were observed (AUER *et al.*, 2004; DUBASOV *et al.*, 2004).

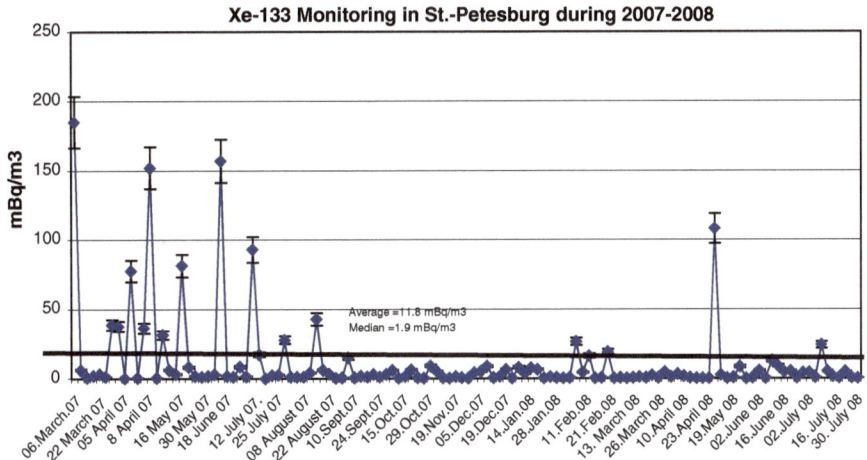

Figure 3

[133]Xe concentration activity in St. Petersburg air during March 2007–July 2008; *black bold line*—mean concentration activity

Table 2

[133]Xe and [135]Xe concentration activity in the atmospheric air of St. Petersburg

Sampling date (07:00 UTC), T_0	Xe volume in measuring ampoule, cm^3	Processed air volume, m^3	Volumetric activity of air at T_0, mBq/m^3		Ratio activity [135]Xe/[133]Xe
			[133]Xe	[135]Xe	
06. Mar. 07	3.75	43.1	185 ± 20	66 ± 7	0.36
16. Apr. 07	3.46	39.8	36.4 ± 3.7	3.4 ± 1.6	0.093
18. Apr. 07	6.0	69.0	152 ± 16	23.9 ± 1.0	0.16
25. Apr. 07	2.5	28.8	31.4 ± 3.2	1.3 ± 0.6	0.041
21. May 07	3.9	44.8	81.5 ± 8.2	4.4 ± 0.5	0.054
23. May 07	2.6	30.0	8.1 ± 0.8	4.2 ± 0.5	0.53
13. June 07	3.5	40.2	157 ± 16	10.8 ± 1.1	0.069
27. June 07	4.5	51.7	8.6 ± 0.9	0.8 ± 0.1	0.093
19. July 07	1.0	11.5	55 ± 6	1.8 ± 0.6	0.033
15. Aug. 07	2.0	23.0	43.1 ± 0.9	7.4 ± 2.1	0.17
05. Sept. 07	4.3	49.4	14.9 ± 0.6	5.0 ± 0.5	0.34
05. Dec. 07	7.8	89.7	5.0 ± 0.4	1.9 ± 0.6	0.38
14. Jan. 08	3.2	36.8	7.7 ± 1.2	6.1 ± 0.9	0.79
16. Jan. 08	7.1	81.6	6.4 ± 0.5	22.7 ± 1.0	3.5
06. Feb. 08	3.3	37.9	26.6 ± 2.7	93 ± 10	3.5
21. Feb. 08	3.8	43.7	19.2 ± 2.0	63 ± 6.0	3.3
10. Apr. 08	5.2	59.8	1.9 ± 0.2	3.7 ± 0.3	1.9
28. Apr. 08	3.8	43.7	108 ± 11	25.0 ± 3.0	0.23
12. May 08	4.2	48.3	2.6 ± 0.3	2.2 ± 0.3	0.84
09. July. 08	2.3	26.4	24 ± 2	2.5 ± 0.4	0.10

[135] Xe MDC is 0.6 mBq/m^3

4. Discussion

The data of Table 2 were analyzed, accounting for the obtained meteorological information on particle transfer and air mass trajectories. It is determined that there is no correlation between an increased value of ^{133}Xe concentration activity and any sector of particle transfer. However, in most cases we succeeded in identifying the sources of ^{133}Xe release into the atmosphere. They were located both in the European territory of the former USSR and in Western Europe.

From the sampling, it appears that ^{133}Xe atmospheric transport is from regions of Kola Peninsula, Ukraine, Smolensk Region, Tver' Region, Lithuania,

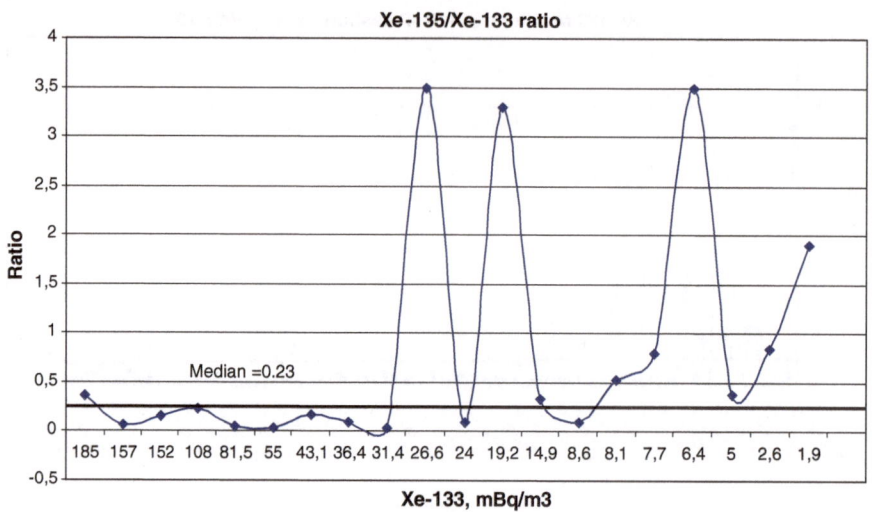

Figure 4
^{135}Xe/^{133}Xe ratio changing depending on ^{133}Xe activity in atmospheric air of St. Petersburg; *black bold line*—median

Scandinavia, and Western Europe. Nevertheless, each specific case needs to be analyzed separately.

4.1. Monitoring in Cherepovets City

On the 27th of September, 2006, an increased concentration of ^{133}Xe (2.5 Bq/m^3) was recorded in the atmospheric air of Cherepovets (59.12 N 37.85 E). The atmoshpheric or air mass transport for two days was from the north of Finland and Karelia, near Peninsula (Russia), where the Kola NPP (67°28′N, 32°28′E) is located and that for three days—from the Barents Sea water area to the west from Spitsbergen. Backward air mass transport trajectories for this date are presented in Fig. 5.

On November 30, 2006 ^{133}Xe concentration in atmospheric air of Cherepovets was equal to 1.62 mBq/m^3. Three-day backward air mass transport trajectories for November 30 start in Belgium (near Mol, 51.2°N, 5.2°E), Denmark, Northern Germany, pass over the southern part of Sweden and then, pass over Latvia and Estonia, and arrive at Cherepovets on November 30 (ref. to Fig. 6).

On December 5, 2006 ^{133}Xe concentration in the atmospheric air of Cherepovets was equal to 1.66 mBq/m^3. The air mass transport is 72 h: from Germany and France (ref. to Fig. 7), through Poland and then through Byelorussia and through Smolensk

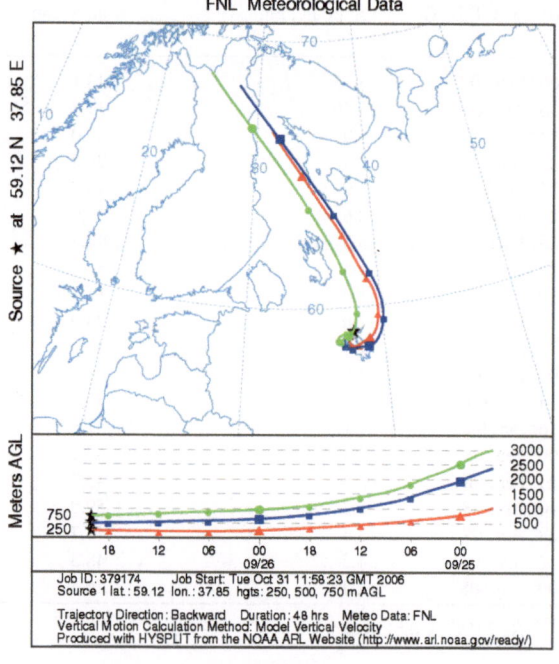

Figure 5
Two-day backward trajectories, on 27 September 2006, Cherepovets control site

NPP region (54°10′N, 33°14′E), towards the northeast to Vologda Province.

Increased ^{133}Xe concentrations in Cherepovets were measured on January 16 and February 10, 2008.

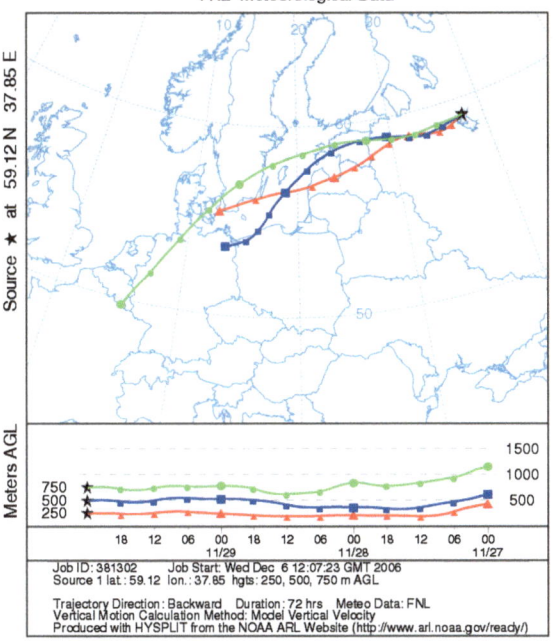

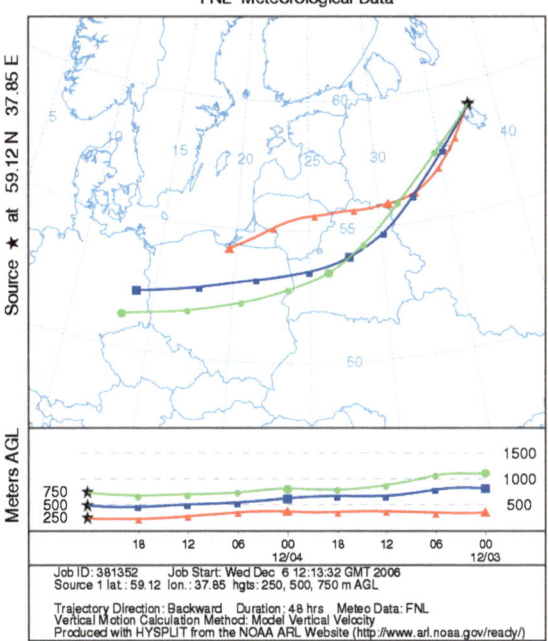

Figure 6
Three-day trajectories on November 30, 2006 at 250, 500, 750 m levels to Cherepovets control site

Figure 7
Three-day trajectories on December 5, 2006 at 250, 500, 750 m levels to Cherepovets control site

The air mass transport from Eastern Europe to Cherepovets took two days. Backward air mass transport trajectory analysis gives evidence to the fact that at 12 UTC on January 15 the air mass, which was detected 15 h earlier near the Ignalina NPP (55°36′N, 26°34′E), arrived at Cherepovets at the height of 800 m. Following more detailed analyses of semidiurnal air mass transport trajectories, the results allow for an assumption that at 12–15 UTC on January 15, it is proposed that the radioisotopes, which arrived at Cherepovets at the height of 800 m, within the air mass which had passed over the Ignalina NPP, and at 18–24 UTC on 16 January, 2008 mixed with the air mass that had passed over the region of Kalinin NPP (57°54′N, 035°02′E) (ref. to Fig. 8). To the height of 500 and 300–100 m the ^{133}Xe (2.02 mBq/m^3) and ^{85}Kr (1.76 Bq/m^3) radionuclides could have only been derived from Kalinin NPP area on January 16 and 15, accordingly.

The concentration of ^{133}Xe averaged for three days, determined at the sampling midpoint, i.e., February 10, 2008 had a value of 1.32 mBq/m^3.

Diurnal backward air mass transport trajectories (ref. to Fig. 9) for the moment of 01 UTC on February 10, 2008 were passing at an altitude of 800 m in direct proximity to the Loviisa NPP with light ascending motions. During the first half of the day on February 10 backward air mass transport direction changed and air masses moved to Cherepovets from the southwest. Backward air mass transport trajectory, which had passed over the Kalinin NPP, descended from about 1,000 m to the level of 500 m. The total air mass transport pattern described in the preceding paragraph suggests a conclusion: That at levels above 500 m, the air mass transported to the Cherepovets region on February 9–11, 2008 could have come from the Loviisa, Leningrad, and Kalinin NPPs.

The period from December 2007 to February 2008 was characterized by increased ^{133}Xe concentration activity in Cherepovets' atmosphere, notably from 0.42 to 2 mBq/m^3, i.e., from 4 to 22 times above the minimum level. The analysis of these periodic synoptic conditions indicates that air mass transport is from western and southern directions, generally

121

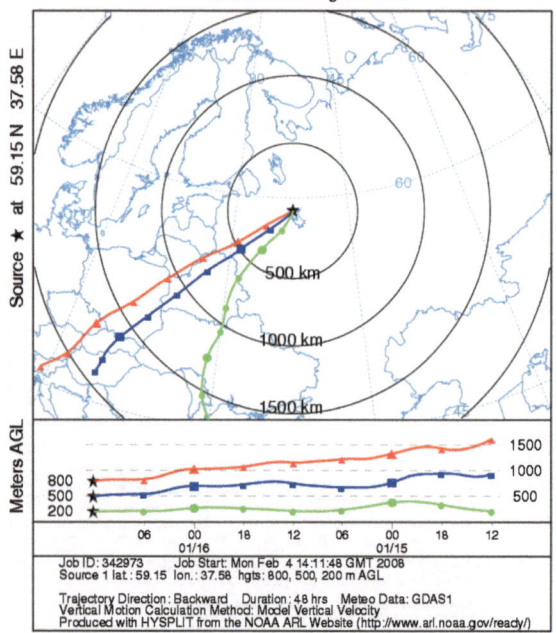

Figure 8
Two-days backward trajectories, which had reached Cherepovets
site at 12 UTC on 16 January 2008

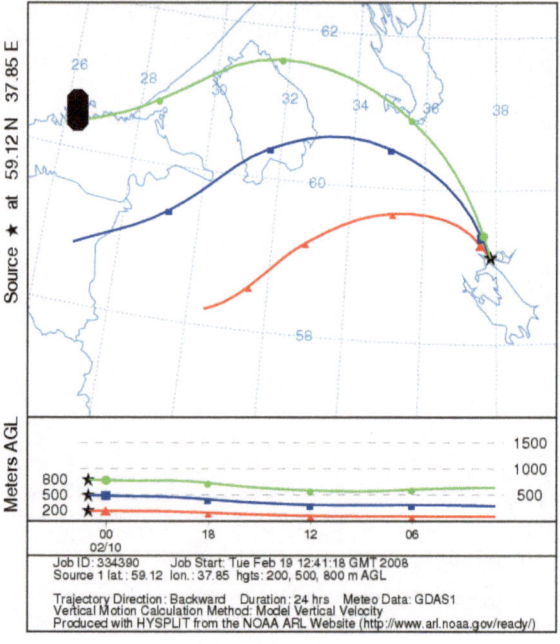

Figure 9
Diurnal backward trajectories to Cherepovets control site on the
levels of 200, 500, and 800 m. 01 UTC on February 10, 2008.
Location of «Loviisa» NPP is marked with *black bar*

caused by cyclones, and was the main particle
transfer sector.

4.2. Monitoring at St. Petersburg

The data of Table 2 show that during the period of
monitoring in St. Petersburg, measurements of ^{133}Xe
and ^{135}Xe occurred in 20% of the samples.

On May 21, 2007, ^{133}Xe concentration activity
was 81.5 mBq/m^3, and the ^{135}Xe/^{133}Xe ratio was
0.054. Air mass transport was from the southwest. As
can be seen from Fig. 10, daily backward trajectories
start in southern Sweden and have very similar
directions along the southern shore of the Gulf of
Finland.

On June 13, 2007, the measurement of the ^{133}Xe
concentration was considerably high (157 ± 16
mBq/m^3), however, the ^{135}Xe/^{133}Xe ratio was lower
in comparison with that of equilibrium one (0.24–
0.33). This can be related either to transport from a
distant region or to air mixture discharged from air
purification system after long-term storage. As shown

in Fig. 11, transport from the eastern shore of Sweden
lasted for no less than 24 h.

Considering those events, the ^{135}Xe/^{133}Xe ratio
was essentially higher than that at equilibrium for the
reactor zone.

So, on January 16, 2008, the ^{135}Xe/^{133}Xe ratio
was 3.5, which suggests a gas mixture release from
reactor with fresh fuel or any other source, for
instance, isotope producing facilities. The backward
air mass transport trajectories for the period from
11:00 UTC on January 15 to 11:00 UTC on January
16 are shown in Fig. 12. The trajectory at 800 m
altitude passes to the north of the trajectory of that at
the 500-m level, whose trajectory passes over north
Poland and then over Latvia and Lithuania, and that
on the level of 200 m—the trajectory is further east,
over Byelorussia. The point of the Ignalina NPP is
situated between these trajectories of 500-m and 200-
m levels. The air mass transport trajectory was
passing over Ignalina on January 15 at approximately
16:00 UTC on the level of 400 m.

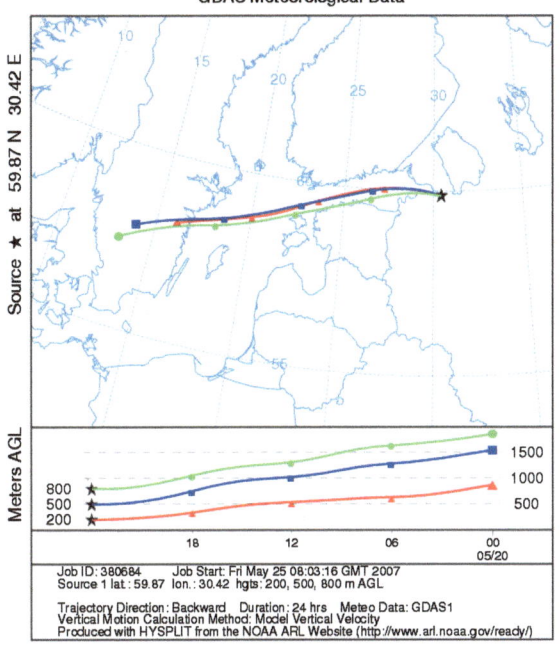

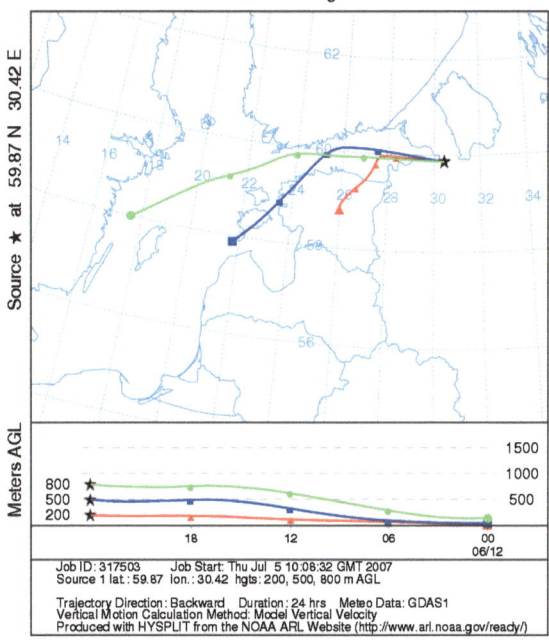

Figure 10
One-day backward trajectories on heights of 200, 500, and 800 m in St. Petersburg, on May 21, 2007

Figure 11
Daily backward trajectories of air transfer to St. Petersburg, on June 13, 2007

On February 21, 2008 the air mass on heights of 500 and 800 m reached St. Petersburg from Finland, being driven by a pressure ridge from the northwest to the southeast. The air mass crossed the Gulf of Finland and the southwest area of Leningrad Province. It is hypothesized, that this was related to an effect from the Loviisa and Leningrad NPPs. In actuality, the trajectories drawn up for 09:00 and 10:00 UTC were passing at the aforementioned altitudes in the immediate vicinity of those NPPs (ref. to Fig. 13). Thus, the effect from the Loviisa and Leningrad NPPs during the period of sampling (from 9:30 to 10:30 UTC) can be considered only for the altitudes above 500 m.

Analysis of backward air mass trajectories on 12 May, 2008, as shown in Fig. 14, indicates that the air mass within the layer from 500 to 800 m, when moving towards St. Petersburg from the southern regions of Sweden, passed in immediate proximity to the Ringhals (57.75°N, 12.00°E) and the Oskarshamn (57.25°N, 16.50°E) NPPs (Sweden). However, analysis of the vertical air currents at the 975-hPa level

indicates that during this period active descending vertical currents were observed over the Oskarshamn NPP. Therefore, it is of low probability that any releases within those two NPP areas could be transported from the ground level to altitudes of 500–800 m and entrained into the air flow.

Modeling of vertical currents at an altitude of 975 hPa, 65 h prior to sampling over the southwestern part of Finland, indicating a period of stable descending currents of about 1.0–1.5 mbar/h were observed in the region of the Olkiluoto NPP (61.5°N, 21.5°E). Hence, the probability of any release capture by air flow in this region is also low.

However, in the region of the Ringhals NPP at altitudes of 975 and 950 hPa, just three days prior to sampling, there were recorded ascending vertical currents of about 1.5 mbar/h. That is the reason that any release (impurity) is the most probable source captured by air flow from this NPP, since the ascending vertical currents were recorded only within the layer of 200–500 m at this location. If it is true, the air mass would have taken 84 h from the

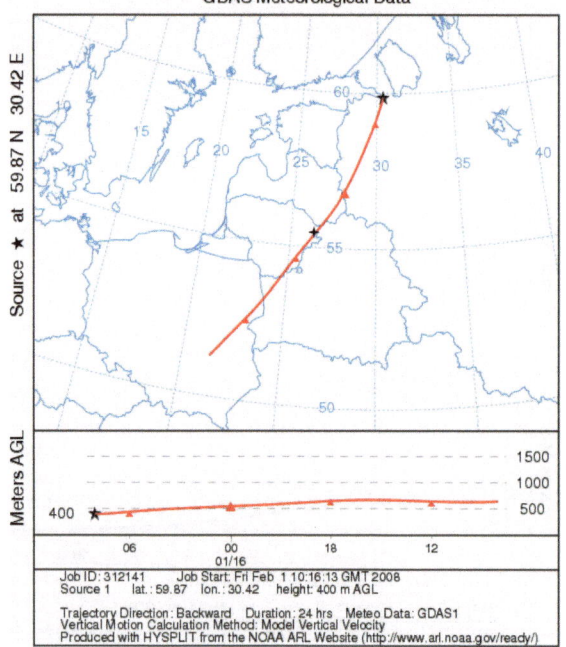

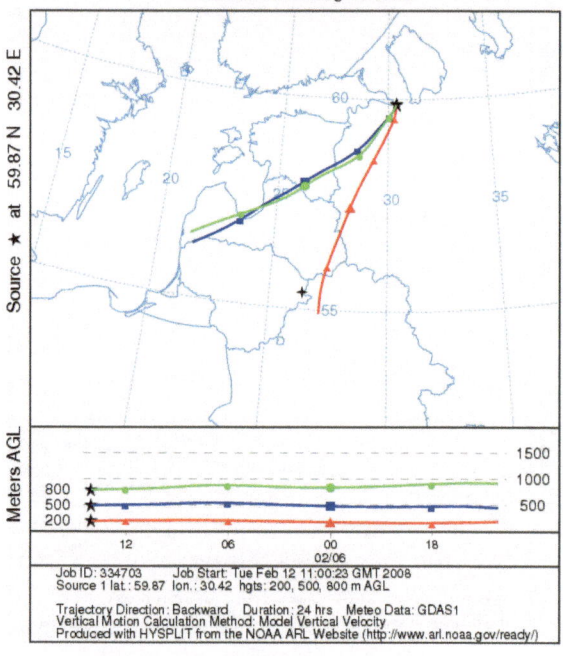

Figure 12

Daily backward trajectory of air mass transfer to St. Petersburg on the level of 400 m. 08:00 UTC on 16.01.2008. Ignalina location is marked with *black asterisk* (55.6°N, 26.56°E)

Figure 13

Two-day backward trajectories of air mass transfer to St. Petersburg at 200, 500, and 800 m levels. 14:00 UTC, February 6, 2008. Ignalina location is marked with *black asterisk*

radionuclide release to reach the sampling location and the radionuclide $^{135}Xe/^{133}Xe$ ratio would have decreased by 0.84 due to decay. Hence, when precise backward trajectories of air mass transfer are available, it is possible to identify the source location, of radioactive noble gas release by the radionuclide chronometry method with the aid of xenon radionuclides ratios.

On December 13, 24, 2007, January 21, 28, 30, and February 18, 20, 26 2008, the ^{133}Xe concentration activity within St. Petersburg's air was either minimum or at the MDC level and did not exceed 0.2–0.6 mBq/m^3. Analysis of calculations of backward air mass trajectories for the period of 1–3 days before the time of sampling indicated that: an air particle within the layer between 200 and 1,000 m was moving to St. Petersburg from the central regions of the Barents Sea and northern regions of Scandinavian, Peninsula (December 13), and there was stable western and south western air mass transport trajectory which passed 15 km southward of the

Leningrad NPP (December 24). This air mass transport was from the south-eastern region of Sweden (January 21, 2008); the diurnal trajectory passed over the northern regions of Finland and Karelia, moving towards St. Petersburg from north to south (January 28); the air mass transport trajectories start over the North Sea and pass at the 200 m level through Sweden, the Gulf of Bothnia, the southern end of Finland, and the eastern part of the Gulf of Finland; trajectories on 500 m and 800 m levels pass further north, through central regions of Scandinavia and Finland (January 30); stable northwestern air mass transport, was observed that began to the north of Finland (levels of 500 m and 800 m) and to the north of Sweden, on 200 m (February 18, 2008); the air mass arrived at St. Petersburg (from 12.30 to 13.30 on February 20) and passed near the Loviisa NPP; air mass transport trajectories (February 26) on levels of 200 and 500 m are located not far from Loviisa, Olkiluoto, and Leningrad NPPs.

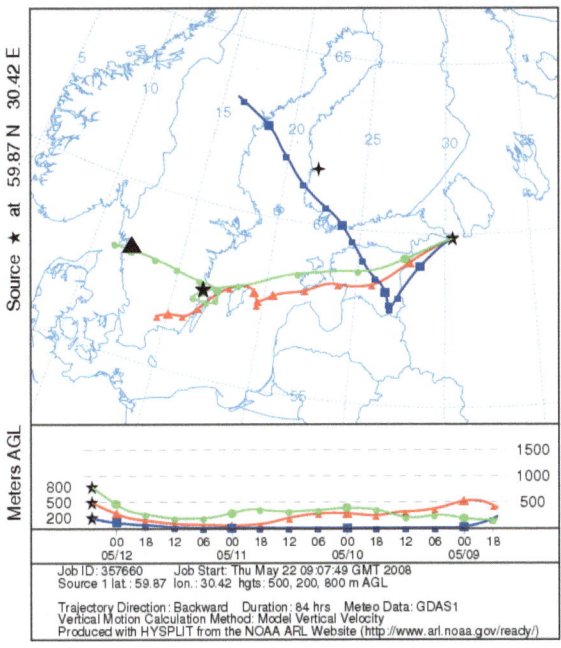

NOAA HYSPLIT MODEL
Backward trajectories ending at 05 UTC 12 May 08
GDAS Meteorological Data

Figure 14

84-h backward trajectories on the heights of 200, 500, and 800 m to St. Petersburg. 05:00 UTC, May 12, 2008 (Start of sampling). Location of Ringhals NPP is marked by *black triangle*, Oskarshamn NPP is marked by *black star*, and Olkiluoto NPP is marked by *black asterisk*

Therefore from the presented data, it can be discerned that the minimum (background) concentrations of ^{133}Xe in St. Petersburg were observed as a rule of either that the air mass trajectory did not pass over the regions of NPPs location or any releases from those NPPs were at a minimum during these periods.

5. Conclusion

Monitoring of [85]Kr in the Russian Federation, which was terminated in 1994, has resumed within the framework of ISTC Project #2133. [85]Kr mean concentration activity in the northwest region of the Russian Federation has increased by 1.5 times and reached the value of 1.55 ± 0.13 Bq/m^3 in 2006–2008, with a range from 1.32 to 1.79 Bq/m^3. This mean value correlates with the similar data for Western Europe and Japan (HIROTA *et al.*, 2004; SMITH *et al.* 2005; WINGER *et al.*, 2005).

Xe radionuclides monitoring in the northwest region of the Russian Federation was accomplished at St. Petersburg and in Cherepovets of Vologda Province. Thanks to the application of high sensitivity measurement techniques developed by our efforts, it has been determined that the minimal ^{133}Xe concentration activity in the air of Cherepovets City was 0.09 mBq/m^3, and that the highest concentration reached 2.5 mBq/m^3 with a mean concentration equal to 0.79 mBq/m^3.

The level of ^{133}Xe concentrations in St. Petersburg, which is 80 km distant from the Leningrad NPP, is substantially higher and when averaged over more than 120 analyses, the mean concentration reaches 12 mBq/m^3, and the median value was 1.9 mBq/m^3, varying from 0.2 to 180 mBq/m^3. In contrast to Cherepovets, two xenon radionuclides ^{133}Xe and [135]Xe are sometimes recorded here, and their ratio varies at that within wide limits, thereby pointing to an effect from sources other than the Leningrad NPP. The fact is because of western air mass transport trajectories, St. Petersburg falls within the effects of Swedish and Finnish NPPs. This coupled with the lack of Xe radionuclides concentration activity standardization and that the concentrations are substantially lower in comparison with Rn concentration, defines for us the necessity to continue such manner of monitoring. This will provide verification of safety of the present day nuclear-power engineering, especially with regard for its further development in the nearest prospect that is necessary for public opinion.

While the change in Xe radionuclides concentration can be explained by means of designing backward trajectories of air mass transfer, it does not work in the case of [85]Kr.

Therefore, noble gas radionuclide monitoring allows both obtainment of the data on man-made radioactivity of atmosphere in points of monitoring and extrapolation of conditions in release sources or region.

Acknowledgments

The authors kindly thank the employees of V.G. Khlopin Radium Institute: A.V. Malyshenkov, V.A.

Knyazev, S.I. Malimonova, S.P. Erofeev, V.N. Selifonov, who actively participated in implementation of this work, and also Dr. Yu. M. Liberman with the employees of the North-West Meteorology Department for realization of designs of backward air mass transfer trajectories. We kindly thank Dr. Bayron Ristvet for assistance in editing the English version of this manuscript.

REFERENCES

AUER, M., AXELSSON, A., BLANCHARD, X. et al. (2004), Intercomparison experiments of systems for the measurement of Xenon radionuclides in the atmosphere. J. Appl. Radiat. Isot. 60, 863–877.

BOWYER, T. W., PERKINS, R.W., ABEL, K. H. et al. (1998), Xenon radionuclides, atmospheric monitoring. In (Myers, R. A., ed), Encyclopedia of Environmental Analysis and Remediation. John Wiley and Sons.

DUBASOV, YU. V., POPOV, and YU. S. (2004), Xe radionuclide monitoring in atmosphere. Proc. Intern. Conf. on Unification and Optimization of Radiation Monitoring on NPP Location Regions, Armenia, Yerevan, September 22–26, 2004, pp. 18–26.

FINKELSTEIN, Y. (2001) Fission product isotope ratios as event characterization tools—Part II: Radioxenon isotopic activity ratios, Kerntechnik 66 (5–6), 229–236.

HIROTA, M., NEMOTO, K., WADA, A. et al. (2004), Spatial and temporal variations of atmospheric ^{85}Kr observed during 1995–2001 in Japan: Estimation of Atmospheric ^{85}Kr Inventory in the Northern Hemisphere, J. Radiat. Res. 45, 405–414.

MAKHON'KO, K.P. (ed.) (1990), Radiation conditions in the territory of the USSR in 1989, Annual. Obninsk, RPA "Typhoon". (In Russian)

MAKHON'KO, K.P. (ed.) (1992), Radiation conditions in 1991 in the territory of Russia and neighbouring states, Annual. Obninsk, RPA "Typhoon". (In Russian)

PAKHOMOV, S.A., KRIVOKHATSKY, A.S., SOKOLOV, I.A., et al. (1990), The method to determine Krypton-85 content in atmosphere, based on application of "Beta-2" installation. Radiochemistry 6, 100–103. (In Russian)

PAKHOMOV, S.A., SOKOLOV, I.A., and KRIVOKHATSKY, A.S. (1991), Experience in Krypton-85 Monitoring System Operation in the USSR Territory, Radiochemistry 6, 131–137. (In Russian)

SMITH, K., MURRAY, M., WONG, J. et al. (2005), Krypton-85 and other radioactivity measurements throughout Ireland, Radioprotection, Supp.1, 40, S457–S463.

WINGER, K., FEICHTER, J., KALINOVSKI, M. B. et al. (2005), A new compilation of the atmospheric ^{85}Kr inventories from 1945 to 2000 and its evaluation in a global transport model. J. Environ. Radioactivity 80, 183–215.

(Received October 29, 2008, revised July 27, 2009, accepted July 30, 2009, Published online December 16, 2009)

Pure Appl. Geophys. 167 (2010), 499–515
© 2010 Birkhäuser Verlag, Basel/Switzerland
DOI 10.1007/s00024-009-0034-z

Environmental Radioxenon Levels in Europe: a Comprehensive Overview

PAUL R. J. SAEY,[1,7] CLEMENS SCHLOSSER,[2] PASCAL ACHIM,[4] MATTHIAS AUER,[2] ANDERS AXELSSON,[5,8]
ANDREAS BECKER,[1] XAVIER BLANCHARD,[4] GUY BRACHET,[4] LUIS CELLA,[1] LARS-ERIK DE GEER,[5]
MARTIN B. KALINOWSKI,[3] GILBERT LE PETIT,[4] JENNY PETERSON,[5] VLADIMIR POPOV,[6] YURY POPOV,[6]
ANDERS RINGBOM,[5] HARTMUT SARTORIUS,[2] THOMAS TAFFARY,[4] and MATTHIAS ZÄHRINGER[1]

Abstract—Activity concentration data from ambient radioxenon measurements in ground level air, which were carried out in Europe in the framework of the International Noble Gas Experiment (INGE) in support of the development and build-up of a radioxenon monitoring network for the Comprehensive Nuclear-Test-Ban Treaty verification regime are presented and discussed. Six measurement stations provided data from 5 years of measurements performed between 2003 and 2008: Longyearbyen (Spitsbergen, Norway), Stockholm (Sweden), Dubna (Russian Federation), Schauinsland Mountain (Germany), Bruyères-le-Châtel and Marseille (both France). The noble gas systems used within the INGE are designed to continuously measure low concentrations of the four radioxenon isotopes which are most relevant for detection of nuclear explosions: ^{131m}Xe, ^{133m}Xe, ^{133}Xe and ^{135}Xe with a time resolution less than or equal to 24 h and a minimum detectable concentration of ^{133}Xe less than 1 mBq/m^3. This European cluster of six stations is particularly interesting because it

is highly influenced by a high density of nuclear power reactors and some radiopharmaceutical production facilities. The activity concentrations at the European INGE stations are studied to characterise the influence of civilian releases, to be able to distinguish them from possible nuclear explosions. It was found that the mean activity concentration of the most frequently detected isotope, ^{133}Xe, was 5–20 mBq/m^3 within Central Europe where most nuclear installations are situated (Bruyères-le-Châtel and Schauinsland), 1.4–2.4 mBq/m^3 just outside that region (Stockholm, Dubna and Marseille) and 0.2 mBq/m^3 in the remote polar station of Spitsbergen. No seasonal trends could be observed from the data. Two interesting events have been examined and their source regions have been identified using atmospheric backtracking methods that deploy Lagrangian particle dispersion modelling and inversion techniques. The results are consistent with known releases of a radiopharmaceutical facility.

Key words: Comprehensive Nuclear-Test-Ban Treaty, low-level environmental radioactivity measurements, noble gas, European air, radioxenon, nuclear verification.

The views expressed in this publication are those of the authors and do not necessarily reflect the views of the CTBTO Preparatory Commission or any of the participating institutions.

[1] Preparatory Commission for the Comprehensive Nuclear-Test-Ban Treaty Organisation, Provisional Technical Secretariat, P.O. Box 1200, 1400 Vienna, Austria. E-mail: paul.saey@ati.ac.at
[2] Bundesamt für Strahlenschutz (BfS), Rosastr. 9, 79098 Freiburg, Germany.
[3] Carl Friedrich von Weizsäcker Center for Science and Peace Research, University of Hamburg, c/o Fachbereich Chemie Martin-Luther-King-Platz 6, 20146 Hamburg, Germany.
[4] Commissariat à l'Energie Atomique (CEA), CEA DAM-DIF, 91297 Arpajon, France.
[5] Swedish Defence Research Agency (FOI), Defence and Security, Systems and Technology Division, 172 90 Stockholm, Sweden.
[6] V. G. Khlopin Radium Institute (KRI), 2nd Murinsky Ave. 28, 194021 St. Petersburg, Russian Federation.
[7] *Present Address:*
Vienna University of Technology, Atominstitut, Stadionallee 2, 1020 Vienna, Austria.
[8] *Present Address:*
IAEA, Wagramer Straße 5, 1400 Vienna, Austria.

1. Introduction

The Comprehensive Nuclear-Test-Ban Treaty (CTBT), which was opened for signature in 1996, is a key element in the non-proliferation of nuclear weapons and a crucial basis for the pursuit of nuclear disarmament as it bans any kind of nuclear explosion.

For its verification several techniques are provided. Beside waveform measurements (seismic, hydro-acoustic and infra sound), radionuclides are measured (radioactive particulates and noble gases) and atmospheric transport models are used to identify the possible source regions of radioactive particulate or noble gas detections.

When the International Monitoring System (IMS) network is completed there will be 321 stations worldwide; among them 80 permanently sampling and measuring radioactive particles. At least 40 of these will also be equipped with noble gas measurement systems (UNGA, 1996; DAHLMAN and MYKKELTVEIT, 2009).

Noble gas monitoring is a fundamental and highly sensitive technique for the detection of underground or underwater nuclear explosions. Of all the verification technologies, it is, together with radionuclide particulate monitoring, the only technique that has the potential to provide unmistakable proof of a nuclear explosion (DE GEER, 1996).

To establish this global noble gas network, fully automated radioxenon measurement systems had to be developed, as no commercial systems were available when the Treaty was opened for signature (AUER et al., 2004). Four countries, France, Russia, Sweden and USA, all with experience of atmospheric xenon measurements, offered to develop such systems. With the Provisional Technical Secretariat (PTS) for the CTBT Organisation (CTBTO) and the German Federal Office for Radiation Protection (Bundesamt für Strahlenschutz, BfS), they participate in the International Noble Gas Experiment (INGE) project (AUER et al., 2004).

Under the auspices of the INGE, these systems are now undergoing tests at worldwide locations and send their results to the International Data Centre (IDC) in Vienna for processing and analysis (SAEY and DE GEER, 2005). All four systems that have been developed are able to measure the four radioxenons of interest for CTBT verification: ^{131m}Xe, ^{133m}Xe, ^{133}Xe and ^{135}Xe with the half-lives of 11.9 days, 2.19 days, 5.24 days and 9.14 h, respectively.

The design criterion for all of the equipment is that the minimum detectable concentration (MDC) of ^{133}Xe should be 1 mBq/m^3 or less for a 24 h sampling period. Supporting this, internal documentation of the PTS prescribe an average air flow above 0.4 m^3/h, where the actual sampling rate should never fall outside ±30% of the average during the sampling period. The total air volume sampled should be at least 10 m^3 per day. The 1 mBq/m^3 MDC for ^{133}Xe was state-of-the-art in the mid-1990s when the requirement was set, but has since then improved by almost a factor of ten.

For this study data from three different noble gas systems were used, the Russian ARIX-II (Analyzer of Radioactive Isotopes of Xenon), based on a beta-gated gamma nuclear measurement system (DUBASOV et al.,

2005), the French SPALAX (Système de Prélèvement d'air Automatique en Ligne avec l'Analyse des radio-Xénons), which is based on high-resolution gamma spectroscopy (FONTAINE et al., 2004) and the Swedish SAUNA-II (Swedish Automatic Unit for Noble gas Acquisition), which uses beta-gamma coincidence spectrometry (RINGBOM et al., 2003).

2. Observation of Environmental Radioxenon

2.1. European Radioxenon Measurements in the past: Some History

To put the new, high time resolution measurements of this paper in relation to other radioxenon measurements performed in Europe in the past, we give in the following an overview about older published measurements of atmospheric radioxenon.

2.1.1 World War II

The first atmospheric radioxenon measurements date back to the fall of 1944, at the end of the Second World War and only 2 years after the first successful test of a nuclear reactor in December 1942. Douglas A-26 medium bomber airplanes from the American Ninth Air Force flew low over Germany and sampled air to search for radioxenon fingerprints of possible nuclear reactors operated in Germany and a related weapons programme (ZIEGLER and JACOBSON, 1995).

This operation was set up by a special intelligence unit from the Manhattan Engineer District. The scientific idea of collecting environmental noble gases in the air using large charcoal tubes in the bomb bay of the airplane and then measure the xenon-133 back home in the laboratory, was developed by later Nobel Prize laureate Luis Alvarez. He became the first scientist to develop a radiological air sampling method of overhead reconnaissance. No radioxenon was detected in these campaigns as there were no reactors or nuclear weapons in Germany at that time.

2.1.2 Germany

In the 1960s, radioxenon measurements were reported from the University of Heidelberg. Their goal was to measure ^{85}Kr and ^{133}Xe released by the ongoing nuclear weapon tests in the atmosphere. EHHALT et al., published in 1963 results from a campaign where they indeed observed clear signals from Russian atmospheric tests carried out in the autumn of 1961 (EHHALT et al., 1963). SCHÖLCH et al. (1966) suggested that a peak of about 10 mBq/m^3 detected in a 10-day air collection sample in early June 1965 might have been due to the second Chinese atmospheric test, which had been performed about 3 weeks earlier (on 14 May 1965, a nuclear explosion with a yield of 35 kt TNT equivalent in 500 m above ground; MIKHAILOV et al., 1999). From a few ^{133}Xe measurements in late 1964 and early 1965 they also reported a general background of about 5 mBq/m^3 presumably from European reactors.

In the early 1970s the BfS started to continuously monitor the atmospheric ^{133}Xe and ^{85}Kr activity concentrations in Freiburg, Germany (STOCKBURGER et al., 1977). At present the atmospheric activity concentration of ^{133}Xe in ground level air is continuously monitored at seven sites in Germany as part of the "Integrated Monitoring and Information System" (IMIS) (WEISS and LEEB, 1993), the German surveillance program for radioactivity in the environment. In addition, samples are taken at other stations around the globe, e.g., at the Meteorological Research Institute (MRI) in Tsukuba, Japan (IGARASHI et al., 2000b; SCHLOSSER et al., 2003).

The pre-enriched samples taken at the sites are sent in a 1 L aluminium vessel to the central noble gas laboratory at BfS in Freiburg for analysis. The collection time during routine operation is seven days. The total volume sampled is around 10 m^3 of air. The procedures for sampling, enrichment and purification of the noble gas fractions are all manual (IGARASHI et al., 2000a). The integral beta activity of the samples is measured in proportional counters using methane as an additional gas component. This integral counting method gives the total activity of all radioxenons but a separation of the components can be done by decay analysis.

Besides ^{133}Xe, the most abundant radioxenon isotopes observed in environmental samples, contributions of ^{131m}Xe and ^{135}Xe could be determined to a few percent of the total beta-activity. The MDC for ^{133}Xe in routine samples is about 1 mBq/m^3. The longest time series available is from the station in Freiburg (Fig. 1).

During recent years the average ^{133}Xe activity concentration measured in weekly samples at German stations is around 6 mBq/m^3 with large variations between 1 and 100 mBq/m^3 (BMU, 2007). The maximum activity concentration of 106 Bq/m^3 was measured in the daily sample of 1 May 1986 taken at BfS in Freiburg. It originated from the Chernobyl reactor accident.[1]

In the years between 1987 and 1995 the atmospheric ^{133}Xe activity concentration decreased by a factor of around 20. This behaviour is consistent with reported noble gas release data from nuclear power plants in Germany and could be explained by improvements of the nuclear fuel rod cladding and reactor containment systems as well as with longer delay times before the release of noble gases in the atmosphere (BIERINGER and SCHLOSSER, 2004).

2.1.3 Sweden

A system for sampling and analysis of small amounts of radioxenon in ambient air was developed around 1980 by the Swedish Defence Research Agency (FOI, formerly FOA). This was a forerunner to the SAUNA system but used at that time charcoal adsorption at −80°C and high-resolution gamma spectroscopy for detection. During the development phase, 2–3 days samples were periodically taken in Stockholm and during 1982 also at a satellite station at Ljungbyhed in southern Sweden (BERNSTRÖM and DE GEER, 1983). Initially the average detection limit was 1.8 mBq/m^3, which was later reduced by a factor of three by chromatographic suppression of ^{222}Rn. The reported average activity concentration of ^{133}Xe was 8.6 mBq/m^3 at the Stockholm station and 22.6 mBq/m^3 at the satellite station. These were quite high values and occasionally extreme values were measured up to around 250 mBq/m^3. The reason for the high values was excessive emission from a boiling water reactor near Gothenburg, some 420 km southwest of Stockholm and 150 km north of Ljungbyhed. Up to 26 fuel elements exhibited cracks and these elements were

---

[1] For a long time only Germany was known to have measured radioxenon from the Chernobyl accident. Recently, however, ^{133}Xe concentrations around 1.5 Bq/m^3 were reported from late April 1986 in Cherepovetz, Russia (PAKHOMOV and DUBASOV, 2008).

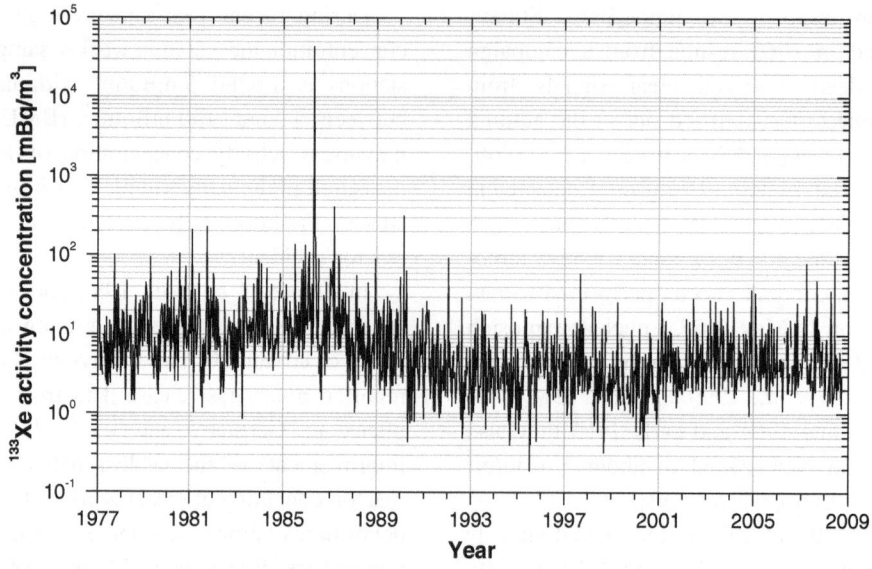

Weekly ^{133}Xe activity concentration of ^{133}Xe in ground level air at Freiburg, Germany

exchanged during the 1981 summer revision. In 1982 the levels then went down to typically below the detection criterion. At the end of October 1980 roundly 100 mBq/m^3 were observed in Stockholm. At that time there were no excessive emissions from Swedish reactors and the signal was interpreted to be due to the last atmospheric nuclear test conducted 2 weeks earlier on 16 October 1980 at Lop Nor, China, with a yield between 150 to 1,500 kt (MIKHAILOV et al., 1999).

The Swedish xenon system was mothballed between 1983 and 1989 but from October 1990 until about 1998 there was a series of measurements from Stockholm and for part of that time also from two satellite systems, one at Ljungbyhed in the south and one in Kiruna in the north. At the satellites pre-processed gas samples were collected and then sent by post in a gas-proof plastic bag to the laboratory in Stockholm for analysis. In October 1990 a 24 mBq/m^3 peak in the xenon activity concentration was observed in Stockholm that, based on careful meteorological backtracking analyses, was believed to be due to a small leak from the last Soviet nuclear test on 24 October 1990 (an eight device shot of in total 40 to 380 kt at Novaya Zemlya, about 2,100 km NE of Stockholm) (MIKHAILOV et al., 1999).

During the period a slowly decreasing trend was observed with an average of 3.0 mBq/m^3 in Stockholm at 59°N, some 0.7 mBq/m^3 in Kiruna at 68°N and some 0.8 mBq/m^3 in Ljungbyhed at 56°N. In late March 1992, a peak at 24 mBq/m^3 was observed in Stockholm. This large activity concentration originated from an accident/incident at the Sosnovy Bor nuclear power plant 100 km west of St. Petersburg and 600 km east of Stockholm. With time the more routine operations of the Swedish xenon stations were phasing out as work increasingly was concentrated on developing the SAUNA system. In preparation for Phase 2 of INGE, when all four systems were to be run in parallel in Freiburg, Germany (see Sect. 2.5), a surveillance period in Stockholm in September 2000 exhibited a mean activity concentration of ^{133}Xe of about 1 mBq/m^3 RINGBOM et al. (2003).

2.1.4 The IMS Noble Gas Network

There are many significant improvements of the new systems set-up for CTBT verification as compared to the older ones just described:

• Beside the long time series of the BfS, measurements in the past were often performed during short, well-defined time periods. The data presented in this paper are collected over several years and the measurements are ongoing;

- The sampling time during these campaigns was often of the order of several days and upto 1 week or even more. In the INGE network, the collection time resolution is between 8 and 24 h;
- In most of the older measurements, the method was optimized for the detection of ^{133}Xe. Due to shorter cycle times, higher processing capabilities and improved counting technologies, the new INGE systems can detect four different radioxenon isotopes simultaneously with a high sensitivity;
- These systems have a very strong detection capability, with ^{133}Xe MDC to 0.1 mBq/m^3 for 12-h samples;

All these improvements make it possible to perform more accurate isotope ratio analysis for source discrimination, better source identification and meteorological geo-location analyses.

2.2. Current Set-up of the Measurement Stations in Europe

For this paper, we used continuous radioxenon data with a high time resolution from six different European locations.

At present, four systems of the INGE network are installed at European IMS stations (Spitsbergen in the Norwegian archipelago Svalbard; Stockholm, Sweden; Dubna, Russia and on the Schauinsland Mountain, near Freiburg, Germany). In Bruyères-le-Châtel, 30 km south of Paris, a system is operated by the French Atomic Energy Commission (Commissariat à l'Énergie Atomique, CEA). The measurements at the Marseille site were performed by Environment S.A (the manufacturer of the French

SPALAX system) with two IMS systems purchased by Health Canada, which collected data prior to shipment to their end destinations in Canada (IMS stations in Yellowknife and St. John's). The geographical and sampling system specifications of the six different sites are listed in Table 1.

The Swedish and Norwegian stations are equipped with SAUNA-II systems, the French and German with SPALAX devices and the Russian with ARIX-II equipment.

The SAUNA-II system samples air in 12-h cycles. Then the collected xenon fraction is purified and concentrated during some 7 h before it is counted with the (plastic and NaI) beta-gamma coincidence detector for approximately 11 h. These spectra have been analysed with PTS developed software, based on a net-count method.

The SPALAX system continuously samples air for 24 h per cycle. At the end of such a collection cycle the final purification and transfer into the counting system needs about one more hour before the start of counting on a broad energy high-purity germanium gamma-ray detector for around 23 h. The spectra from Schauinsland were analysed and reviewed with the PTS developed radionuclide analysis and evaluation software AATAMI (2003), those from Bruyères-le-Châtel and Marseille with Canberra's software Genie 2000. The xenon isotopes analysed with Genie 2000 are based on the gamma ray peak information only—the Aatami software also uses additional information from the xenon and caesium X-rays in the 30 keV region.

The ARIX-II system collects air in 12-h cycles. Then the air is purified and concentrated during approximately 4 h before it is counted with the (plastic and NaI) beta-gated gamma detector for

Table 1

Location of the radioxenon measurement systems considered in this study, listed from north to south

Station	Host country	Latitude	Longitude	Altitude (m)	System
Spitsbergen	Norway	78.1°N	15.2°E	220	SAUNA-II
Stockholm	Sweden	59.4°N	18.0°E	40	SAUNA-II
Dubna	Russian Federation	56.4°N	37.2°E	120	ARIX-II
Bruyères-le-Châtel	France	48.9°N	2.3°E	150	SPALAX
Schauinsland	Germany	47.9°N	7.9°E	1,208	SPALAX
Marseille	France	43.2°N	5.2°E	43	SPALAX

Table 2

Overview of the different measurement periods covered in this study

Station	Operational period	Number of valid measurements
Spitsbergen	13 April 2003–31 August 2008	1,798
Stockholm	22 August 2005–31 August 2008	1,540
Dubna	12 November 2006–31 August 2008	1,146
Schauinsland	20 February 2004–31 August 2008	1,165
Bruyères-le-Châtel	8 August 2003–4 July 2005	391
Marseille	29 September 2004–23 March 2005	176

Values below Lc were not reported for the Marseille station

around 18 h. These spectra have been analysed with software from the system developer KRI.

2.3. Time Series Analysis for ^{133}Xe

The data considered in this study were sampled at different time periods, between August 2003 and August 2008. Table 2 gives an overview of the different measurement periods. During this reporting period, not all systems were transmitting data continuously, some of them being used for training operators or undergoing upgrades. The systems in Marseille were only operated for specific factory acceptance test periods.

The xenon isotope most commonly seen in ambient air samples is ^{133}Xe. Its half-life of 5.24 days and its large cumulative fission yields of 6.7 to 7.0% are factors contributing to its high detectability. It is often observed at locations downwind from nuclear power plants and is almost

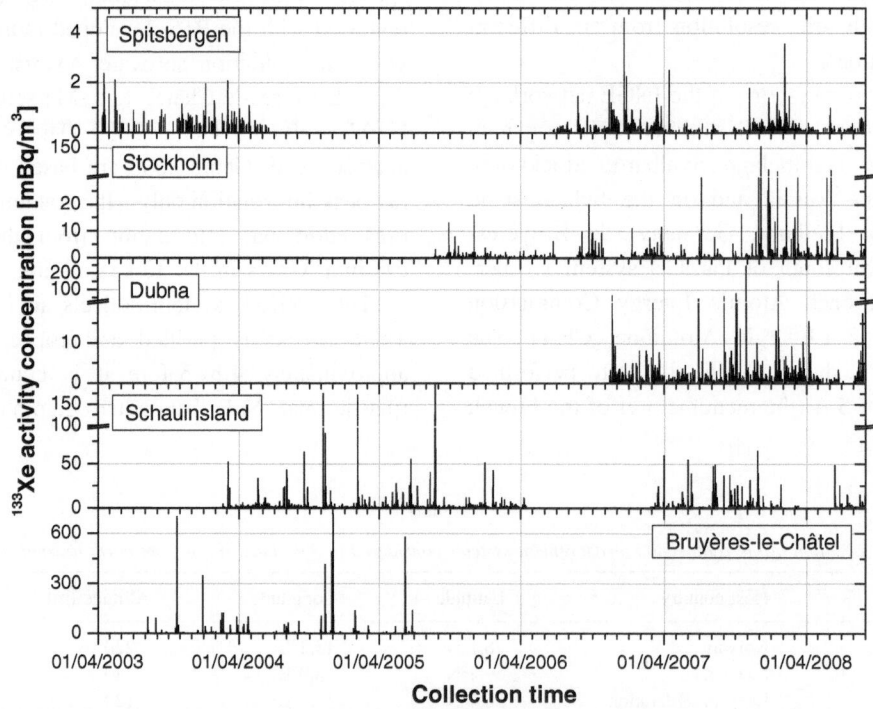

Figure 2
Activity concentration of the ^{133}Xe isotope at the five sites with long time series available. Please note that the plots have different activity concentration scales

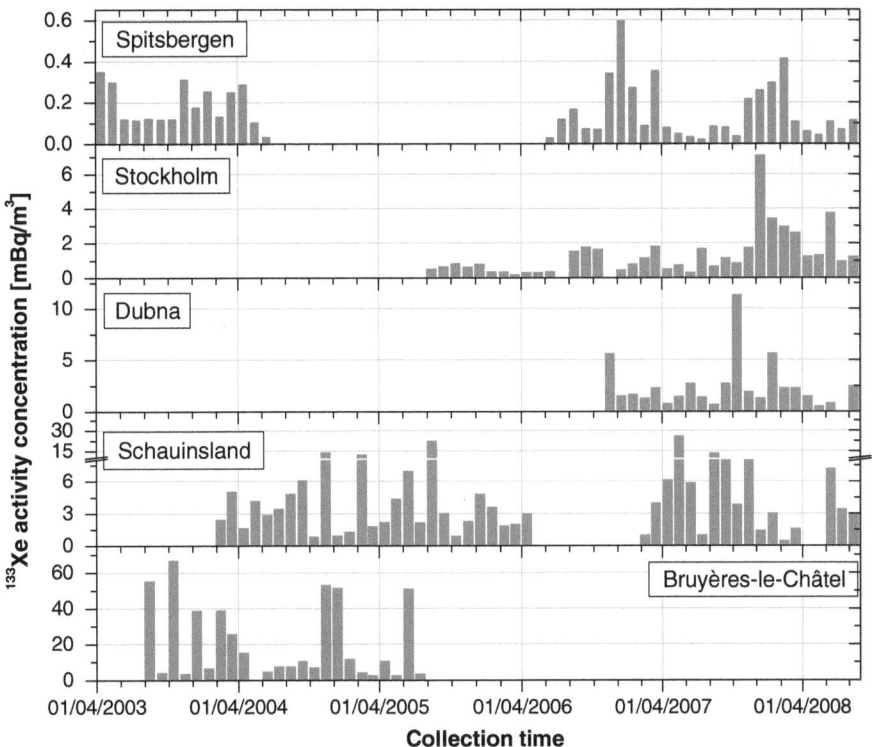

Figure 3

Monthly activity concentrations of ^{133}Xe at five different stations. Please note that the plots have different activity concentration scales

continuously present in the air of the nuclear power dense regions of the globe. Therefore, the main focus of this paper is on this isotope.

Figure 2 presents the activity concentration of all measurements of ^{133}Xe at the five different stations with longer time series. The monthly averages of the activity concentrations are shown in Fig. 3.

At this point it might be useful to note that prevailing activity concentrations of a few mBq/m^3 of ^{133}Xe, as observed in Europe, yields low doses to human beings or other species in the biosphere. The dose rate factor of ^{133}Xe is about 5×10^{-8} mSv/year per mBq/m^3 (KOCHER, 1980). This renders an annual dose to humans of many orders of magnitude less than what is caused by its more natural "cousins" Radon-220 and Radon-222 and their progeny.

The average MDCs of the four different radioxenon isotopes at the six different sites are shown in Table 3. At Marseille, only the MDC from ^{133}Xe was calculated, as the system was under factory

Table 3

The Minimum Detectable Concentration (MDC) and the counting time of the samples at the six stations for the four relevant radioxenon isotopes

Station	^{131m}Xe (mBq/m^3)	^{133m}Xe (mBq/m^3)	^{133}Xe (mBq/m^3)	^{135}Xe (mBq/m^3)	Counting time (h)
Spitsbergen	0.12	0.11	0.19	0.73	11
Stockholm	0.15	0.17	0.23	0.82	11
Dubna	0.3	0.4	0.2	0.9	18
Schauinsland	0.33	1.6	0.12	0.55	23
Bruyères-le-Châtel	6.0	2.0	0.35	0.88	23
Marseille	n.a.	n.a.	0.34	n.a.	23

The MDCs from Bruyères-le-Châtel and Marseille are based on the main gamma peak of the nuclide

133

acceptance testing. One should note the different counting times of the gas in the detector, which depends on the measurement system used (see Table 1). A shorter counting time increases the time resolution of the sampling but it also decreases the sensitivity. All are, however, still well within the minimum specifications of the IMS (1 mBq/m^3 of ^{133}Xe).

In contrast to observations at the Spitsbergen station at 78°N, that do show higher values in winter than in summer (SAEY et al., 2006) there are apparently no such correlations to the season at any of the other stations within the European reactor region. This was also confirmed by Fast Fourier Analyses of the data. We therefore conclude that the release of radioxenon at nuclear facilities in Europe is spread throughout the year whereas Spitsbergen is located remotely enough from xenon emission areas in North America and Europe that long-range atmospheric transport patterns govern the measurements there. These patterns are mainly controlled by the North Atlantic Oscillation that features a pronounced seasonal variability (SAEY et al., 2006).

2.4. Background Determination

The distribution of the different activity concentrations of ^{133}Xe at all six different sites is shown in Fig. 4. Measurements below the critical limit (Lc, the value which is used to concretely decide whether a signal is present or not in an actual measurement, with a confidence level of 99.5%) are not plotted but considered as an offset on the probability scale reflecting the fact that a total of 22% of the data were below Lc. The plot suggests that the data are log-normally distributed, which is typical for environmental atmospheric data.

To confirm this distribution, the D'Agostino test (D'AGOSTINO and STEPHENS, 1986) was performed on the data series of the six different stations. This test can be performed if the data set is larger than 50 and smaller than 1,000 measurements. The non-detects and the extreme values (see Sect. 2.7) were not considered and the significance level α was 0.05. The test was carried out with the values of the activity concentrations and with the logarithmic values of the activity concentrations. Only the test on the logarithmic values confirmed, for all stations except Marseille, the hypothesis of normal distribution which means that the activity concentrations are log-normally distributed.

The frequency distributions, the mean values and the medians of the data sets of ^{133}Xe at the six different sites are shown in Fig. 5.

As five of the data sets were shown to be log-normally distributed, logarithmic scales are used for the activity concentrations. From the sixth, Marseille, very few data points were available. The inflection points in the plots on the higher side indicate optically the change from background to extreme

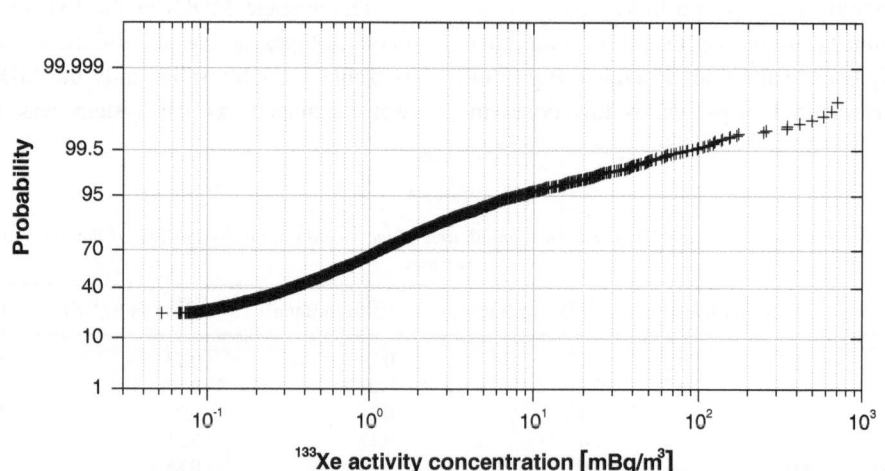

Figure 4
Log-normal Probability Plot of all measurements above the detection limit for all six stations

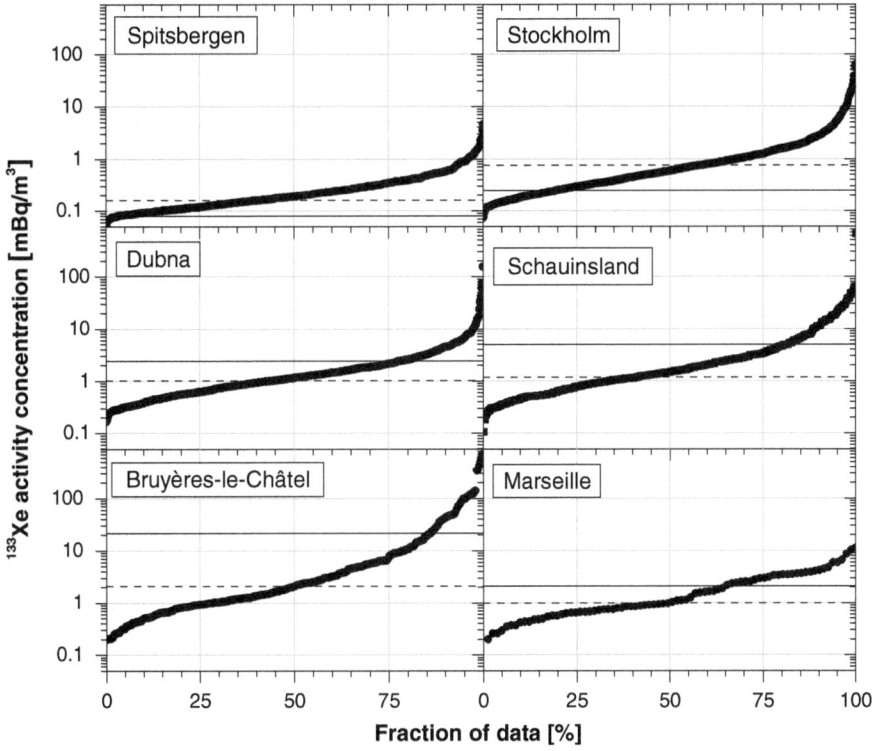

Figure 5

The frequency distribution of ^{133}Xe at the six different sites. The *solid horizontal line* shows the mean value and the *dashed line* the median of the data sets. Data below the critical limit Lc were not plotted

Table 4

Summary of the statistics of the daily ^{133}Xe measurements at the six radioxenon measurement stations presented

Station	Measurements below Lc (%)	Max. (mBq/m^3)	Mean (mBq/m^3)	Median (mBq/m^3)	IQR (mBq/m^3)	75% percentile (mBq/m^3)	95% percentile (mBq/m^3)
Spitsbergen	49	4.61	0.16	0.08	0.25	0.20	0.59
Stockholm	20	155.6	1.39	0.40	0.86	1.01	4.24
Dubna	8.7	249.0	2.43	1.03	1.5	1.97	6.17
Schauinsland	0.18	257	4.3	1.2	2.2	2.8	18.2
Bruyères-le-Châtel	0	717.5	21.4	2.1	5.7	8.0	98.2
Marseille	n.a.	11.1	2.1	1.0	2.3	3.0	5.9

values. Figure 5 shows this in more detail. The Box-and-Whisker diagram of the data set is discussed in Sect. 2.7.

A summary of some statistical parameters is given in Table 4. The values were calculated from all data sets including measurements below Lc. The percentage of measurements below Lc ranges between 0.18% in Schauinsland and 49% in Spitsbergen. For the latter site this means that the median is given by the typical Lc value. The mean has been calculated assuming a zero true value for all measurements below Lc. It should be pointed out that this leads to a systematic underestimation of the true mean, particularly for sites with a high percentage of data points below Lc. For measurements from the Marseille site, no data below Lc were available. In this case the values refer to data above Lc only and both mean and median may be systematically overestimated.

2.5. Comparison with a Previous Measurement Campaign in Germany

In the year 2000, an intercomparison exercise with the four different INGE radioxenon measurement systems took place at the BfS in Freiburg, Germany. The goal was to demonstrate the available capabilities and determine the technical characteristics of these new technologies in an independent laboratory away from the developers but still with experienced staff present. The BfS in Freiburg operates a noble gas laboratory and has the required experience in noble gas monitoring to validate the systems by performing re-analysis of the archived samples. This intercomparison exercise is thoroughly described in BOWYER *et al.* (2002) and AUER *et al.* (2004).

The ARSA (Automated Radioxenon Sampler and Analyzer) system from the Pacific Northwest National Laboratory (PNNL, USA), which, like the SAUNA system, was based on beta-gamma coincidence spectrometry, measured the most complete

time series of the exercise. Therefore, these data were used to compare the results from the year 2000 with those from 2004 to 2008. The data from 2000 for the isotope ^{133}Xe are presented in Fig. 6. They are based on 874 eight-hour measurements over a period of 340 days (coverage of 86%).

The distribution of this isotope measured in the year 2000 in Freiburg as compared to nearby Schauinsland in the current study, illustrates that there was little change: the most frequently measured activity concentration interval was in both periods the one between 0.4 and 0.8 mBq/m^3.

The frequency distribution has the same shape for both periods. However, the mean and median in the year 2000 (2.31 and 0.80 mBq/m^3, respectively) were lower than in the period 2004–2008 (4.3 and 1.23 mBq/m^3, respectively). This is in contrast to the published data of releases of noble gases from nuclear power plants within the European Community (VAN DER STRICHT and JANSSENS, 2005): in the year 2000 a total airborne release of 9.7 PBq is

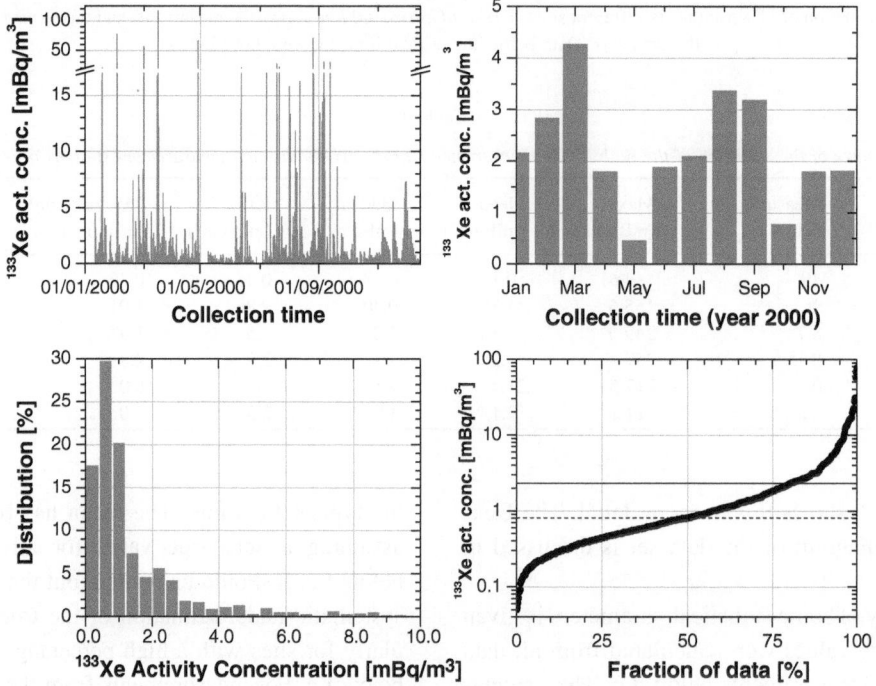

Figure 6

^{133}Xe data from the ARSA system in Freiburg, Germany in 2000. The *upper left graph* shows the activity concentrations of ^{133}Xe. The monthly average activity concentrations are shown in the *upper right graph*. The lower left graph shows the distribution of the activity concentration of ^{133}Xe. The frequency distribution of ^{133}Xe is depicted in the *lower right graph* (the *solid horizontal line* shows the mean value and the *dashed line* the median of the data sets)

reported and 4.7 PBq in 2003, with a generally decreasing trend.

A plausible explanation for the higher activity concentrations measured during 2004 to 2008 is an increased production of the radiopharmaceutical facility in Fleurus, South Belgium (AUER et al., 2010). This facility is remote enough from Freiburg and Schauinsland that emissions from there have already been vented into the free troposphere during their transport to the station. The facility in Fleurus is considered to be a major contributor of environmental radioxenon in Europe (SAEY, 2009). The radiopharmaceutical isotope production facility at Chalk River, Canada (ACHIM et al., 2007) could also impact the detection level in Western Europe.

2.6. Comparison with a Field Campaign in Austria

During July to September 2006 a comprehensive test was conducted in Seibersdorf, Austria, focusing on equipment and procedures to collect and analyse radioxenon samples from the atmosphere and from sub-surface gas. During the test mobile versions of the SAUNA and ARIX systems were deployed. Though the test focussed on operational issues and logistics, 16 atmospheric samples and five sub-

surface gas samples were collected and analysed. The activity concentration of ^{133}Xe in most of the atmospheric samples was between 0.3 and 2.4 mBq/m^3, though on one occasion ^{133}Xe activity concentrations of (17 ± 1) and (51 ± 3) mBq/m^3 were reported by the SAUNA and the ARIX system, respectively (AXELSSON, 2007).

2.7. Tests for Extreme Values

Figure 7 shows the Box-and-Whisker diagram of ^{133}Xe measurements at the six different sites. The length of the central boxes indicates the spread of the central 50% of the data. This interquartile range (IQR) in which 50% of the ranked data are found, describes the dispersion of the measurements. It refers only to the data set with data above Lc. Therefore the values differ from those given in Table 4.

The length of the whiskers indicates the extent that the measurements are spread out below and above the central 50% box. The upper whisker extends to the 95th percentile. The − symbol indicates the largest observations and the × symbols show the 99th percentiles. Data between the − and the × are considered to be extreme values. The □ symbol

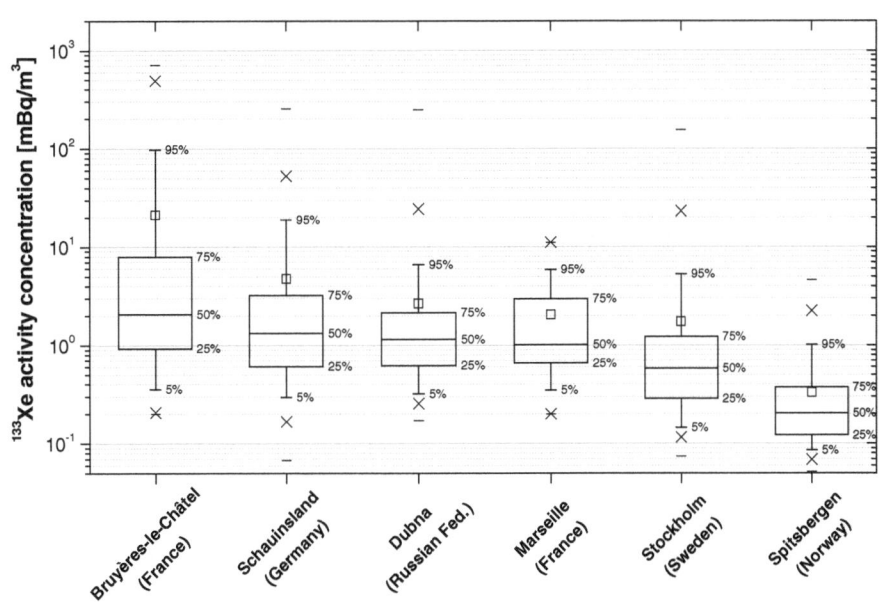

Figure 7

Box-and-Whisker plot of ^{133}Xe at the different sites—values below Lc were not considered here. For details, see text

shows the mean value of the measurements. The lower – symbol indicates the average Lc. In this diagram we can see that the data from the more central European stations Bruyères-le-Châtel and Schauinsland measure generally higher ^{133}Xe activity concentrations than the stations situated on the fringe of the main nuclear power plant region.

The extreme values can be indications of erroneous analyses, but are most probably just the results of extreme emissions at a certain reactor or at a radiopharmaceutical isotope production facility (SAEY, 2009). In the framework of the CTBT verification, an extreme value could trigger a special meteorological analysis to identify its possible source region.

2.8. Atmospheric Transport Modelling

2.8.1 Assumption of Sources

The station of Bruyères-le-Châtel is situated in a region with a high density of nuclear power plants—there are roundly 60 operational reactors within a 500-km range. In a 500 km radius surrounding Schauinsland, there are approximately 70 reactors. France produced in the year 2000 half of all nuclear-generated electricity in Europe—Germany almost 20% and Sweden 8% (Energy Information Administration, Office of Coal, Nuclear, Electric and Alternate Fuels, International Nuclear Model, PC Version—May, 2001). Both stations also have Fleurus, the world's third largest radiopharmaceutical isotope production factory (after Chalk River in Canada and Petten in the Netherlands) within the 500-km range. Petten, the world's second biggest producer, has long delay lines installed, which reduce their emission several orders of magnitudes (SAEY,

Table 5

Order of magnitude of releases of radioxenon at different nuclear facilities

Type of release	Typical order of magnitude of radioxenon release
Hospitals	$\sim 10^6$ Bq/d
Nuclear power plants	$\sim 10^9 - 10^{11}$ Bq/d
Radiopharmaceutical facilities	$\sim 10^{11} - 10^{13}$ Bq/d
1 kton nuclear explosion underground	$0 - 10^{15}$ Bq
1 kton nuclear explosion atmospheric	$\sim 10^{16}$ Bq

2009). The other two stations have fewer reactors within a 500-km radius: Marseille 15, Stockholm 12 and Spitsbergen none.

Table 5 presents an overview of the order of magnitude of radioxenon releases at different types of nuclear installations (UNSCEAR, 2000; VAN DER STRICHT and JANSSENS, 2005; KALINOWSKI and PISTNER, 2006; SAEY, 2007, 2009; KALINOWSKI and TUMA, 2009).

Some studies concerning the identifications and contributions of the main radioxenon sources in Western Europe (ACHIM et al., 2007; LE PETIT et al., 2008) showed that most of the observed detections at Marseilles, Bruyères-le-Châtel and Schauinsland could be for a first order of approximation, accounted as releases from the European nuclear power plants and from the radioisotope production facility in Fleurus, Belgium.

2.8.2 Modelling of Two Selected Events with the Web-Grape Software

The PTS utilizes Atmospheric Transport Modelling (ATM) to estimate in which regions of the globe a surface level emission into the atmosphere would possibly result in radionuclide detection at one of the sampling stations. Based on the routine global wind field analysis assimilated by the European Centre for Medium range Weather Forecasts (ECMWF), backward atmospheric transport calculations are done for every station and every measurement taken in the network (WOTAWA et al., 2003; BECKER et al., 2007). The system is organized in a four-layered workflow where the results from the basic ATM calculations are stored as so-called Source Receptor Sensitivity (SRS) fields at a suitable geo-temporal resolution (currently 1°; 3 h) half-way to the workflow.

These fields are then available for post-processing, both in automated mode and in a more flexible interactive mode. In automated mode standard map products called Field-of-Regards (FOR) are generated and included in the daily issued reviewed radionuclide reports. For the interactive mode the client software, Web-Grape, has been developed (BECKER, 2006). One source region estimation product that can be generated by Web-Grape in addition to the FOR is the Possible Source Region (PSR), which is a map that depicts the area or areas that best fit detections in several samples that can be assumed to belong to the same event. In contrast to the measurement specific FOR, the PSR is event-specific and belongs to a scenario of measurements. Web-Grape also features an event calculator (ECAL) with which the user can test different source hypotheses and see at what stations and in what time periods these emissions would lead to detections in the PTS radionuclide network.

Two interesting events were selected as examples to show how a source of an anomalous detection can be located. These examples demonstrate that a regular (daily) sensitivity calculation for each station (i.e., how sensitive a station on a certain day is towards known radioxenon emitting sources), as proposed by UNGAR et al. (2007), might help in automatically screening out certain events measured at noble gas stations, i.e., classifying certain high values as resulting from civil sources, as the possible source might be known.

2.8.2.1 One station measurements that could point to the release of a radiopharmaceutical isotope production facility, using one radioxenon isotope The Schauinsland mountain station measures a few times per year activity concentrations above 50 mBq/m^3 of ^{133}Xe, which can be considered as extreme values. Most backtracking calculations with Web-Grape

could show that the air that contained those higher radioxenon activity concentrations had passed the south of Belgium 1 or 2 days before. This is an area that hosts the "Institut National des Radioéléments" (IRE) which produces ^{99}Mo for ^{99m}Tc generators and sells its by-product ^{133}Xe to "MDS Nordion SA" located on the same premises near the little town of Fleurus some 40 km south of Brussels. A certain amount of the radioxenon activities produced in this process is released into the atmosphere (SAEY, 2009). As shown in Table 5, these releases can be 10^{11}–10^{13} Bq/day.

Figure 8 displays a backtracking example. The lower left image shows the FOR belonging to the detection of 15 mBq/m^3 of ^{133}Xe in air collected between 14 July 2008 0600 hours and 15 July 2008 0600 hours at the station Schauinsland in Germany (all times in this paper are in UTC). The sensitivity of this measurement (receptor) to all sources on 13 July between 9–12 h is colour-coded and indicates the area where the air sampled during this single measurement at Schauinsland was located at this time. The lower right image shows the PSR of the four measurements taken at the same station between collection start 13 July and collection stop 17 July 2008. There is a confined area where the PSR exhibits values between 0.9 and 1.0. These correlation coefficient values indicate a high consistency of a singular

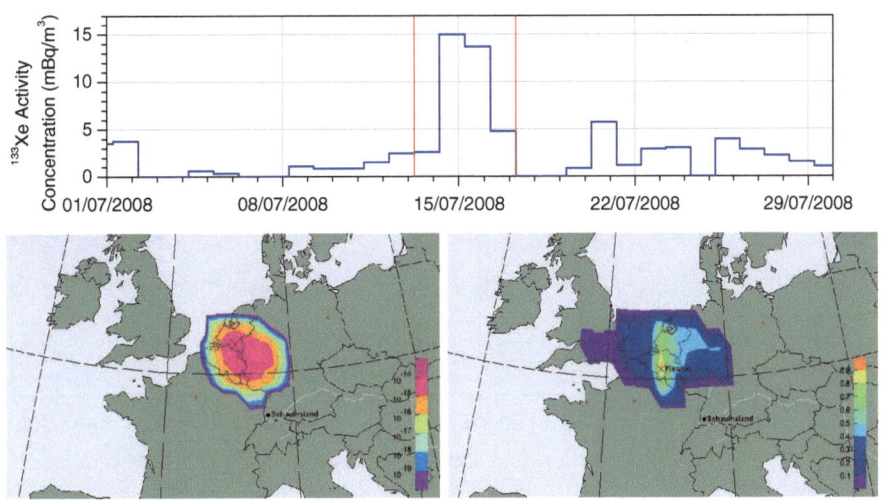

Figure 8

Time series (*upper graph*) of ^{133}Xe for this station for the considered time period. The values between the *red vertical lines* are those used for the PSR calculations. FOR (*lower left*) and PSR (*lower right*) for an interesting measurement of ^{133}Xe at the INGE station on Schauinsland Mountain, Germany

source from there with the four measurements encountered. It is worth noting that the PSR pattern, which utilises more data than the FOR, is more confined than the FOR and corresponds to the region around Fleurus in the south of Belgium. Applying the simple source-receptor relationship equation, Source strength [Bq] × Sensitivity [1/m^3] = Observation [Bq/m^3], the FOR across that area reveals information on the strength of the release: The relation to the release activity is for this area 10^{-15}. The measured activity was 15 mBq/m^3 of ^{133}Xe. Thus it can be concluded that the source should have released around 15×10^{12} Bq on 13 July 2008 between 0900 and 1200 hours. This estimation is consistent with the daily releases from this facility.

2.8.2.2 Two station measurements that could indicate different releases at one radiopharmaceutical facility, using two isotopes

Another case was studied in which during the same period (mid-November 2004)

in both the Schauinsland Mountain station and the Bruyères-le-Châtel station extreme ^{133}Xe values were measured, 83 and 400 mBq/m^3, respectively (see Fig. 9). The ^{135}Xe isotope was also present in these samples.

Figure 9 shows the isotopic activity ratio of ^{135}Xe/^{133}Xe of the samples collected between 11 and 13 November 2004. They are positioned on one of the parallel lines which has the slope of the radioactive decay of the ratio. This could indicate the hypothesis that they are related to the same emission.

The sample with collection stop on 6 November 2004 is clearly related to another source and the 13 November sample is not fully consistent with the cluster of four samples and will, therefore, not be considered further.

Initial atmospheric transport modelling indicates that the samples measured in Bruyères-le-Châtel and Schauinsland might originate from the same sources: the radiopharmaceutical isotope production facility in Fleurus. However, meteorological backtracking analysis reveals that both stations did not see the same

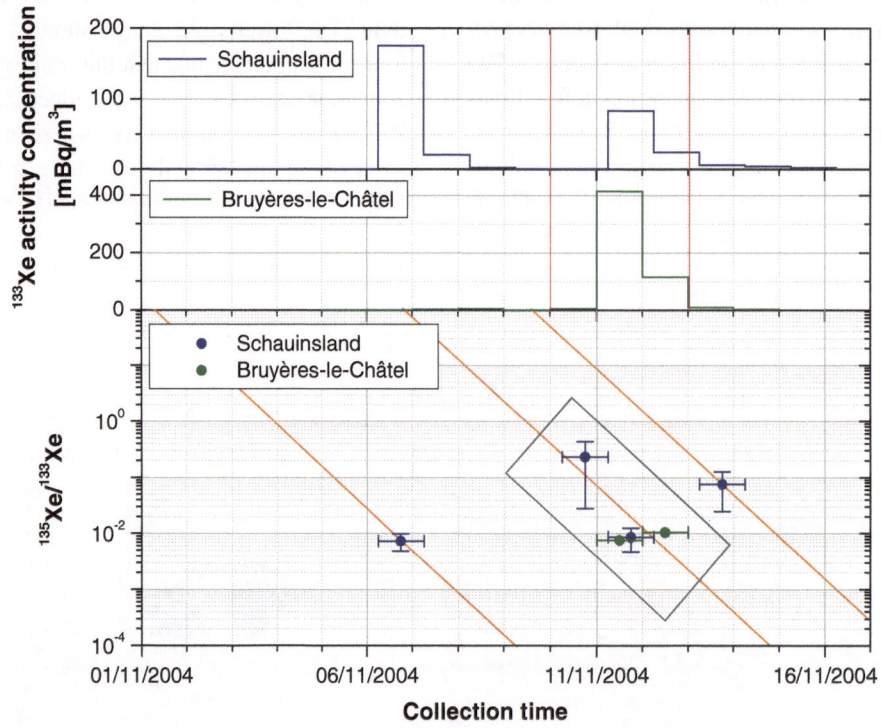

Figure 9

Time series (upper two graphs) of ^{133}Xe for Schauinsland and Bruyères-le-Châtel for the considered time period. The values between the *red vertical lines* are those used for the PSR calculations. The *lower graph* shows the ^{135}Xe/^{133}Xe ratio in early November 2004. The *orange lines* have slopes corresponding to the ratio decay. The four samples in the *box* may be associated with the same event in accordance with radioactive decay, but only ATM shows that it is not the case. The *horizontal bars* indicate the 24-h sampling time

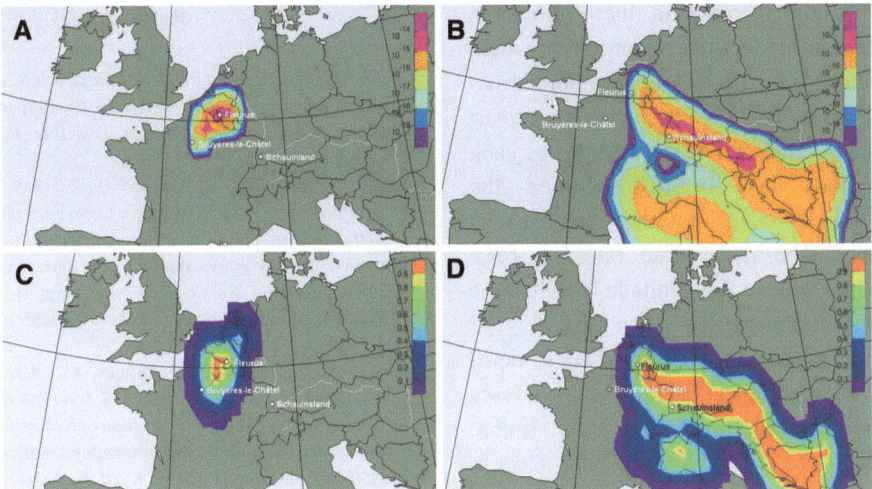

Figure 10

FOR (*upper two images*) and PSR (*lower two images*) for an extreme value measurement of ^{133}Xe at the INGE station in Bruyères-le-Châtel, France (*two left images*) and on Schauinsland Mountain, Germany (*two right images*), for the time period indicated in Fig. 9—A detailed description is given in the text

release. The facility in Fleurus released in that period daily. The samples measured in Schauinsland originate from a release that took place on 9 November (see right-hand side FOR and PSR plots for 3–6 UTC of Fig. 10) while the samples measured in Bruyères-le-Châtel most likely originate from a release 2 days later, on 11 November, 3–6 UTC (see left-hand side FOR and PSR plots of Fig. 10). This result has been verified by forward atmospheric modelling, assuming these release times at Fleurus yield a very good correlation between the observed and modelled concentrations at Bruyères-le-Châtel and Schauinsland of 0.94 and 0.96, respectively, however the latter only while applying a 1-d phase correction in view of the excessively flat representation of the topography in the wind-field model running at 1° horizontal resolution.

Additionally other sources were checked. During that period, there was only one nuclear power plant that went through a shut-down or start-up operation, according to reports from the IAEA Power Reactor Information System (PRIS). This was a reactor start-up operation at Gravelines, France, situated at the sea side near the Belgium border, after a refuelling on 8 November. According to the ATM calculations, the release should have been six orders of magnitude above the average nuclear power plant releases to explain the measurements and can, therefore, be excluded.

This example illustrates that simultaneous detections at different stations and measurements of multiple isotopes can point to a possible source. It also demonstrates the crucial role of ATM in source identification.

3. Summary and Outlook

Long-term continuous data from four radioxenon isotopes at six different European radioxenon stations have been analysed:

- Three stations are situated in the middle of the European nuclear facilities (nuclear power plants and two radiopharmaceutical isotope production facilities), two are just outside this zone and one is a remote, polar station. The measurements show that the releases of radioxenon at nuclear facilities in Europe are spread throughout the year;
- The radio xenon background at several stations in the European network, including variations in time and geographical location, was well characterized during recent years. This is a very valuable data set for further developing a categorisation scheme to discriminate releases from civil sources against releases from nuclear tests;
- A comparison between ^{133}Xe data from the year 2000 and the years 2003 to 2008 has shown that

there was a slight increase in the radioxenon activity concentrations in the Freiburg area, which might originate from an increase in radiopharmaceutical production of ^{99}Mo of around 5% per year;

- The background values for these nuclides at each of these stations have been determined—the atmospheric activity concentrations were log-normal distributed. The mean dose rate has been calculated to be 9 orders of magnitude less than the one from the common present ^{222}Rn gas;

- Some selected examples have been used to demonstrate how meteorological backtracking analyses can be used to indicate the possible source locations and source strengths that would account for the observations.

To reach a routine for an extreme value flagging system which could be used to automatically screen out extreme values, the following topics would need more careful study:

- release scenarios from different types of nuclear facilities (theoretical and experimental);

- automatic sensitivity calculations of the INGE stations towards all known major radioxenon emitters;

- field measurements around these facilities to better understand and quantify the initial dilution of batch releases.

An early information system of the nuclear facilities themselves regarding significant releases would help build confidence in the measurements.

Acknowledgments

The presented data would not be available without the conscientiousness of the local station operators, the crew from CEA in Bruyères-le-Châtel, from KRI in Dubna, from SFI in Marseille, from the BfS on the Schauinsland Mountain, from SvalSat in Spitsbergen and from FOI and Gammadata in Stockholm. We are grateful to Dr. Kurt Ungar (Health Canada) for giving permission to use the data acquired by their equipment in Marseille. We would also like to thank all colleagues within the INGE collaboration for fruitful work and discussions. The authors are also grateful to Dr. Gerald Kirchner (BfS, Germany) for his critical review and valuable comments on the manuscript.

REFERENCES

AATAMI (2003), *User Manual of Radionuclide Analysis and Evaluation Software Aatami*, Version 3.15. Preparatory Commission of the Comprehensive Nuclear Test Ban Treaty Organisation, Vienna, Austria.

ACHIM, P., BELLIVIER, A., ARMAND, P., TAFFARY, T., FONTAINE, J.P., and PIWOWARCZYK, J., C. (2007), *Contributions of xenon releases in the atmosphere from radionuclide production facilities and nuclear nuclear power plants to the detection of ^{133}Xe by SPALAX systems in werstern Europe*, Proc. 11th Internat. Conf. Harmonisation within Atmospheric Dispersion Modelling for Regulatory Purposes.

AUER, M., KUMBERG, T., SARTORIUS, H., WERNSPERGER, B., and SCHLOSSER, C. (2010), *Ten years of development of equipment for measurement of atmospheric radioactive xenon for the verification of the CTBT*, Pure Appl. Geophys. *167*, 4–5.

AUER, M., AXELSSON, A., BLANCHARD, X., BOWYER, T.W., BRACHET, G., BULOWSKI, I., DUBASOV, Y., ELMGREN, K., FONTAINE, J.P., HARMS, W., HAYES, J.C., HEIMBIGNER, T.R., McINTYRE, J.I., PANISKO, M.E., POPOV, Y., RINGBOM, A., SARTORIUS, H., SCHMID, S., SCHULZE, J., SCHLOSSER, C., TAFFARY, T., WEISS, W., and WERNSPERGER, B. (2004), *Intercomparison experiments of systems for the measurement of xenon radionuclides in the atmosphere*, Appl. Radiation and Isotopes *60*, 6, 863–877.

AXELSSON, A. (2007), *Development, Demonstration, Testing and Evaluation of On-Site Inspection Equipment for Xenon Sampling, Separation and Measurement*. CTBTO Technical Report, CTBT/PTS/TR/2007-1, CTBTO, Vienna.

BECKER, A. (2006), *A new tool for NDC analysis of atmospheric transport calculations*, CTBTO Spectrum 7, 19–24.

BECKER, A., WOTAWA, G., DE GEER, L.-E., SEIBERT, P., DRAXLER, R.R., SLOAN, C., D'AMOURS, R., HORT, M., GLAAB, H., HEINRICH, P., GRILLON, Y., SHERSHAKOV, V., KATAYAMA, K., ZHANG, Y., STEWART, P., HIRTL, M., JEAN, M., and CHEN, P. (2007), *Global backtracking of anthropogenic radionuclides by means of a receptor oriented ensemble dispersion modelling system in support of Nuclear-Test-Ban Treaty verification*, Atmos. Environ. *41*, 21, 4520–4534.

BERNSTRÖM, B. and DE GEER, L.-E. (1983), *Mätning av Små Mängder Xenon-133 I Luft*, FOA-rapport ISSN 0347-3694, C 20515-A1, FOA, Stockholm, Sweden.

BIERINGER, J. and SCHLOSSER, C. (2004), *Monitoring ground-level air for trace analysis: Methods and results*, Analyt. Bioanalyt. Chem. *379*, 2, 234–241.

BMU (2007), *Umweltradioaktivität und Strahlenbelastung—Jahresbericht 2006* (in German), Reaktorsicherheit, Bundesministerium für Umwelt, Naturschutz und Reaktorsicherheit, pp. 100–104.

BOWYER, T.W., SCHLOSSER, C., ABEL, K.H., AUER, M., HAYES, J.C., HEIMBIGNER, T.R., McINTYRE, J.I., PANISKO, M.E., REEDER, P.L., SATORIUS, H., SCHULZE, J., and WEISS, W. (2002), *Detection and analysis of xenon isotopes for the comprehensive nuclear-test-ban treaty international monitoring system*, J. Environ. Radioact. *59*, 2, 139–151.

D'AGOSTINO, R.B. and STEPHENS, M.A. (1986), *Goodness-of-Fit Techniques*, (Marcel Dekker) 0824774876.

DAHLMAN, O., MYKKELTVEIT, S., and HAAK H. (2009), *Nuclear Test Ban: Converting Political Visions to Reality* (Springer, 1st edition, 2009), ISBN-13: 978-1402068836.

DE GEER, L.-E. (1996), *Atmospheric radionuclide monitoring: A Swedish perspective*. In Huseby, E.S. and Dainty, A.M., eds., *Monitoring a Comprehensive Nuclear Test Ban Treaty* (pp. 157–177) (The Netherlands: Kluwer Academic Publishers, 1996).

DUBASOV, Y.V., POPOV, Y.S., PRELOVSKII, V.V., DONETS, A.Y., KAZARINOV, N.M., V. V. MISHURINSKII, POPOV, V.Y., RYKOV, Y.M., and SKIRDA, N.V. (2005), *The АРИКС-01 Automatic facility for measuring concentrations of radioactive xenon isotopes in the atmosphere*, Instrum. Experim. Techn. *48*, 3, 373–379.

EHHALT, D., MUENNICH, K.O., ROETHER, W., SCHOELCH, J., and STICH, W. (1963), *Artificially produced radioactive noble gases in the atmosphere*, J. Geophys. Res. *68*, 13, 3817–3821.

FONTAINE, J.P., POINTURIER, F., BLANCHARD, X., and TAFFARY, T. (2004), *Atmospheric xenon radioactive isotope monitoring*, J. Environ. Radioact. *72*, 1–2, 129–135.

IGARASHI, Y., MIYAO, T., AOYAMA, M., HIROSE, K., SARTORIUS, H., and WEISS, W. (2000a), *Radioactive noble gases in surface air monitored at MRI, Tsukuba, before and after the JCO accident*, J. Environ. Radioact. *50*, 1–2, 107–118.

IGARASHI, Y., SARTORIUS, H., MIYAO, T., WEISS, W., FUSHIMI, K., AOYAMA, M., HIROSE, K., and INOUE, H.Y. (2000b), *^{85}Kr and ^{133}Xe monitoring at MRI, Tsukuba and its importance*, J. Environ. Radioact. *48*, 2, 191–202.

KALINOWSKI, M.B. and PISTNER, C. (2006), *Isotopic signature of atmospheric xenon released from light water reactors*, J. Environ. Radioact. *88*, 3, 215–235.

KALINOWSKI, M.B. and TUMA, M.P. (2009), *Global radioxenon emission inventory based on nuclear power reactor reports*, J. Environ. Radioact. *100*, 58–70.

KALINOWSKI, M.B., AXELSSON, A., BEAN, M., BLANCHARD, X., BOWYER, T.W., BRACHET, G., MCINTYRE, J.I., PETERS, J., PISTNER, C., RAITH, M., RINGBOM, A., SAEY, P.R.J., SCHLOSSER, C., STOCKI, T.J., TAFFARY, T., and UNGAR, R.K. (2010), *Discrimination of nuclear explosions against civilian sources based on atmospheric xenon isotopic activity ratios*, Pure Appl. Geophys. *167*, 4–5.

KOCHER, D.C. (1980), *Dose-rate conversion factors for external exposure to photon and electron radiation from radionuclides occurring in routine releases from nuclear fuel cycle facilities*, Health Phys. 38, 543.

LE PETIT, G., ARMAND, P., BRACHET, G., TAFFARY, T., FONTAINE, J., P., ACHIM, P., BLANCHARD, X., PIWOWARCZYK, J. C., and POINTURIER, F. (2008), *Contribution to the development of atmospheric radioxenon monitoring*, J. Radioanalyt. Nuclear Chem. *276*, 2, 391–398.

MIKHAILOV, V., DUBASOV, Y.V., and MATUSHENKO, A.M. (1999), *Nuclear Explosions in the USSR, The North Test Site, Reference Material on nuclear explosions, radiology, radiation safety (ver. 2)*. Tests, Interagency Expert Commission on Assessment of Radiation and Seismic Safety of Underground Nuclear, V.G. Khlopin Radium Institute, Russian Nuclear Society, St. Petersburg, Russian Federation.

PAKHOMOV, S.A. and DUBASOV, Y.V. (2008). *Estimation of explosion energy yield at Chernobyl NPP accident*, Geophys. Res. Abstr. *10*, A-09280.

RINGBOM, A., LARSON, T., AXELSSON, A., ELMGREN, K., and JOHANSSON, C. (2003), *SAUNA–a system for automatic sampling, processing, and analysis of radioactive xenon*, Nuclear Instruments and Methods in Physics Research Section A: Accelerators, Spectrometers, Detectors and Associated Equipment, *508*, 3, 542–553.

SAEY, P.R.J. (2007), *Ultra-low-level measurements of argon, krypton and radioxenon for treaty verification purposes*, ESARDA Bull. *36*, July 2007, 42–55.

SAEY, P.R.J. (2009), *The influence of radiopharmaceutical isotope production on the global radioxenon background*, J. Environ. Radioact. *100*, 5, 396–406.

SAEY, P.R.J. and DE GEER, L.-E. (2005), *Notes on radioxenon measurements for CTBT verification purposes*, Appl. Radiation Isotopes *63*, 5–6, 765–773.

SAEY, P.R.J., WOTAWA, G., DE GEER, L.-E., AXELSSON, A., BEAN, M., d'AMOURS, R., ELMGREN, K., PETERSON, J., RINGBOM, A., STOCKI, T.J., and UNGAR, R.K. (2006), *Radioxenon background at high northern latitudes*, J. Geophys. Res. *111*, D17306.

SCHLOSSER, C., SARTORIUS, H., and SCHMID, S. (2003), *Status of the SPALAX at the RN-33 station*, International Workshop on Atmospheric Radioxenon Measurements, Ottawa, Canada, 29 September–2 October, 2003.

SCHÖLCH, J., STICH, W., and MÜNNICH, K.O. (1966), *Measurements of radioactive xenon in the atmosphere*, Tellus *XVIII*, 2, 298–300.

STOCKBURGER, H., SARTORIUS, H., and SITTKUS, A. (1977), *Messung der Krypton-85- und Xenon-133-Aktivität der atmosphärischen Luft.*, Z. Naturforsch. *32a*, 1249–1253.

UNGA (1996), *United Nations General Assembly Resolution Number 50/245*. In U.N. New York, U.S.A.

UNGAR, K., HOFFMAN, I., YI, J., STOCKI, T., DE GEER, L.-E., WALTERS, M., and BECKER, A. (2007), *Use of existing IDC routine numerical calculations as a prototype standard radionuclide numerical ATM product for operational categorization and categorization development*. In *International Noble Gas Experiment, Workshop*, Las Vegas, NV, U.S.A.

UNSCEAR (2000), *Sources and Effects of Ionizing Radiation*, Report to the General Assembly by the United Nations Scientific Committee on the Effects of Atomic Radiation. 1, Nations, United, UNSCEAR, New York.

VAN DER STRICHT, S., and JANSSENS, A. (2005), *Radioactive effluents from nuclear power stations and nuclear fuel reprocessing plants in the European Union, 1999–2003*. Radiation Protection 143, Directorate-General Environment.

WEISS, W. and LEEB, H. (1993), IMIS—The German integrated radioactivity information and decision support system, Radiat. Prot. Dosimetry *50*, 2–4, 163–170.

WOTAWA, G., DE GEER, L.-E., DENIER, P., KALINOWSKI, M., TOIVONEN, H., D'AMOURS, R., DESIATO, F., ISSARTEL, J.-P., LANGER, M., SEIBERT, P., FRANK, A., SLOAN, C., and YAMAZAWA, H. (2003), *Atmospheric transport modelling in support of CTBT verification—Overview and basic concepts*, Atmos. Environ., *37*, 18, 2529–2537.

ZIEGLER, C.A. and JACOBSON, D. (1995), *Spying Without Spies: Origins of America's Secret Nuclear Surveillance System*, Westport, CT, USA (Praeger Publishers 1995), ISBN-13: 978-0275950491.

(Received November 21, 2008, revised March 24, 2009, accepted April 3, 2009, Published online January 14, 2010)

Reprinted from the journal

Pure Appl. Geophys. 167 (2010), 517–539
© 2010 Birkhäuser Verlag, Basel/Switzerland
DOI 10.1007/s00024-009-0032-1

Discrimination of Nuclear Explosions against Civilian Sources Based on Atmospheric Xenon Isotopic Activity Ratios

Martin B. Kalinowski,[1] Anders Axelsson,[2] Marc Bean,[3] Xavier Blanchard,[4] Theodore W. Bowyer,[5] Guy Brachet,[4] Simon Hebel,[1] Justin I. McIntyre,[4] Jana Peters,[1] Christoph Pistner,[6] Maria Raith,[7] Anders Ringbom,[8] Paul R. J. Saey,[9] Clemens Schlosser,[10] Trevor J. Stocki,[3] Thomas Taffary,[4] and R. Kurt Ungar[3]

Abstract—A global monitoring system for atmospheric xenon radioactivity is being established as part of the International Monitoring System that will verify compliance with the Comprehensive Nuclear-Test-Ban Treaty (CTBT) once the treaty has entered into force. This paper studies isotopic activity ratios to support the interpretation of observed atmospheric concentrations of ^{135}Xe, ^{133m}Xe, ^{133}Xe and ^{131m}Xe. The goal is to distinguish nuclear explosion sources from civilian releases. Simulations of nuclear explosions and reactors, empirical data for both test and reactor releases as well as observations by measurement stations of the International Noble Gas Experiment (INGE) are used to provide a proof of concept for the isotopic ratio based method for source discrimination.

Key words: CTBT, environmental monitoring, international monitoring system, isotope activity ratios, noble gas, radioactivity monitoring, radioxenon, source discrimination, test ban, xenon.

[1] Carl Friedrich von Weizsäcker Center for Science and Peace Research (ZNF), Beim Schlump 83, 20144 Hamburg, Germany. E-mail: Martin.Kalinowski@uni-hamburg.de

[2] International Atomic Energy Agency, P.O. Box 100, Wagramer Strasse 5, 1400 Vienna, Austria.

[3] Radiation Protection Bureau, 775 Brookfield Rd., A.L. 6302D1, Ottawa, ON K1A 1C1, Canada.

[4] CEA, DAM, DIF, 91297 Arpajon, France.

[5] Pacific Northwest National Laboratory, P.O. Box 999, Richland, WA 99352, USA.

[6] Öko-Institut e.V., Rheinstraße 95, 64295 Darmstadt, Germany.

[7] Austrian Research Centers, Seibersdorf Research GmbH, 2444 Seibersdorf, Austria.

[8] Swedish Defence Research Agency (FOI), 172 90 Stockholm, Sweden.

[9] Preparatory Commission for the Comprehensive Nuclear-Test-Ban Treaty Organization (CTBTO), Provisional Technical Secretariat, Vienna International Centre, P.O. Box 1200, 1400 Vienna, Austria.

[10] Federal Office for Radiation Protection (BfS), Rosastraße 9, 79098 Freiburg, Germany.

1. Introduction

Xenon isotopes and their isomers are the most likely observable radioactive signatures of underground nuclear explosions at IMS stations. However, radioactive xenon is released during normal operations of nuclear facilities. So, at stations located downwind to nuclear facilities, these isotopes can frequently be detected and be unrelated to nuclear testing. Therefore, proper source characterization is important for determining whether an event is possibly a nuclear explosion.

Xenon samples will be collected and measured on a daily basis at 40 radionuclide stations of the International Monitoring System (IMS) that is currently being established (see e.g., Bowyer *et al.*, 1998; Hoffmann *et al.*, 1999). Due to their half-lives and fission yields the xenon radionuclides ^{135}Xe, ^{133m}Xe, ^{133}Xe and ^{131m}Xe are relevant for detecting a nuclear explosion (De Geer, 2001). Though two of these are metastable isomers, for convenience, this paper refers to the entities of this quartet as the four relevant xenon isotopes.

The measurements will be analyzed at the International Data Centre (IDC) in Vienna as well as at National Data Centers (NDC). The goal of the analysis is to characterize the samples and screen them according to their relevance to the signature of a nuclear explosion, though no judgment would be made by the IDC. Appropriate criteria are needed to characterize the samples as either "definitely not a nuclear explosion (i.e., screened out)" or "consistent with a nuclear explosion or inconclusive due to insufficient data (i.e., not screened out)". The latter would be flagged for further study. The screening-out procedure has to be done based on scientifically proven and robust criteria without any prejudice, and the final judgement is left to the member states of the CTBT. The goal is to screen out as many events as reasonably possible and to keep the number of events flagged as low as possible in order to minimize costs for additional analysis. This also allows states to concentrate on fewer important events.

Several methods for categorizing radioxenon measurements are under investigation. High concentrations of one xenon isotope (e.g., ^{133}Xe) detected distantly from nuclear reactors at locations with historically low concentrations of xenon isotopes, and no other exceptional release possibilities, could serve as a criterion. In addition, the combination with other data such as relevant anthropogenic radioactivity in air filter samples or a seismic signal (HOFFMANN et al., 1999), could increase confidence that a suspected detection results from a nuclear detonation. Information on releases from reactors or isotope production facilities as part of confidence building measures will help to clarify if the source is of a civilian nature. Determination of the specific sensitivity to known sources based on atmospheric transport simulation appears to be particularly helpful for categorization (WOTAWA et al., 2010). Another approach is the Bayesian analysis of radio xenon samples (ZÄHRINGER and KIRCHNER, 2008).

In the past it was suggested that single xenon activity ratios would be useful both for source discrimination as well as for event timing (see e.g., BOWYER et al., 1998; FINKELSTEIN, 2001; CARMAN et al., 2002). A positive conclusion on whether a measured activity ratio is consistent with a nuclear explosion has been based on assumptions that did not take into account the full range of possible isotopic activity ratios.

The main assumptions, in previous work, have been that the extended irradiation time in a reactor leads to equilibrium operational characteristics of steady power production and no change to routine procedures for off-gas treatment and release, whereas a nuclear explosion is an instantaneous process with a very short duration of a fast neutron flux resulting in different isotopic ratios. However, initial data have

shown that the theoretically derived isotopic activity ratios for reactors out of equilibrium are not uncommon (AUER et al., 2004; BOWYER et al., 2002; STOCKI et al., 2005) and therefore new tools for discrimination of nuclear explosions and reactor operations are being developed.

A robust screening method is required for IDC sample categorization. To support this, new methods for source discrimination based on the relationship of two different isotopic activity ratios are investigated in this paper. It is applicable if several of the xenon isotopes are measured in a sample and if the measured ratios originate from a single release source. It requires three or four xenon isotopes to be detected and quantified. If they are not detected they can be, in certain cases, replaced by an upper limit that is related to the detection limit. The most important feature of the screening approach suggested here is the robustness of the method, independent of decay of the species. The advantage of this time-invariance is that the age of the xenon plume that intercepts the sampling site does not need to be known.

In this paper, the combinations of isotopic ratios of various calculated and empirical data sets as well as atmospheric observations are used to validate the new screening method. These data sets are:

1. Reported gaseous effluents from light water reactors (LWRs)
2. Simulations for LWR operational cycles
3. Simulations of mixing between fresh and aged air masses
4. Other civilian sources (accidental releases from low burn-up fuel, medical isotope production)
5. Simulations of various nuclear explosion scenarios
6. Empirical release data of underground nuclear tests
7. Short-time irradiated HEU targets
8. Observations of atmospheric xenon concentrations

In the first section of this paper, all these data sets are introduced. The second section uses the data of items 1–7 to demonstrate the useability of isotopic activity ratios for source discrimination. The screening method based on the relationship of different xenon isotopic activity ratios is validated against real atmospheric measurements (data set number 8) in the third section.

2. Radioxenon Releases, Observations and Simulations

2.1. Reported Light Water Reactor Release Data

A large part of the data set on radioxenon releases from nuclear power plants is taken from the reports on activity releases to the environment of radioactive materials in airborne and liquid effluents during the period 1995–1999 (VAN DER STRICHT and JANSSENS, 2001) and 2000–2003 (EC, 2004). It covers discharges from operational nuclear power stations of capacity greater than 50 MW (electric power) in the European Union (EU). During the time covered by the first report, 73 nuclear power stations totaling 148 reactors were operational and are spread over 64 different sites within the territory of the European Union. The annual xenon release data are available for 66 nuclear reactors at 55 sites in Finland, France, Germany, Spain and Sweden. In some cases, the discharge values for noble gases are reported to be below the detection limit. In addition, release data from the four Swiss nuclear power plants are included (BAG, 2000, 2001 and 2002). Another large portion of reactor release data is taken from 41 quarterly reports of nuclear power plants in the USA. The North American Technical Center (NATC) Public Radiation Safety Research Program at the University of Illinois at Urbana-Champaign compiled these data for the US effluent database (NATC, 1999). This database is the official US database provided to the United Nations (UNSCEAR) for their global report on radiation dose to man. The special feature of this database is to report continuous and batch releases separately. Distinct differences are noted between the isotopic activity ratios appearing in these two different release modes.

It should be noted that the list of individual nuclides routinely measured might vary from one installation to another. Also, the way the corresponding activity is measured or estimated, as well as measurement accuracy and precision, can be different at each installation location or in terms of the requirements of each State. The data are cumulative annual figures and, therefore, not overly sensitive for possible outliers in single measurements. Nevertheless, single discharge events can still have isotopic

ratios outside the ranges presented here, for example, when new nuclear technologies are implemented.

Figure 1 shows the distributions of six different xenon isotopic activity ratios with the isotopes ^{135}Xe, ^{133m}Xe, ^{133}Xe and ^{131m}Xe derived from all reported nuclear reactor xenon releases as described above. In total there are 340 data sets, each of which contains the activities of up to four xenon isotopes. The activity ratios are sorted by increasing value. The first sorting numbers are reserved for those data sets in which both numerator and denominator are not reported and hence no ratio is available. By this way, the plot can show the number of samples in which none of the two required activities is reported. For example, the activities of both isomers ^{131m}Xe and ^{133m}Xe are missing in 140 data sets and accordingly, in Fig. 1, the entries for the ^{133m}Xe/^{131m}Xe activity ratio begin at sorting number 141. The highest sorting numbers are used for cases in which only one of the required activities is given. Accordingly, the entries end at lower sorting numbers if more data sets with one out of the two required activities are available. In particular, it can be seen from Fig. 1 that ^{133}Xe and ^{135}Xe are available in most reports and only for a handful of data sets is the activity of exactly one of these two isotopes reported.

The reported data shown in Fig. 1 exhibit a broad range of ratios spreading over several orders of magnitude. The logarithmic mean ratio is given for each isotope. Part of the spread is likely due to radioactive decay after different residence times within the reactor containment. Other effluent data clearly indicate that the LWRs were in many cases not in equilibrium that is reached after power and neutron flux are maintained constant for several weeks.

On the other hand, previous atmospheric measurements at some distance from nuclear reactors as published by AUER et al. (2004), BOWYER et al. (2002), LE PETIT et al. (2008), STOCKI et al. (2005) and SAEY et al. (2010) suggest that remote measurement of xenon is consistent with equilibrium operation in many cases. This apparent contradiction could be explained by the fact that when measurements are made closer to a reactor, it is much more likely that all four isotopes will be at high enough concentrations so as to have a higher sensitivity to

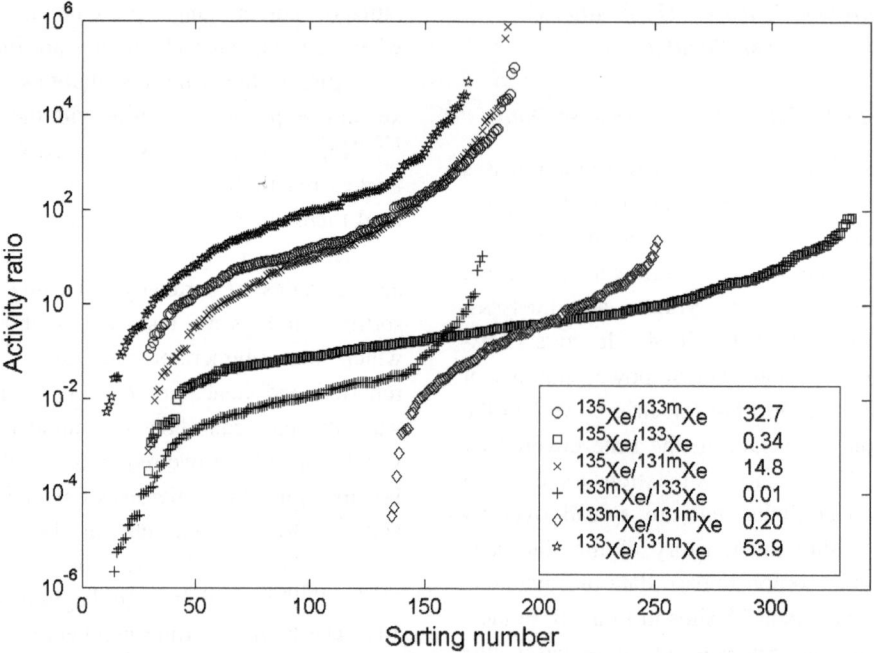

Figure 1
Distribution of released xenon isotopic activity ratios as reported from nuclear power sites. The logarithmic mean ratios are given in the text box

determining the equilibrium state of the reactor. In fact, in most cases where the measurements were made remote from reactors, only ^{133}Xe was detected, and when other isotopes were measured, usually only 2 or 3 isotopes were at high enough concentrations to allow a ratio to be formed.

A more detailed consideration on power reactor emissions can be found in KALINOWSKI and TUMA (2009).

2.2. Simulations for Light Water Reactors

The reasons for, and the potential extent of the variability of reactor releases under different operational conditions need to be understood to make progress on the screening of data collected for nuclear explosion monitoring. Figure 2 shows results of LWRs fuel burn-up simulations covering the whole operational time with the typical three-one-year power cycles with 3.2% enrichment in ^{235}U (KALINOWSKI and PISTNER, 2006).

The isotopic activity ratios depend significantly on the neutron flux and can vary over several orders of magnitude within a few days of release. Any power change and especially shut-down and start-up of a reactor (BOWYER et al., 2002) causes a deviation from the initial values. Effluent treatment can have an impact, particularly if this leads to an extended residence time in the reactor because the isotopic ratios change by decay. Uranium enrichment, fuel burn-up and diffusion through the fuel matrix or effluent treatment filters have a comparatively low impact on the xenon ratios. Diffusion of fission gases from the fuel matrix into the free volume causes fractionation of the isotopes (see Table 4, KALINOWSKI and PISTNER, 2006).

2.3. Simulations of the Effect of Mixing between Fresh and Aged Air Masses

The process of mixing air volumes that emanate from nuclear reactors can lead to isotopic activity ratios different from any of those in the contributing air parcels. This is true even for a mixture of two parcels of air that shared the same starting ratios, but are of different ages. Mixing of this sort can occur within the contained spaces of the facility or in the atmosphere.

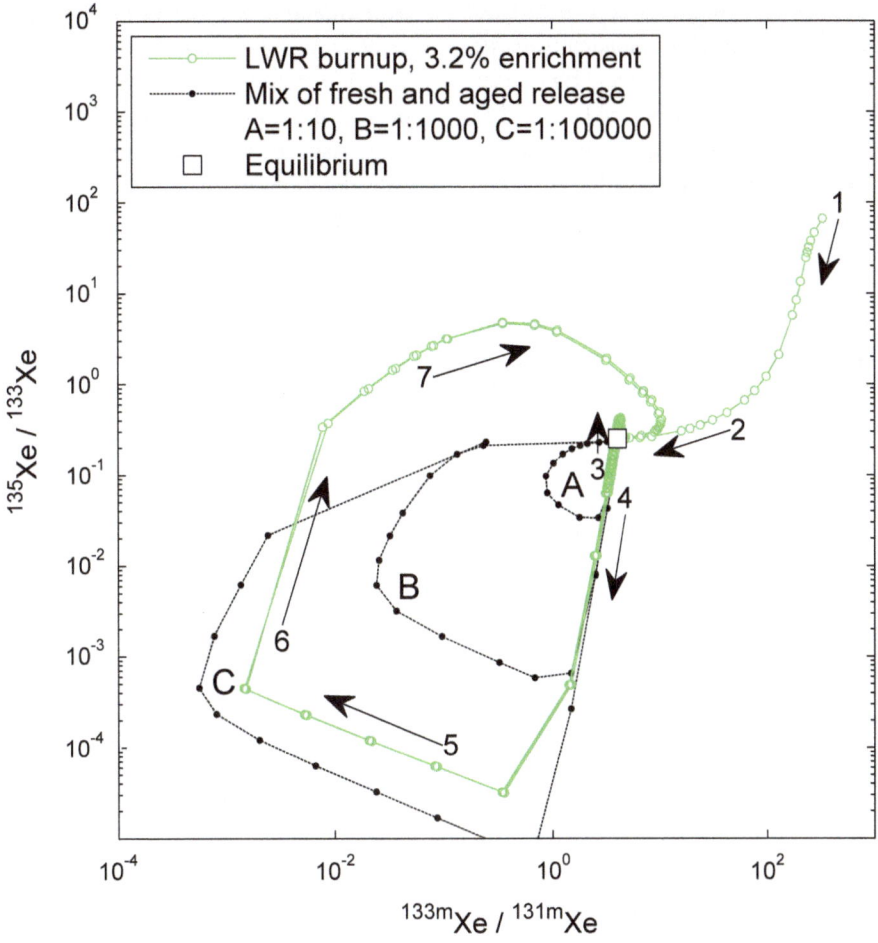

Figure 2
Reactor ratio trajectory of the first one-year power cycle of fuel plus a 30-day revision period with 3.2% enrichment in [235]U (KALINOWSKI and PISTNER, 2006). The equilibrium, reached after a few weeks, is marked with a large square. Description of the isotopic ratio track during the first reactor power cycle: 1. Start-up with fresh fuel only 2. Approaching equilibrium 3. Short-time increase of [135]Xe after power shut-down 4. Radioactive decay 5. Low neutron flux during revision period 6. Restart of reactor to full power 7. Circling back to equilibrium. The second and third power cycle repeat the track from part 3–7 very close to the first cycle. The most significant difference can be seen along part 7 of the track

In order to study the impact of mixing on the isotopic activity ratios, a number of mixing scenarios is set up. A fresh and an aged air mass are mixed in three volume ratios. Both start from the equilibrium activities in fuel with 3.2% enrichment as reached at the end of the third cycle (KALINOWSKI and PISTNER, 2006). The fresh part is kept at time zero, while the other portion is assumed to have decayed for a certain number of days. Figure 2 shows the trajectories for the three mixing scenarios. All three are similar to the no-mixing trajectory.

2.4. Other Civilian Sources

Besides discharges from nuclear reactors during normal operation and the regular large releases from medical isotope production facilities, other less frequent sources have to be considered. These include spent fuel reprocessing, medical isotope usage and disposal, and exceptional accidental releases from nuclear reactors. The latter case is of particular interest with the scenario of a short irradiation reactor, i.e., less than 7 days (FINKELSTEIN, 2001).

Normally, nuclear fuel is reprocessed with at least a year cooling time after the fuel is unloaded from the reactor core. At that time, only ^{131m}Xe and tiny traces of ^{133}Xe remain at measurable activity levels. The ^{133}Xe/^{131m}Xe isotopic ratio might still be in the order of 10^{-5} (Bowyer et al., 1998) but could be significantly less due to the radioactive decay. This is already more than two orders of magnitude below the minimum found in the reactor releases (see Fig. 1). Unless the detection is very close to the reprocessing activity, or reprocessing is undertaken with an unusually short decay time, the activities will be extremely low, consequently this activity will likely have no impact on source discrimination. The equilibrium activity of ^{133}Xe reached with a burn-up of 30 GWd/MTU (Giga Watt days per metric ton of uranium) at the time of unloading the fuel is 6×10^{18} Bq/MTU, but it will be decayed to 6×10^{-3} Bq/MTU after 1 year. After that time, even the full release of all fission noble gases will most certainly not allow detection of ^{133}Xe further away than meters from an atmospheric source.

The main source of radioactive xenon isotopes for medical applications is separation of fission gases from irradiated highly enriched uranium (HEU) with a low exposure of only normally 2–11 days. In fact, radioactive xenon often is harvested as a by-product of breeding ^{99}Mo in HEU targets. Table 1 shows isotopic activity ratios for various irradiation times and days of decay. For data based on other assumptions see Saey (2009). Some of these ratios happen to be close to the respective logarithmic mean ratio for power plant releases as shown in Fig. 1. However, other low-exposure ratios are indeed similar to explosion ratios, as found in calculations and

experiments described in Sects. 1.5 and 1.7, below, respectively (Kalinowski and Pistner, 2006).

As a result, isotope production from HEU can pose a difficulty to source discrimination. It could give rise to a false alarm and requires special consideration. A worst case estimate has been created by assuming total release from a large 1 kg target exposed for 5 days with a total neutron flux of 3.5×10^{14} s^{-1}cm^{-2} following 1 day of decay. The ^{133}Xe content at the time of the release is 10^{16} Bq. Under the assumption of a 1-h release duration and a stack air volume rate of 10,000 m^3 h^{-1}, the air concentration at the point of release is 10^{12} Bq m^{-3}. This is certainly detectable at some distance (Stocki et al., 2005). The number of isotope production facilities in the world is limited, and even if the number increases their releases can be taken into consideration though it complicates the test ban verification and makes it less confident. Further information on radiopharmaceutical production facilities can be found in Saey (2009).

Radioactive xenon isotopes for medical applications are in general not produced by exposing natural xenon to a high thermal neutron flux. Though ^{133}Xe might be produced by neutron absorption in ^{132}Xe, the content of other radioactive xenon isotopes, especially ^{125}Xe would result in an excessively high dose rate. However, the production of ^{125}I, another isotope important for medical applications, is based on the irradiation of natural xenon gas. After the nuclear reaction ^{124}Xe(n,γ)^{125}Xe, the product decays to ^{125}I. Therefore, one needs to take into account the possibility of detecting radioactive xenon resulting from this source. As an example, the isotopic composition of 5 ml natural xenon after a 6 h irradiation in a neutron flux of 7×10^{11} s^{-1} cm^{-2} is considered (Keller, 2004). The related isotopic activity ratios are found in the nuclear explosion domain only for one of the xenon ratio relationship plots. This is the one that contains all isotopes with the exception of ^{131m}Xe.

A possible source of pure ^{131m}Xe is the medical application of its precursor, ^{131m}I, which has a half-life of 8.02 days. The ^{131m}Xe frequently observed in Ottawa, Canada originates from ^{131}I production at Chalk River Nuclear Laboratory (Stocki et al., 2005). The precursor of this iodine isotope can be generated

Table 1

Isotopic activity ratios for 93% enrichment in ^{235}U and an irradiation time of 10 and 5 days of decay and for an irradiation time of 5 and 2 days of decay

Isotopic activity ratios	Irradiation time = 10 days, decay time = 5 days	Irradiation time = 5 days decay time = 2 days
^{133m}Xe/^{133}Xe	0.02	0.04
^{133m}Xe/^{131m}Xe	6	36
^{133}Xe/^{131m}Xe	330	890

with a high purity by neutron capture: ^{130}Te(n,γ) ^{131}Te.

Another scenario that could result in isotopic ratios in the explosion domain would be an emission from a short-irradiation reactor accident. In this case, the source term could be significantly larger than in the case of medical isotope production. After an irradiation time of 15 days, the isotopic ratios are still found in the area between the reactor and the test domain. At that time, the ^{133}Xe content in one fuel element with a weight of 1 MTU results in the activity release of 10^{18} Bq. If this would be fully released within 1 h through an air volume rate of 100,000 m^3 h^{-1}, the air concentration at the stack is 10^{13} Bq m^{-3}. The simultaneous release of radioactive aerosols would certainly help to relate this kind of event to a reactor accident.

2.5. Simulations of Various Nuclear Explosion Scenarios

A nuclear explosion takes place in a very short time and the nuclear chain reaction ends in less than a second. Two extreme scenarios are possible, one in which the Xe is removed from their precursors immediately and allowed to decay, and another in which all fission products are held together, perhaps in an underground cavity, such that in-growth from precursors is allowed and the full cumulative yield of Xe is produced.

Three different neutron sources are studied with regard to the fissile material used and the neutron energy. Fission of ^{235}U and ^{239}Pu is simulated at fission neutron energies and ^{238}U with high-energy neutrons using the Bateman equations (BATEMAN, 1910) implemented in Matlab (KALINOWSKI, 2010a, b).

2.6. Empirical Release Data of Underground Nuclear Tests

SCHOENGOLD et al. (1996) report detailed atmospheric radioactivity release information for 433 nuclear tests conducted on the Nevada Test Site (NTS) from 15 September 1961 through 23 September 1992. An analysis of these data can be found in KALINOWSKI (2010a, b). Figure 3 shows the distribution of xenon isotopic activity ratios that are calculated from the reported activity releases of these

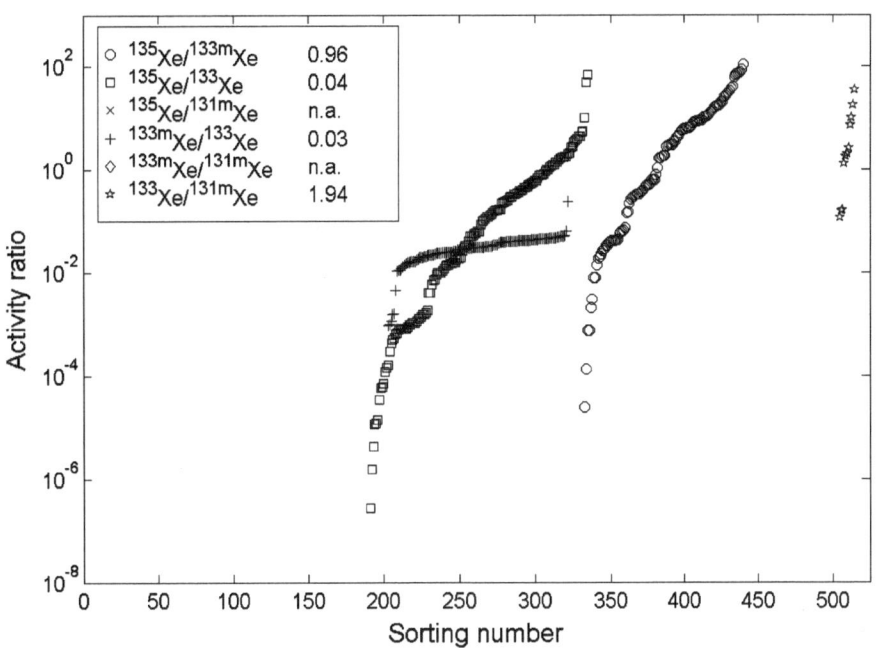

Figure 3
Distribution of released xenon isotopic activity ratios as reported for underground nuclear explosions at Nevada. The logarithmic mean ratios are given in the text box, n.a. not applicable

underground nuclear explosions. The data were reported at two significant figures accuracy. The spread of the activity ratios over several orders of magnitude can be explained by the delay and duration of the releases.

Only 102 of the 433 releases are considered. The selection criterion used was to pick only those events for which single release events are reported, and therefore excludes aggregated activities that are summed up from different release events at different times originating from the same test explosion.

All these 102 cases are operational releases resulting from the purging of tunnels or sometimes shafts to minimize the exposure to personnel, from drill-back operations to recover samples for diagnostic purposes, from gas sampling or from sealing the drill hole with a plug and cementing it to the surface. These releases had delays ranging between 31.5 h and 24.5 days with durations of at least 2 min and at most 5 days.

In most of the cases in which the activities of three xenon isotopes were measured and reported, the activity of ^{131m}Xe was not included. That isotope is reported only for a very few late operational and seepage release cases. The likely reason for this isomer not being measured is that its independent fission yield is 10^{-6}% or less depending on the fission reaction. In addition, the half-life of the immediate precursor is 8 days, too long for a significant build-up through the decay chain shortly after the explosion. The only exception is a late-time seepage release that occurred after operations in the relevant test area ceased with a delay of 9.5 days and a duration of 30 days. All isotopes but ^{135}Xe were measured in that case.

The two plots on top of Fig. 4 show the calculated xenon isotopic ratios that are possible for nuclear detonations using different scales; the right one being zoomed in. The change in the isotopic ratios is shown for complete fractionation involving only the independent yields (pure xenon curve) and without any fractionation converging within 1 day close to the cumulative yields (unfractionated decay chain curve). Depending on the amount of fractionation of the xenon isotopes from their parents, the actual ratios will fall between the two or precisely on one of the lines. Also shown in the figure are data points from the Nevada Test Site (NTS) (SCHOENGOLD et al.,

1996). It is notable that in all of these examples from the NTS, the xenon parents were allowed to come into equilibrium with the xenon isotopes before the measurement, as can be concluded from the fact that all points lie quite close to the unfractionated decay chain curve. This seems consistent with the fact that all samples were collected at least 31.5 h after the explosion.

Figure 4, bottom left, shows the single reported seepage release in the isotopic activity ratio relationship plot for the three isotopes excluding ^{135}Xe. No plots are shown with other ratio pairs because there are no data available with the required isotope combinations.

The agreement between the NTS measurements and the model is good, and therefore engenders confidence that the model at least bounds (see unfractionated decay chain curve in Fig. 4) the correct answer for the "worst case." Further, the comparison of data from the Nevada Test Site with the calculations illustrates that the xenon isotopes and their precursors remained basically non-fractionated before the releases took place. In other words, irrespective of the release scenario for an operational release, there is high confidence that the airborne xenon activity ratio will exhibit negligible fractionation and lie close to the unfractionated decay chain curve. This may be different for rapid uncontrolled venting but cannot be demonstrated here due to the lack of data.

2.7. Short-time Irradiated HEU Targets

RAITH (2006) reports samples containing 1 μg of 90% enriched ^{235}U were irradiated at the Training, Research, Isotopes, General Atomic (TRIGA) Mark II research reactor at the Atomic Institute of the Austrian Universities in Vienna. Figure 5 shows short-time irradiated 90% enriched ^{235}U measurements. The measurements lie near the unfractionated decay chain curves and are consistent with the results of Sect. 2.5 (compare with Fig. 4).

2.8. Observations of Atmospheric Xenon

The first reported regular global xenon observations of all four relevant xenon isotopes began with the International Noble Gas Experiment (INGE), initiated

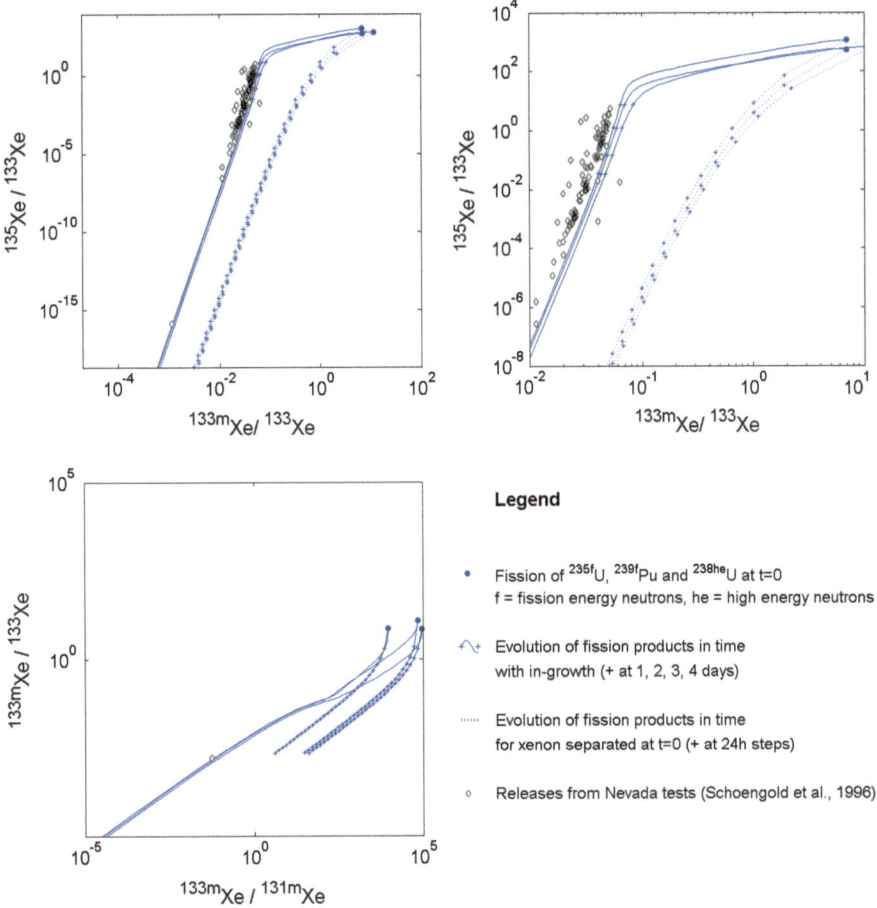

Figure 4

Comparison of simulated and reported xenon isotopic activity ratios released from nuclear tests (102 selected). The expected isotopic ratios present at $t = 0$ are denoted with the *filled circles*, and *lines* and "+" symbols are used to show the change in the isotopic ratios in time steps of full days

and headed by the CTBTO PrepCom Preparatory Commission (BOWYER *et al.*, 2002; AUER *et al.*, 2004, 2010; and SAEY and DE GEER, 2005). The study reported here uses the concentrations measured in more than 1,800 xenon samples taken at five different locations. Data analyzed with two different measurement systems are used here. These are the Système de Prélèvement d'Air Automatique en Ligne avec l'Analyse des radioXénons atmosphériques (SPALAX) from the French Atomic Energy Commission (CEA), France (FONTAINE *et al.*, 2004), and the Swedish Automatic Unit for Noble Gas Acquisition (SAUNA) developed by the Swedish Defence Research Agency (FOI), Sweden (RINGBOM *et al.*, 2003).

SPALAX measurements have collection times of 24 h and were taken at Ottawa and at Yellowknife, both in Canada (STOCKI *et al.*, 2004, 2005). Other sets of SPALAX data have been taken on Schauinsland, a mountain in nearby Freiburg (AUER et al.; 2010), Germany, as well as at Bruyères-le-Châtel, France. Samples taken in Stockholm have collection times of 12 h and were measured with the SAUNA system.

Apart from the samples taken in Ottawa, most of the daily samples have no radioxenon or solely ^{133}Xe, only a few of these samples have high ^{133}Xe concentration and also include ^{135}Xe and/or ^{133m}Xe. ^{131m}Xe is detected with the lowest frequency.

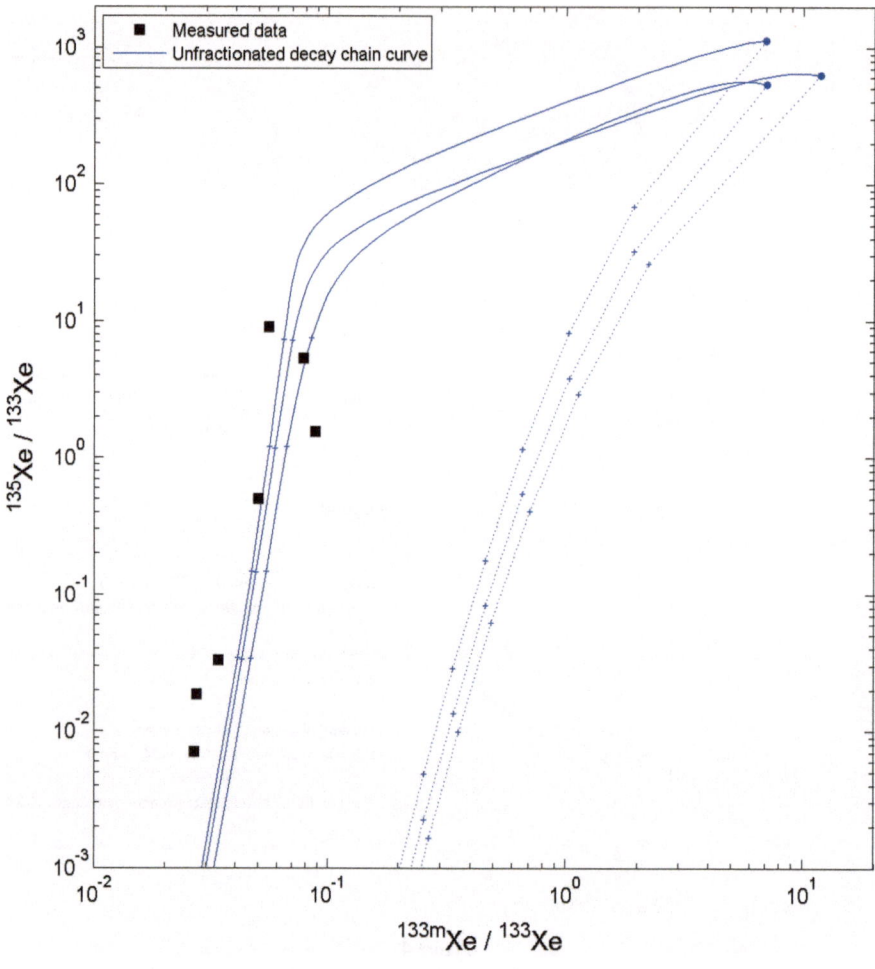

Figure 5
Measured data for very short-term irradiation of HEU targets (RAITH, 2006)

Figure 6 shows the distribution of all measured xenon isotopic activity ratios. They are spread over several orders of magnitude.

The Indian and Pakistani nuclear tests in May 1998 could not be detected by INGE stations because they were not operative. Regarding the claimed nuclear test in North Korea in October 2006, measurements were taken in South Korea by RINGBOM et al. (2007), RINGBOM et al. (2009) and backtracked to the test location by BECKER et al. (2010). Earlier SAEY et al. (2007) presented a study relating a radioxenon detection at the IMS station in Yellow-knife, Canada to the same event.

3. Source Discrimination Based on Single Xenon Isotopic Activity Ratio Relationship

3.1. Isotopic Activity Ratios as a Function of Time

The iodine activity released by nuclear reactors is by at least three orders of magnitude smaller than the discharged xenon activity (see e.g., VAN DER STRICHT and JANSSENS, 2001). Therefore, the xenon isotope concentrations are governed by decay with negligible in-growth.

Figure 7 depicts the change over time of xenon isotopic activity ratios simulated for LWRs. The reactor equilibrium and the simulation maximum are

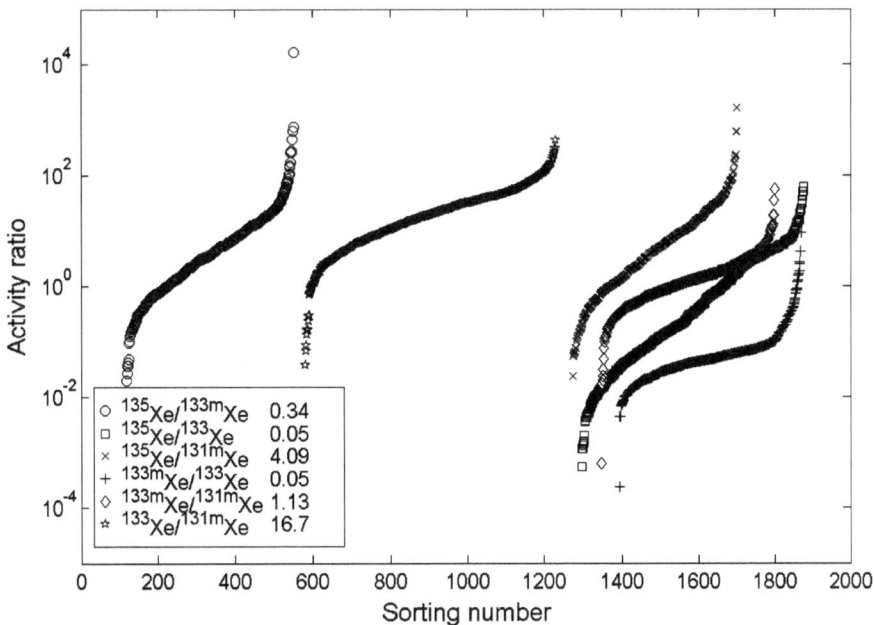

Figure 6
Distribution of atmospheric xenon isotopic activity ratios as measured at different sites of the International Noble Gas Experiment (INGE) in Europe and North America. The logarithmic mean ratios are given in the text box

displayed to mark the band of ratios that can be expected as a result of normal operational releases. This has to be compared to the isotopic ratios resulting from a nuclear explosion.

For fully fractionated releases from underground explosions the evolution in time follows an exponential decrease and will show as a straight line unless ^{133}Xe is involved. For that one of the precursor ^{133m}Xe is present. In this case a bent curve can be seen in the semi-logarithmic graph (see Fig. 7).

3.2. Window of Opportunity for Source Discrimination

For comparison, the change over time of xenon isotopic activity ratios is shown in Fig. 7 for both reactor simulations and releases of underground nuclear explosions. In addition, measured releases at the Nevada Test Site are marked with the reported start time of the release. In general, the largest amount of activity is released shortly after the emission begins. The emission rate typically declines steeply and remains at a low level before it is reported to end.

All isotopic ratios used here have the shorter-lived isotope in the numerator so as to have them decreasing with progressing time with their maximum at time zero. Discrimination of nuclear test releases against reactor discharges in the sense of a test being the sole explanation requires that the isotopic ratio remains higher than the maximum of the reactor discharge band at time zero, because in general the age of the sampled plume is not known. This requirement can be relaxed if the suspected detonation time is known from seismic analysis.

Therefore, a margin for source discrimination is defined by the gap between the lowest possible activity ratio for nuclear explosions and the maximum of the reactor emission simulation. If it exists at all, it not only has a certain width (height of the gap) but also a limited duration, since it takes a certain time until this gap vanishes due to the radioactive decay.

Besides isotopic activity ratios, there are other screening methods that may be based on absolute concentrations and may involve an outlier analysis to classify a measurement as anomalous in comparison to typical atmospheric background at that detector

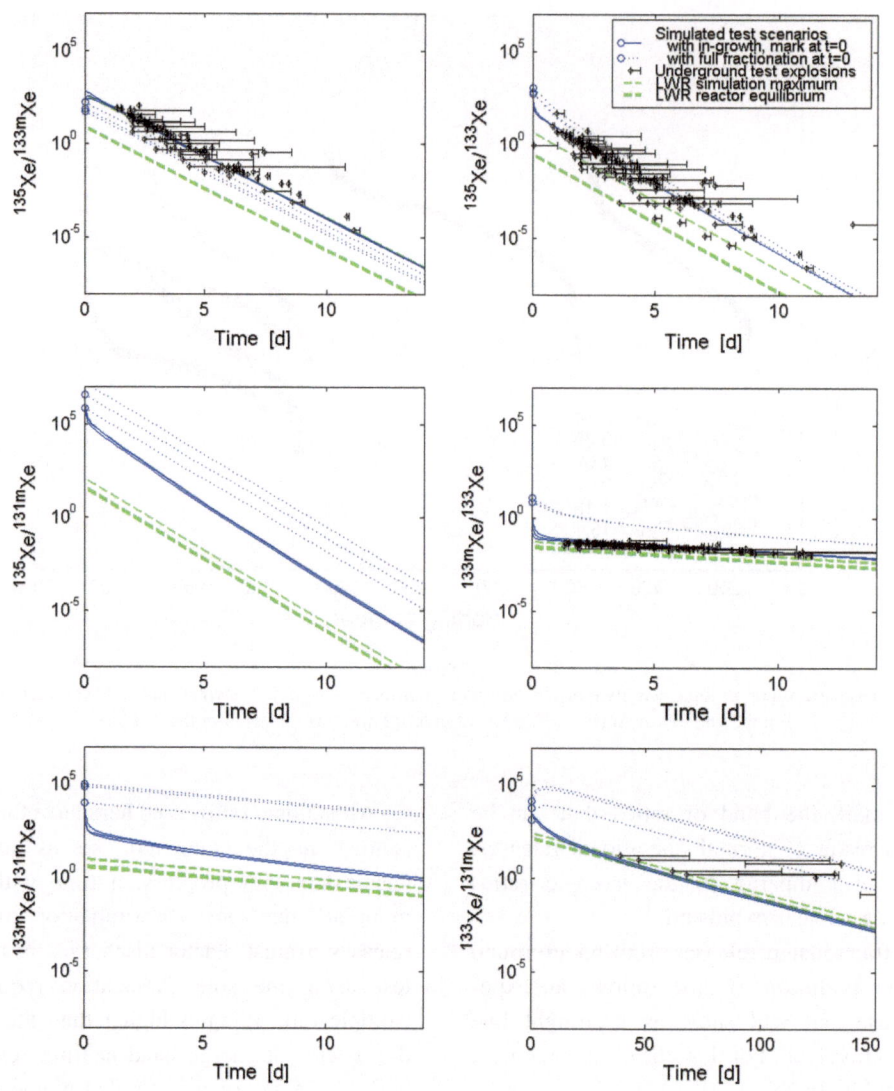

Figure 7
Change over time of isotopic activity ratios of reactor emission simulations compared with nuclear explosion scenarios and Nevada test release data. Note different time scale in the last subplot. The *horizontal bars* indicate the reported duration of the release

site. This is of particular relevance on the southern and in remote areas of the Northern Hemisphere. However, source discrimination may be less complicated and more robust if three or four isotopes are measured and used for source discrimination as described in the following section.

3.3. Single Isotopic Ratios

At the time of a nuclear explosion all combinations of isotopes as shown in Fig. 7 have ratios that are above typical reactor emissions as indicated by the equilibrium of light water reactors (LWRs). Information relative to the time evolution of the four xenon isotope activities for a 1 kt nuclear test can be found in KALINOWSKI (2010a). For all combinations of isotopes with ^{135}Xe in the numerator it takes less than 5 days and for the activity ratio ^{133m}Xe/^{133}Xe 6 days, before the non-fractionated release from nuclear explosions reaches the reactor equilibrium. For the remaining two plots shown in Fig. 7 it takes significantly longer. The margin of discrimination would be even better, if fractionation would occur, because in general this results in activity ratios that decrease more slowly. However, reactor emissions may have xenon isotopic activity ratios that are found above the equilibrium. If the reactor simulation maximum is considered, the discrimination method based on single isotopic ratios is less powerful.

Since the ratios decrease with time, any observed isotopic activity ratio could indicate a nuclear explosion. Once a ratio that is specific for a nuclear explosion has declined below the maximum possible for reactor emissions, it loses its unambiguous character. A nuclear reactor emission is never unambiguous. It can always be suspected to be an aged nuclear explosion signature. Therefore, the ratios do not strictly allow for screening out irrelevant cases in general.

4. Source Discrimination Based on Xenon Isotopic Activity Ratio Relationship

4.1. Multiple Isotopic Ratios

To overcome these shortcomings of combinations of two xenon isotopes regarding source discrimination, more information is necessary. The xenon activity concentrations themselves are not widely applicable parameters for this purpose, since they change by dilution during the transport through the atmosphere. However, all isotopic ratios are independent of this effect. They change only by radioactive decay, unless precursors are carried along in the same plume at relevant concentration. Therefore, a third or fourth measured xenon isotope could carry in its ratio to another measured isotope the additional information required for source discrimination. The analysis method proposed here is to use a

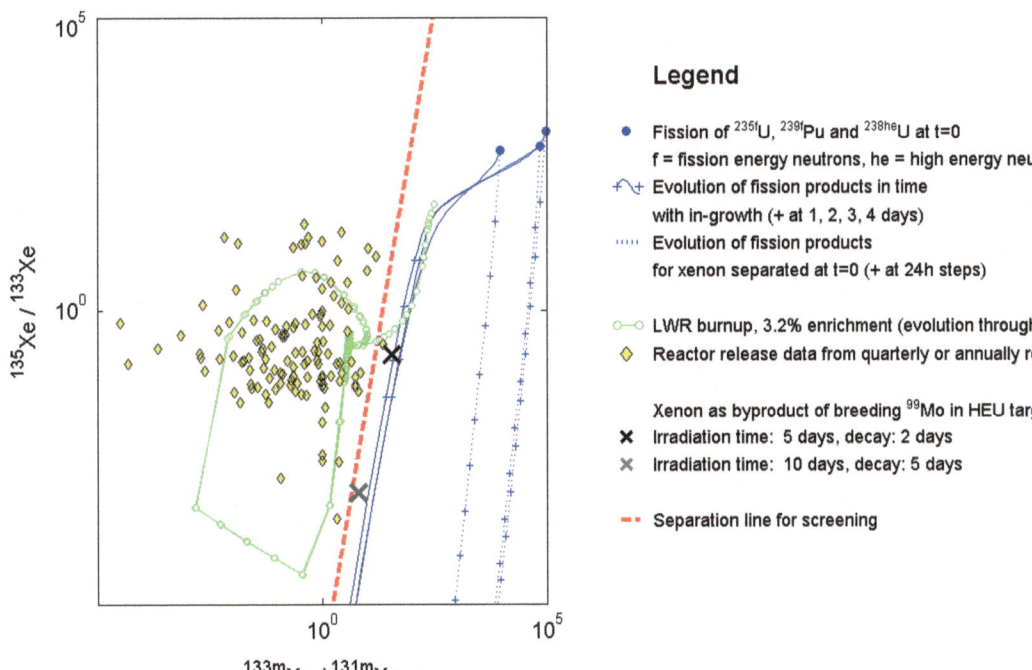

Figure 8

Time-invariant source discrimination based on xenon isotopic activity ratio relationship with reactor emission data for the case in which all four isotopes are measured. The *dashed line* marks the time-invariant screening separation. All isotope ratio relations found above (i.e., left to) this line can be screened out, i.e. the related samples are definitely irrelevant for CTBT verification purposes, because they cannot be explained by a nuclear explosion. All samples that have ratio dependencies found below (i.e., right to) the line might be relevant for CTBT monitoring purposes. The exact location of the separation line is subject to further studies

157

set of plots that presents the relationship of one isotopic activity ratio to another one. The activity ratios change due to decay and with time they move along a defined line that is straight on a log–log plot in case no precursors are present in the plume. If ^{133}Xe is involved the presence of its precursor ^{133m}Xe causes the line to be slightly bent. It should be noted that this paper assumes that pure air masses from

single sources, and in particular no mixing between releases from reactors and with nuclear weapons test emissions.

In Fig. 8, the proposed method is shown for the case that all four xenon isotopes are measured and made use of. Two isotopes are used on the abscissa; the other two are taken for the activity ratio on the ordinate. The legend shown in Fig. 8 also applies to

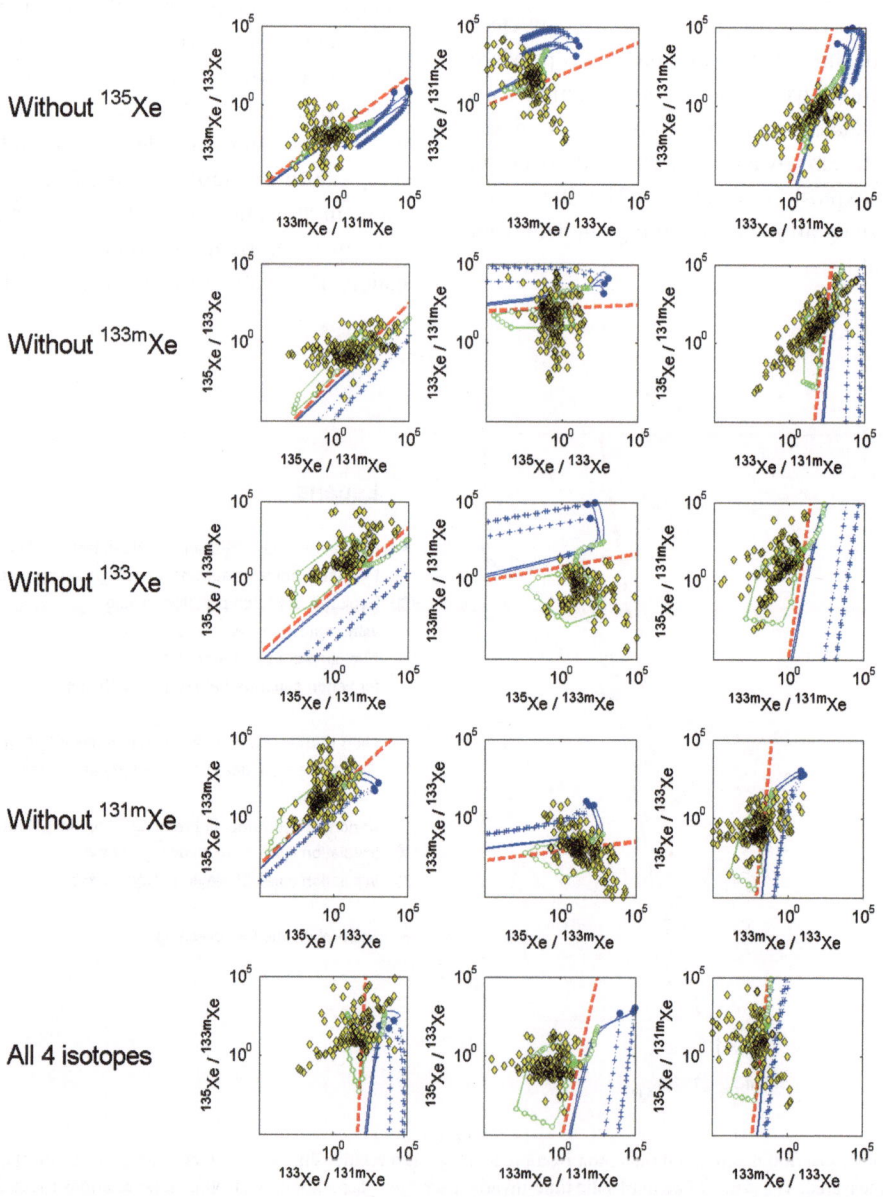

Figure 9

All 15 possible combinations of two xenon isotopic activity ratios without exchange of axes and restricted to those that have the shorter-lived isotope in the numerator. Reported annual reactor emission data are shown together with simulated curves

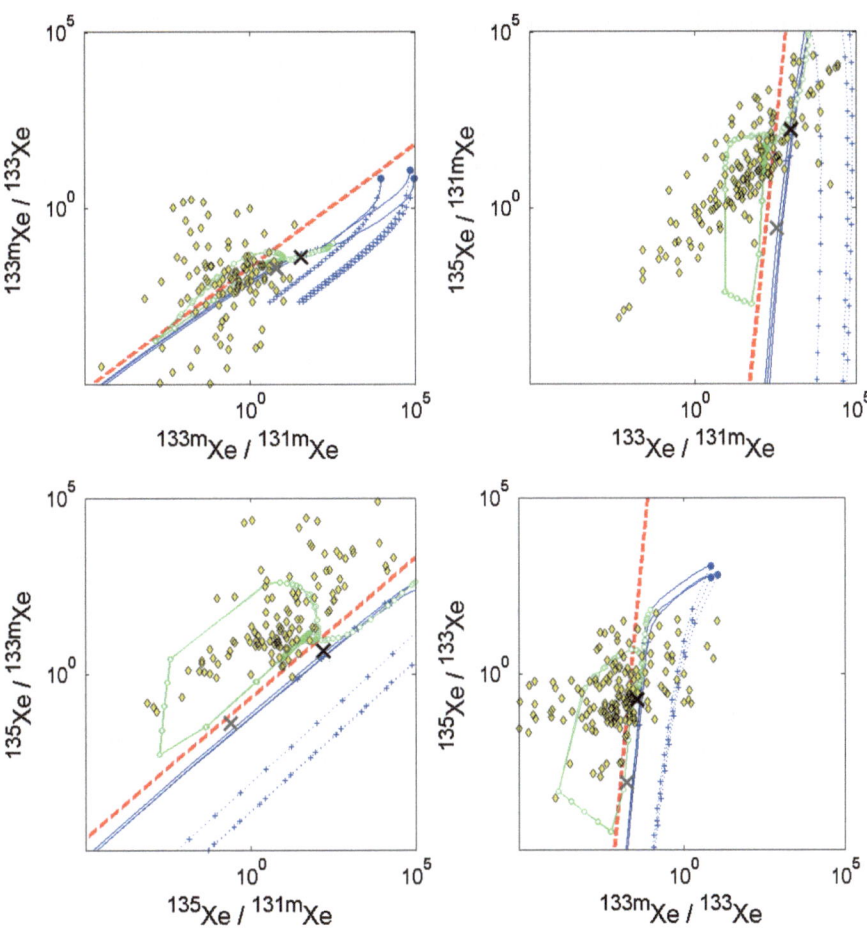

Figure 10
Time-invariant source discrimination based on xenon isotopic activity ratio relationship with reactor emission data for the case in which three out of four isotopes are measured

the following figures. The simulation curves entered here separate the plot area into two distinctive domains. The trajectories for three operational cycles of a nuclear reactor follow a circular pattern in the left half of the plot (KALINOWSKI and PISTNER, 2006). The simulation curves for nuclear explosion scenarios remain in the right half of the plane.

Figure 9 shows all 15 possible xenon isotopic activity ratio combinations without permutation of axes and is restricted to those that have the shorter-lived isotope in the numerator. Each row of sub-plots applies to a different set of xenon isotopes. Most of the 15 plots exhibit overlapping regions for nuclear reactor and explosion scenarios.

Since each row of three plots in Fig. 9 is a different projection of the same three-dimensional

space, the plots in one row are not independent from each other. Therefore, it is sufficient to consider only one representative plot for each set of three or four isotopes. The most discriminatory plot with all four isotopes is the one shown in Fig. 8. The selection of best plots for each of the four combinations of three different isotopes is shown in Fig. 10. All further discussions refer to these one plus four plots.

The method for source discrimination proposed here is to define a screening condition as a function of three or four xenon isotopes and to keep it independent from radioactive decay. Two different isotopic ratios are shown on the abscissa and ordinate. Since the axes are on the logarithmic scale, the decay causes the point representing a certain isotopic starting mixture to move on a straight diagonal line towards minus infinity in both dimensions. In the case of xenon precursors being present in the plume, this would be a bent curve that converges with the straight line defined by decay. The direction of this straight line is determined by the decay constant of the three or four isotopes under consideration.

In each plot a separation line is drawn between the reactor and the explosion domain. Any isotopic activity ratio combination on the separation line moves down along that line with the radioactive decay. Any isotopic activity ratio combination off that line remains on the same side of the line forever. The only exception is ^{133m}Xe/^{133}Xe that could cross the line from a starting point on the nuclear test domain close to the separation line. Radioactive decay changes the isotopic activity ratios along a line that is parallel to the separation line.

This decay invariance offers a new and important quality for source discrimination. It allows for screening out all irrelevant detections because the separation between explosion and non-explosion scenarios is independent from the age of the detected release, i.e., neither the delay before release nor the time for transport through the atmosphere needs to be known.

4.2. Validation of the Multiple Isotopic Relationship Screening Method with Atmospheric Observations Data

The isotopic activity ratio relationship screening method can be validated with atmospheric observation data described in Sect. 2.8. Figure 11 presents the isotopic ratio relations for all air samples in which all four xenon isotopes were measured. Figure 12 shows the same data in the four plots with three out of four isotopes measured, i.e., all samples in which one isotope was not detected are included as well. The error bars shown are statistical contributions only. According to AUER et al. (2004), the systematic error is estimated to be on the order of ±10%. Table 2 shows the screening statistics for all air samples. The screening method works exceptionally well if all four isotopes are measured. Among the cases where three out of four isotopes are measured the case without ^{131m}Xe is the worst. If this isotope is not present or not considered because of a possible background due to its long half-life, the screening method no longer works. If only ^{135}Xe is missing three quarters can be screened out. The screening method is almost 100% successful in the cases without ^{133m}Xe and without ^{133}Xe.

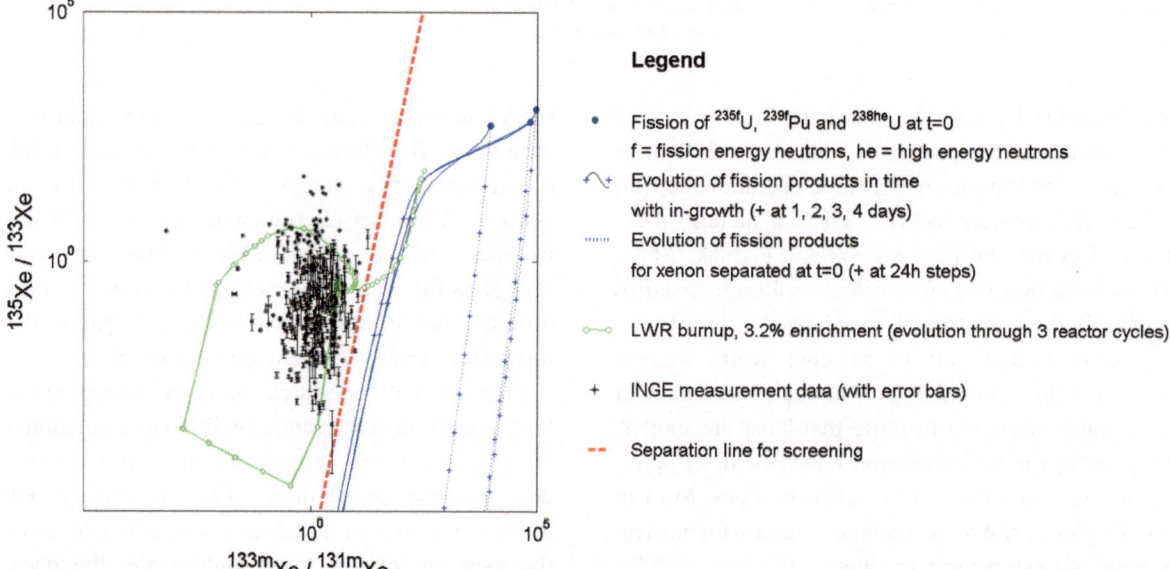

Figure 11
Time-invariant source discrimination for atmospheric measurements with all four isotopes, based on xenon isotopic activity ratio relationship

The station CAX05 in Ottawa is close to one of the isotope production facilities (Chalk River Nuclear Laboratory). Emissions after short-time irradiation of HEU targets have ratios below the screening line are illustrated in Figs. 8 and 10. Therefore one could expect at CAX05 to detect isotopic ratios below the screening line. However, as it can be seen in Fig. 12 this does not happen. The reason for this is twofold. The ^{131m}Xe, due to its long half-life and on-site storage of fission wastes tends to remain in the atmosphere as background from the waste silos and from past emissions. Furthermore, ^{131m}Xe is formed independently at CRL as a by-product of ^{131}I production via irradiation of ^{130}Te targets. Mixing this background into a fresh released plume from isotope production shifts all ratios with ^{131m}Xe in the denominator to lower

values and presumably above the screening line. Therefore it is vital to know the typical local background of ^{131m}Xe at the given station when using the multiple isotopic ratio plot method for screening.

4.3. Quantifying the Separation Line

The optimum position of the separation line for operational use is subject to further studies and depends on user-specific decision criteria. Therefore, an indicative line is constructed which has an optimized slope and a free parameter that allows it to be moved closer or further away from the nuclear test domain.

The straight lines marking the separation between explosive and power reactor scenarios in the plots of

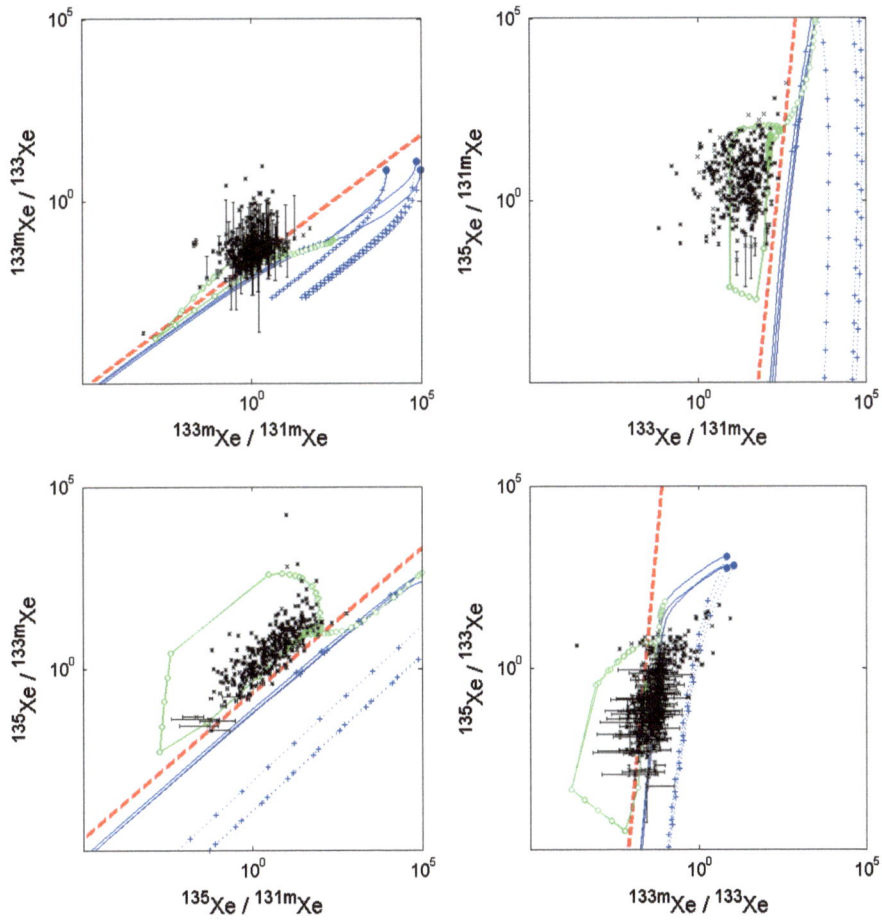

Figure 12
Time-invariant source discrimination for atmospheric measurement data with three out of four isotopes measured based on xenon isotopic activity ratio relationship

Table 2

Screening statistics for atmospheric observation data

Measured isotope	Total	Above (%) (reactor domain)	Below (%) (test domain)
All four isotopes	286	285 (99.7)	1 (0.3)
All or without ^{135}Xe	361	270 (74.8)	91 (25.2)
All or without ^{133m}Xe	305	305 (100)	0 (0)
All or without ^{133}Xe	288	281 (97.6)	7 (2.4)
All or without ^{131m}Xe	310	39 (12.6)	271 (87.4)

1,875 measurements were made. The column "total" shows how many measurements reported all four or three out of four isotopes. The columns "above" and "below" show how many of them lie above or below the screening line

Figs. 8, 9, 10, 11 and 12 are parallel to a linear least-squares fit to a lower part of the unfractionated decay chain curve. The slope of the separation line is denoted by $m_{a,b,c,d}$ where a, b, c and d indicate xenon isotopes (any of ^{135}Xe, ^{133m}Xe, ^{133}Xe and ^{131m}Xe), a stands for the numerator and b for the denominator of the ratio along the ordinate, c and d for the ratio along the abscissa.

A sample can be screened out, if the entry is found above (or to the left of) this screening line. This can be expressed as the following condition:

$$\log(R_{a,b} - \sigma_{a,b}) > m_{a,b,c,d} \cdot \log(R_{c,d} + \sigma_{c,d}) + B_{a,b,c,d}, \tag{1}$$

where the isotopic activity ratio for the xenon isotopes a and b is

$$R_{a,b} = \frac{C_a}{C_b} \tag{2}$$

and the variance of the isotopic activity ratio for the xenon isotopes a and b is

$$\sigma_{a,b}^2 = \left(\frac{\sigma_a}{C_a}\right) + \left(\frac{\sigma_b}{C_b}\right) - 2\frac{\sigma_a \sigma_b}{C_a C_b}\rho \tag{3}$$

with C_a and C_b being the measured atmospheric concentrations, σ_a and σ_b are their measurement uncertainties and ρ_{ab} is the correlation coefficient. Most data considered here satisfy $\rho_{ab} = 0$ as they are analyzed by gamma-spectrometry. The error is overestimated in the few cases where the X-ray lines are involved. In certain cases, one or two concentrations might be replaced by the Minimum Detectable Concentration (MDC, see below). The constant $B_{a,b,c,d}$ is a discrimination margin that can be used to position the screening line at the desired distance to the two separate clusters. The exact position has yet to be determined. Moving the screening line away from the nuclear test domain would render the discrimination more conservative.

Raising both sides of the screening condition to the power of 10 leads to the following relation:

$$R_{a,b} - \sigma_{a,b} > K_{a,b,c,d} \cdot (R_{c,d} + \sigma_{c,d})^{m_{a,b,c,d}} \tag{4}$$

with the constant $K_{a,b,c,d}$ being derived from one selected point on the screening line denoted by the index 0:

Table 3

Screening parameters K and m for all cases with three or four xenon isotopes

Measured isotope	Ordinate: a/b abscissa: c/d	Separation parameter K (tentative)	Slope m of the decay line	Number of cases in atmospheric observation data
All four isotopes	^{135}Xe/^{133}Xe ^{133m}Xe/^{131m}Xe	1.00×10^{-6}	4.4388	286
All but ^{135}Xe	^{133m}Xe/^{133}Xe ^{133m}Xe/^{131m}Xe	2.00×10^{-2}	0.6972	75
All but ^{133m}Xe	^{135}Xe/^{131m}Xe ^{133}Xe/^{131m}Xe	3.78×10^{-22}	9.0774	19
All but ^{133}Xe	^{135}Xe/^{133m}Xe ^{135}Xe/^{131m}Xe	2.00×10^{-1}	0.7996	2
All but ^{131m}Xe	^{135}Xe/^{133}Xe ^{133m}Xe/^{133}Xe	3.27×10^{15}	9.7847	24

$$K_{a,b,c,d} = 10^{B_{a,b,c,d}} = \frac{R_{a,b,0}}{R_{c,d,0}^{m_{a,b,c,d}}}. \qquad (5)$$

The separation parameter K is related to the constant B and defines where the screening line is positioned. It is placed above the nuclear test domain in the ratio versus ratio relationship plot. Increasing K moves the screening line up and renders the source discrimination to become more conservative, i.e. it increases the detection probability at the cost of raising the false alarm rate at the same time.

The screening criteria for all cases with three or four different xenon isotopes are given in Table 3. Thick dashed lines mark them for a tentatively selected discrimination margin B in Figs. 8, 9, 10, 11 and 12.

For example, the following criterion for screening out is found to be applicable, if the atmospheric concentrations of all four isotopes are quantified (see dashed line in Fig. 9, bottom middle sub-plot as well as in Figs. 8 and 11):

$$\frac{C_{Xe-135}}{C_{Xe-133}} - \sigma_{Xe-135,Xe-133} > 1$$
$$\times 10^{-6} \left(\frac{C_{Xe-133m}}{C_{Xe-131m}} + \sigma_{Xe133m,Xe-133} \right)^{4.4388}. \qquad (6)$$

As mentioned above, the exact locations of the screening lines are subject to further studies. Especially, further nuclear reactor source and release scenarios as well as the effect of mixing a weak relevant signal with ambient background have to be studied. The operational screening parameters will be selected according to the wanted detection probability and the acceptable false alarm probability.

The main lesson of these plots is that with the exception of fresh fuel being exposed for only a very short time (less than 20–30 days), there are no conditions during reactor operation that generate xenon isotopic activity ratios in the domain of nuclear test explosions in three out of the five different plots of isotopic activity ratio relationships. The combination of isotopes that provides the least opportunity for discrimination between explosions and reactor sources is the one without ^{131m}Xe. The parent of this isomer, ^{131}I, has a half-life of 8.02 days. The other isotopes have shorter-lived precursors with a maximum of 2.19 days (^{133m}Xe) for ^{133}Xe, 20.8 h (^{133}I)

for ^{133m}Xe and 6.61 hours (^{135}I) for ^{135}Xe. Due to the long half-life of its precursor, the presence of ^{131m}Xe in the set of quantified isotopes provides for good separation capability.

4.4. Detection Limit to Substitute Missing Concentration Measurements

Even in sensitivity measurements of atmospheric xenon activity concentrations in ground level air (AUER et al., 2004; SAEY and DE GEER, 2005; STOCKI et al., 2004), the instrumentation used cannot always detect all of the xenon isotopes because two or more of them are at very low concentrations and thus below the detection limits. Therefore, a robust screening method using xenon ratios should be able to handle the circumstances in which only 2 or 3 of the isotopes are measured, and the others are presumed to be either at or below the detection limit.

In the event that only two or three of the xenon isotopes are detected, the isotopic ratio relationship method might still be applicable. If the isotope in the numerator of an isotopic activity ratio is not quantified, the real ratio is to a high probability lower than one gets by replacing the activity with the minimum detectable concentration (MDC). If the isotope concentration in the denominator is replaced by its MDC, the real ratio is larger than the calculated value. The goal for the MDC set by CTBTO guidelines is 1 mBq/m^3 for ^{133}Xe. SPALAX data analyzed with Aatami achieved an MDC of 0.9 mBq/m^3 for ^{133m}Xe and 0.7 mBq/m^3 for ^{131m}Xe (STOCKI et al., 2004). For ^{135}Xe, an MDC of 0.4 mBq/m^3 can be achieved and 0.2 mBq/m^3 for ^{133}Xe. The application of the Bayesian method (ZÄHRINGER and KIRCHNER, 2008) allows extraction an optimum information for the isomers (VIVIER et al., 2009), especially at very low activities.

If only two xenon isotopes are measurable, the ratio relationship method still appears to work by applying the MDC for a third one with the exception of ^{135}Xe and ^{133m}Xe being the measured isotopes. That case should be rare since these are the two shorter-lived isotopes that are most likely those having decayed below the detection limit at the time of the measurement. Table 4 presents the applicability of screening based on two measured xenon

Table 4

Applicability of isotopic activity ratio relationship method for all cases with two measured xenon isotopes and with the MDC being used for one of the non-detected isotopes

Measured isotope	Using MDC for third isotope	Comment: Screening possible Yes/no	Best option	Number of cases in atmospheric observation data
^{135}Xe, ^{133m}Xe	^{133}Xe	No	None	0
	^{131m}Xe	No		
^{135}Xe, ^{133}Xe	^{133m}Xe	Yes	Without ^{131m}Xe	66
	^{131m}Xe	No		
^{135}Xe, ^{131m}Xe	^{133m}Xe	Yes	Without ^{133}Xe or without ^{133m}Xe	0
	^{133}Xe	Yes		
^{133m}Xe, ^{133}Xe	^{135}Xe	Yes (practically)[a]	Without ^{131m}Xe	41
	^{131m}Xe	No		
^{133m}Xe, ^{131m}Xe	^{135}Xe	Yes (practically)[a]	Without ^{135}Xe	0
	^{133}Xe	Yes		
^{133}Xe, ^{131m}Xe	^{135}Xe	Yes (practically)[a]	Without ^{133m}Xe	127
	^{133m}Xe	Yes (practically)[a]		

[a] Theoretically, time independent screening is not possible. However, if from practical considerations the age of the sample can be limited to a reasonable maximum (e.g., 2 weeks), screening is possible

Table 5

Applicability of isotopic activity ratio relationship method for all cases with one measured xenon isotope only and with the MDC being used for two of the non-detected isotopes

Measured isotope	Using MDC for two isotopes	Comment: Screening possible Yes/no	Best option	Number of cases in atmospheric observation data
^{135}Xe	^{133m}Xe, ^{133}Xe	No	None	28
	^{133m}Xe, ^{131m}Xe	No		
	^{133}Xe, ^{131m}Xe	No		
^{133m}Xe	^{135}Xe, ^{133}Xe	No	None	3
	^{135}Xe, ^{131m}Xe	No		
	^{133}Xe, ^{131m}Xe	No		
^{133}Xe	^{135}Xe, ^{133m}Xe	Yes (practically)[a]	Without ^{131m}Xe	875
	^{135}Xe, ^{131m}Xe	No		
	^{133m}Xe, ^{131m}Xe	No		
^{131m}Xe	^{135}Xe, ^{133m}Xe	Yes (practically)[a]	Without ^{133}Xe	3
	^{135}Xe, ^{133}Xe	Yes (practically)[a]		
	^{133m}Xe, ^{133}Xe	Yes (practically)[a]		

[a] Theoretically, time independent screening is not possible. However, if from practical considerations the age of the sample can be limited to a reasonable maximum (e.g. 2 weeks), screening is possible

isotopes with the MDC being used for one of the non-detected isotopes.

Table 5 lists the applicability of the screening method based on only a single measured xenon isotope and with the MDC being used for two of the non-detected isotopes. This still works, if only ^{133}Xe or only ^{131m}Xe are measured. The first is in fact one of the isotopes being most frequently detected alone, because its share in nuclear reactor releases is the highest and also it has the second longest half-life. Giving a sufficiently large source term, ^{131m}Xe might be the last isotope moving below the detection threshold during the transport through the atmosphere, since it has the longest half-life.

5. Conclusions and Implications for Monitoring Nuclear Explosions

Single xenon isotopic activity ratios may not be as well suited for source discrimination as previously thought, unless the explosions are highly fractionated or no reactor sources are present in the region. Their applicability has been overestimated mainly because the large range of these single ratios in reactor emissions was not adequately considered. There is only one single xenon isotopic activity ratio that can be used alone, namely, $^{133m}Xe/^{131m}Xe$. It can be used with confidence for source discrimination for all nuclear testing scenarios if a single release source is assumed. There is some limited potential for the activity ratios without ^{135}Xe. $^{133m}Xe/^{133}Xe$ is the most relevant example, particularly if early fractionation takes place.

Even if a single xenon isotopic activity ratio is not suitable for source discrimination, it may still be valuable for determining the plume age (KALINOWSKI, 2010a) and tracking it back with atmospheric transport simulation in order to investigate a spatial overlap of the backward plume with waveform location predictions related to a suspected nuclear test. This has to be considered when drawing conclusions on the usability of xenon isotopic ratios for nuclear explosion monitoring.

A new method for source discrimination has been proposed here based on the relationship of two different isotopic activity ratios. This requires three or four xenon isotopes to be quantified. In some cases the method works with the detection limit by substituting a missing concentration value, if certain isotopes are not detected. Hence, it is possible to use this method if only two isotopes are detected. A special advantage of this new method is its independence on the time periods elapsed between generation and release as well as between release and detection, i.e., it is dilution and decay invariant. Quantitative screening conditions are derived for all combinations of three or four xenon isotopes.

A proof of concept for this new screening method is given here by validating it against a comprehensive data set. These data include simulations as well as empirical data of nuclear reactors and explosions as well as observations of atmospheric xenon at different locations in Europe and North America.

In particular the conditions and limitations are:

- Suitable combination of isotopes need to be detected
- Background must be precisely known, in particular ^{131m}Xe
- No mixing in of fresh releases from other sources
- Dependent on decision procedure whether or not a nuclear explosion is likely indicated by the ratio (in this paper simply indicated by separation line)
- Excessively large xenon isotopic activity ratio uncertainties

Further work is required in order to put this method in operation. In particular in the light of the detection goals, various uncertainties and possible mixing of a weak relevant signal with ambient background have to be considered. The limitations posed by the need to use three or more xenon activities calls for additional screening methods for the cases of fewer isotopes being quantified. These may be based on a combination of single ratios as well as absolute concentrations and may involve an outlier analysis to classify a measurement as anomalous in comparison to typical atmospheric background at that detector site. This can be supported by atmospheric transport simulations using information about known civilian sources.

Acknowledgments

Most of this work was carried out while M.K. was a staff member with the Provisional Technical Secretariat of the Preparatory Commission for the Comprehensive Nuclear-Test-Ban Treaty Organisation (CTBTO). Part of the work was accomplished during his stay at the Department of Nuclear, Plasma, and Radiological Engineering (NPRE) and the Program in Arms Control, Disarmament, and International Security (ACDIS) of the University of Illinois at Urbana-Champaign during spring of 2005. At that time, his work was sponsored by the John D. and Catherine T. MacArthur Foundation. The completion of this work was funded by the German Foundation for Peace Research (DSF). The views expressed herein are those of the authors and do not necessarily reflect the views of the CTBTO

Preparatory Commission or any other organisation one of the authors is affiliated with.

REFERENCES

AUER, M., AXELSSON, A., BLANCHARD, X., BOWYER, T.W., BRACHET, G., BULOWSKI, I., DUBASOV, Y., ELMGREN, K., FONTAINE, J.P., HARMS, W., HAYES, J.C., HEIMBIGNER, T.R., MCINTYRE, J.I., PANISKO, M.E., POPOV, Y., RINGBOM, A., SARTORIUS, H., SCHMID, S., SCHULZE, J., SCHLOSSER, C., TAFFARY, T., WEISS, W., and WERNSPERGER, B. (2004), *Intercomparison experiments of systems for the measurement of xenon radionuclides in the atmosphere*, Appl. Rad. Isotopes 60, 863–877.

AUER, M., KUMBERG, T., SARTORIUS, H., WERNSPERGER, B., and SCHLOSSER, C. (2010), *Ten years of development of equipment for measurement of atmospheric radioactive xenon for the verification of the CTBT*, Pure Appl. Geophys. Topical Volume. *Recent Advances in Nuclear Explosion Monitoring*, this volume, 167, 415.

BATEMAN, H. (1910), *The solution of a system of differential equations in the theory of radio-active transformation*, Proc. Cambridge Phil. Soc. (16), p. 423.

BECKER, A., WOTAWA, G., RINGBOM, A., and SAEY, P.R.J. (2010), *Backtracking of noble gas measurements taken in the aftermath of the announced October 2006 event in North Korea by means of PTS methods in nuclear source estimation and reconstruction*, Pure Appl. Geophys. Topical Volume *Recent Advances in Nuclear Explosion Monitoring*, this volume, 4/5.

BOWYER, T.W., PERKINS, R.W., ABEL, K.H., HENSLEY, W.K., HUBBARD, C.W., MCKINNON, A.D., PANISKO, H.E., REEDER, P. L., THOMPSON, R.C., and WARNER, R.A. (1998), *Xenon radionuclides, atmospheric: Monitoring*. In *Encyclopaedia of Environmental Analysis and Remediation* (ed. Meyers R. A.) (Wiley, New York).

BOWYER, T.W., SCHLOSSER, C., ABEL, K.H., AUER, M., HAYES, J.C., HEIMBIGNER, T.R., MCINTYRE, J.I., PANISKO, M.E., REEDER, P.L., SARTORIUS, H., SCHULZE, J., and WEISS, W. (2002), *Detection and analysis of xenon isotopes for the comprehensive NuclearTest-Ban Treaty international monitoring system*, J. Environ. Radioact. 59(2), 139–151.

BAG (Bundesamt für Gesundheit), (2000, 2001 and 2002), *Überwachung der Umweltradioaktivität*, Jahresberichte 2000, 2001 und 2002, Bern, Switzerland, http://www.bag.admin.ch/strahlen/ionisant/download/radio_env/d/re-pdf.php.

CARMAN, A.J., MCINTYRE, J.I., BOWYER, T.W., HAYES, J.C., HEIMBIGNER, T.R., and PANISKO, M.E. (2002), *Discrimination between anthropogenic sources of atmospheric radioxenon*, Trans. Am. Nucl. Soc. 87, 89-90.

DE GEER, L.-E. (2001), *Comprehensive Nuclear-Test-Ban Treaty: Relevant radionuclides*, Kerntechnik 66 (3), 113–120.

EC (2004), *Radioactive effluents from nuclear power stations in the European Union, 2000–2003*, Personal communication with the European Commission radiation protection unit DG TREN H4 on 14 October, 2004.

FINKELSTEIN, Y. (2001), *Fission product isotope ratios as event characterization tools - Part II: Radioxenon isotopic activity ratios*, Kerntechnik 66 (5–6), 229–236.

FONTAINE, J.P., POINTURIER, F., BLANCHARD, X., and TAFFARY, T. (2004), *Atmospheric xenon radioactive isotope monitoring*, J. Environ. Radioact. 72, 129–135.

HOFFMANN, W., KEBEASY, R., and FIRBAS, P. (1999), *Introduction to the verification regime of the Comprehensive Nuclear-Test-Ban Treaty*, Phys. Earth and Planet. Interiors 113, 5–9.

KALINOWSKI, M.B. (2010a), *Characterisation of prompt and delayed atmospheric radioactivity releases from underground nuclear tests at Nevada as a function of release time*, accepted by J. Environ. Radioact.

KALINOWSKI, M.B. (2010b), *Nuclear explosion time assessment based on xenon isotopic activity ratios*, accepted by Appl. Rad. Isotopes.

KALINOWSKI, M.B. and PISTNER, Ch. (2006), *Isotopic signature of atmospheric xenon released from light water reactors*, J. Environ. Radioact. 88 (3), 215–235.

KALINOWSKI, M.B. and TUMA, M.P. (2009), *Global radioxenon emission inventory based on nuclear power reactor reports*, J. Environ. Radioact. 100 (1), 58–70.

KELLER, H. (2004), University Mainz, Germany, Private Communication received on 27 September 2004.

LE PETIT, G., ARMAND, P., BRACHET, G., TAFFARY, T., FONTAINE, J. P., ACHIM, P., BLANCHARD, X., PIWOWARCZYK, J. C., and POINTURIER, F. (2008), *Contribution to the development of atmospheric radioxenon monitoring*, J. Radioanal. Nucl. Chem. 276 (2), 391–398.

NATC (1999), *1999 US Gaseous and Liquid Effluent Reports*. North American Technical Center, University of Illinois at Urbana-Champaign (1999), http://hps.ne.uiuc.edu/natcenviro/planteffluent.htm (latest access 3 August 2004).

RAITH, M. (2006), *Development and quantification of new methods for the determination and preparation of short-lived xenon isotopes for the verification of the Comprehensive Nuclear Test Ban Treaty (CTBT) in a laboratory*, Dissertation, Vienna University of Technology.

RINGBOM, A., ELMGREN, K., and LINDH, K. (2007), *Analysis of radioxenon in ground level air sampled in the Republic of South Korea on October 11–14*, Report FOI-R-2273-SE.

RINGBOM, A., LARSON, T., AXELSSON, A., ELMGREN, K., and JOHANSSON, C. (2003), *SAUNA—A system for automatic sampling, processing, and analysis of radioactive xenon*, Nucl. Instrum. and Methods A 508 (3), 542–553.

RINGBOM, A., ELMGREN, K., LINDH, K., PETERSON, J., BOWYER, T.W., HAYES, J.C., MCINTYRE, J.I., PANISKO, M., and WILLIAMS, R. (2009), *Measurements of radioxenon in ground level air in South Korea following the claimed nuclear test in North Korea on October 9, 2006*, J. Radioanal. Nucl. Chem 282, 773–779.

SAEY, P.R.J. and DE GEER, L.-E. (2005), *Notes on radioxenon measurements for CTBT verification purposes*, Appl. Rad. Isotopes 63, 765–773.

SAEY, P.R.J. (2009), *The influence of radiopharmaceutical isotope production on the global radioxenon background*, J. Environ. Radioact., doi:10.1016/j.jenvrad.2009.01.004.

SAEY, P. R. J., BEAN, M., BECKER, A., COYNE, J., d'AMOURS, R., DE GEER, L.-E., HOGUE, R., STOCKI, T. J., UNGAR, R. K., and WOTAWA, G. (2007), *A long distance measurement of radioxenon in Yellowknife, Canada, in late October 2006*, Geophys. Res. Lett. 34.

SAEY, P.R.J., SCHLOSSER, C., AUER, M., AXELSSON, A., BECKER, A., BLANCHARD, X., BRACHET, G., DE GEER, L.-E., KALINOWSKI, M.B., PETERSON, J., POPOV, V., POPOV, Y., RINGBOM, A., SARTORIUS, H., TAFFARY, T., and ZÄHRINGER, M. (2010), *Environmental radioxenon levels in Europe— A comprehensive overview*, Pure Appl. Geophys. Topical Volume *Recent Advances in Nuclear Explosion Monitoring*, 167, 4/5.

SCHOENGOLD, C.R., DEMARRE, M.E., and KIRKWOOD, E.M. (1996), *Radiological effluents released from U.S. continental tests 1961 through 1992*, United States Department of Energy - Nevada Operations Office, DOE/NV-317 (Rev.1) UC-702, Las Vegas, August 1996.

STOCKI, T.J., BEAN, M., UNGAR, R.K., TOIVONEN, H., ZHANG, W., WHYTE, J., and MEYERHOF, D. (2004), *Low level noble gas measurements in the field and laboratory in support of the Comprehensive NuclearTest-Ban Treaty*, Appl. Rad. Isotopes *61*, 231–235.

STOCKI, T.J., BLANCHARD, X., D'AMOURS, R., UNGAR, R.K., FONTAINE, J.P., SOHIER, M., BEAN, M., TAFFARY, T., RACINE, J., TRACY, B.L., BRACHET, G., JEAN, M., and MEYERHOF, D. (2005), *Automated radioxenon monitoring for the Comprehensive NuclearTest-Ban Treaty in two distinctive locations: Ottawa and Tahiti*, J. Environ. Radioact. *80* (3), 305–326.

VAN DER STRICHT, S. and JANSSENS, A. (2001), *Radioactive effluents from nuclear power stations and nuclear fuel reprocessing plants in the European Union, 1995–1999*, Radiation Protection *127*, European Commission, Luxembourg.

VIVIER, A., LE PETIT, G., PIGEON, B., and BLANCHARD, X. (2009), *Probabilistic assessment for a sample to be radioactive or not: application to radioxenon analysis*, J. Radioanal. Nucl. Chem. *282* (3), 743–748.

WOTAWA, G., BECKER, A., KALINOWSKI, M.B., SAEY, P.R.J., TUMA, M.P., and ZÄHRINGER, M. (2010), *Computation and Analysis of the Global Distribution of the Radioxenon Isotope ^{133}Xe based on Emissions from Nuclear Power Plants and Isotope Production Facilities and its relevance for the Verification of the Nuclear Test–Ban Treaty*, Pure Appl. Geophys. Topical Volume *Recent Advances in Nuclear Explosion Monitoring, 167*, 4/5.

ZÄHRINGER, M. and KIRCHNER, G. (2008), *Nuclide ratios and source identification from high-resolution gamma-ray spectra with Bayesian decision methods*, Nucl. Instrum. and Methods A *594*, 400–406.

(Received December 17, 2008, revised September 7, 2009, accepted September 8, 2009, Published online January 30, 2010)

Pure Appl. Geophys. 167 (2010), 541–557
© 2009 Birkhäuser Verlag, Basel/Switzerland
DOI 10.1007/s00024-009-0033-0

❘ Pure and Applied Geophysics

Computation and Analysis of the Global Distribution of the Radioxenon Isotope ^{133}Xe based on Emissions from Nuclear Power Plants and Radioisotope Production Facilities and its Relevance for the Verification of the Nuclear-Test-Ban Treaty

GERHARD WOTAWA,[1,2] ANDREAS BECKER,[2] MARTIN KALINOWSKI,[3] PAUL SAEY,[2] MATTHIAS TUMA,[3] and MATTHIAS ZÄHRINGER[2]

Abstract—Monitoring of radioactive noble gases, in particular xenon isotopes, is a crucial element of the verification of the Comprehensive Nuclear-Test-Ban Treaty (CTBT). The capability of the noble gas network, which is currently under construction, to detect signals from a nuclear explosion critically depends on the background created by other sources. Therefore, the global distribution of these isotopes based on emissions and transport patterns needs to be understood. A significant xenon background exists in the reactor regions of North America, Europe and Asia. An emission inventory of the four relevant xenon isotopes has recently been created, which specifies source terms for each power plant. As the major emitters of xenon isotopes worldwide, a few medical radioisotope production facilities have been recently identified, in particular the facilities in Chalk River (Canada), Fleurus (Belgium), Pelindaba (South Africa) and Petten (Netherlands). Emissions from these sites are expected to exceed those of the other sources by orders of magnitude. In this study, emphasis is put on ^{133}Xe, which is the most prevalent xenon isotope. First, based on the emissions known, the resulting ^{133}Xe concentration levels at all noble gas stations of the final CTBT verification network were calculated and found to be consistent with observations. Second, it turned out that emissions from the radioisotope facilities can explain a number of observed peaks, meaning that atmospheric transport modelling is an important tool for the categorization of measurements. Third, it became evident that Nuclear Power Plant emissions are more difficult to treat in the models, since their temporal variation is high and not generally reported. Fourth, there are indications that the assumed annual emissions may be underestimated by factors of two to ten, while the general emission patterns seem to be well understood. Finally, it became evident that ^{133}Xe sources mainly influence the sensitivity of the monitoring system in the mid-latitudes, where the network coverage is particularly good.

Key words: Atmospheric transport modelling, radioxenon monitoring, CTBT verification.

1. Introduction

The Comprehensive Nuclear-Test-Ban Treaty (CTBT) was opened for signature in 1996. The Treaty bans all nuclear explosions regardless of purpose and size (CTBT, 1996) and is thus a landmark international agreement in the area of arms control. The backbone of the CTBT verification regime is a global International Monitoring System (IMS) with four different measurement technologies, namely seismic, hydro-acoustic, infrasound and radionuclide monitoring (CTBT, 1996; HOFFMANN et al., 2000). The Provisional Technical Secretariat (PTS) of the Preparatory Commission for the Comprehensive Nuclear-Test-Ban Treaty Organization (CTBTO) was established in Vienna in 1997. The CTBT has not entered into force yet.

As regards as the radionuclide (RN) monitoring subsystem, the treaty mentions 80 stations equipped with aerosol samplers and high-resolution germanium detectors (CTBT, 1996; SCHULZE et al., 2000), 79 of which have already been named (see Fig. 1). Out of the 80 RN locations, 40 will additionally be equipped with xenon detectors (CTBT, 1996; KALINOWSKI et al., 2008), and 39 of those are known to date. The missing RN site was reserved for one of the current non-signatory states. After Entry into Force of the

[1] Central Institute for Meteorology and Geodynamics, Vienna, Austria. E-mail: Gerhard.wotawa@zamg.ac.at

[2] Preparatory Commission for the Comprehensive Nuclear-Test-Ban Treaty Organization, PTS, Vienna, Austria. E-mail: gerhard.wotawa@ctbto.org

[3] Carl Friedrich von Weizsäcker Center for Science and Peace Research, University of Hamburg, Hamburg, Germany.

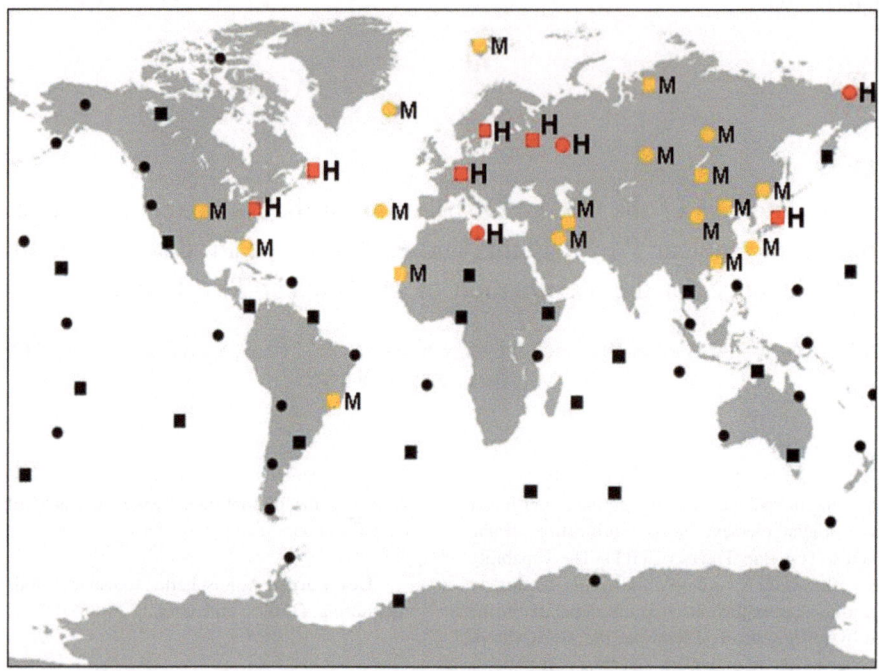

Figure 1
Display of the 79 radionuclide sites of the International Monitoring System as foreseen in the Protocol to the CTBT. *Square symbols* mark sites with additional noble gas capability. The sites marked in *red*/with capital *H* are predicted to be significantly influenced by the known xenon emitters; the sites marked in *orange*/with capital *M* occasionally. A significant influence is assumed if the median value of the model predicted ^{133}Xe concentrations exceeds 0.1 mBqm^{-3}; an occasional influence if the 95% percentile value exceeds 0.1 mBqm^{-3}

treaty, subject to a decision of the Conference of the States Parties, all 80 stations may be equipped with xenon detectors (CTBT, 1996). The xenon systems will measure the isotopes ^{131m}Xe, ^{133m}Xe, ^{133}Xe and ^{135}Xe, with half-lives of 11.93 days, 2.19 days, 5.24 days and 9.14 h, respectively. The xenon network is important to uncover testing scenarios in which no particulate debris is injected into the atmosphere (underground or underwater). Noble gases like radioxenons are created in significant quantities. They are chemically inert, not subject to deposition, and are very difficult to contain, consequently there is an increased probability that they will escape from a clandestine test point (DE GEER, 1996). In case of a 1-kiloton nuclear fission explosion, calculations show that about 10^{16} Bq of ^{133}Xe would be produced in total. Depending on the actual test scenario, it is estimated that between 10^{14} and 10^{15} Bq of ^{133}Xe could be released into the atmosphere during the first 3 h after an underground explosion (PERKINS and CASEY, 1996). In connection with the announced nuclear explosion conducted on the territory of the

Democratic People's Republic of Korea (DPRK) on 9 October 2006, the ^{133}Xe signal measured in Yellowknife (Canada) at the end of October 2006, about two weeks after the event, was found to be consistent with a release of about 10^{15} Bq of ^{133}Xe at the position of the recorded seismic event within the first 3 h, using the same model as applied in this study (SAEY et al., 2007). Measurements of ^{133}Xe in the Republic of Korea after 9 October showed that xenon was also released in the days after the explosion, although an order of magnitude less (RINGBOM et al., 2007; BECKER et al., 2010). The seismic signal recorded by the PTS on 9 October referred to an explosion of somewhat less than 1 kt.

As part of the International Noble Gas Experiment (INGE), noble gas measurement systems have been set up worldwide (AUER et al., 2004; WERNSBERGER and SCHLOSSER, 2004; SAEY and DE GEER, 2005). Four different types of automatic detection systems were developed: the Russian ARIX system (Analyzer for Xenon Measurements, see DUBASOV et al., 2005), the American ARSA system (Automated Radioxenon

Sampler/Analyze, see BOWYER et al., 1999), the Swedish SAUNA system (Swedish Automatic Unit for Noble Gas Acquisition, see RINGBOM et al., 2003) and the French SPALAX system (Système de Prélèvement d'air Automatique en Ligne avec l'Analyse des radio-Xénon, see FONTAINE et al., 2004). In all systems, ambient air is collected and a xenon-enriched gas sample is generated. In a second stage the gas is further concentrated and purified. Measurement of the activity of the final gas sample is done either by high-resolution gamma spectroscopy (SPALAX) or by beta–gamma spectroscopy (ARIX, ARSA and SAUNA)). Between September 2007 and June 2008, data from 13 stations were regularly available (12 IMS sites plus one national contribution station in Ottawa, Canada).

The minimum detection limit requirement for all systems is 1 mBqm^{-3} of ^{133}Xe over a full day sampling period. Current systems achieve, however, a minimum detectable concentration (MDC) between 0.3 and 0.1 mBqm^{-3}, depending on the performance of the system and its capability to suppress radon. The collection times of the measurements are currently either 12 or 24 h, depending on the station and the system used there.

It was found that ambient background levels of ^{133}Xe range over several orders of magnitude. The background is negligible at island locations of the Southern Hemisphere (STOCKI et al., 2005), but evident even in remote locations of the Northern Hemisphere in terms of the isotope ^{133}Xe (SAEY et al., 2006). Close to known emitters, significant concentrations of ^{133}Xe were found, for example in Freiburg, Germany and Ottawa, Canada (see AUER et al., 2004; STOCKI et al., 2005). The other three isotopes of interest are typically only detected on an occasional basis. Since the isotope ^{133}Xe is the most prevalent one globally, this study will focus solely on that isotope.

To interpret the ambient radionuclide concentration measurements, the PTS operates an Atmospheric Transport Modelling (ATM) system based on the Lagrangian Particle Diffusion Model FLEXPART (STOHL et al., 2005). The FLEXPART model is designed to simulate the long-range and mesoscale transport, diffusion, dry and wet deposition, and radioactive decay of tracers released from point, line, area or volume sources. It can be used forward in time to simulate the dispersion of tracers from their sources, or backward in time to determine potential source contributions for given receptors. The PTS uses FLEXPART in backward mode to compute daily Source-Receptor-Sensitivity (SRS) fields for all radionuclide measurement locations (WOTAWA et al., 2003; BECKER et al., 2007). The model is fed with analysed wind data provided by the European Centre for Medium-Range Weather Forecasts (ECMWF). The horizontal resolution of the input data is 1° in both longitude and latitude, the temporal resolution is 3 h. The maximum transport duration (backtracking time) for the SRS computations is set to 14 days. The 14 days have been chosen taking into account the fact that it was one of the design criteria of the IMS RN network (90% detection probability within 14 transport days), and for computational reasons. Equivalent to the input, the SRS fields have a spatial resolution of 1° × 1° and a temporal resolution of 1 h.

2. Model Simulation of ^{133}Xe Emissions and Transport

To facilitate the interpretation of observed xenon concentrations, generic annual radioxenon emission strengths for all operable Nuclear Power Plants (NPPs) were recently estimated (KALINOWSKI and TUMA, 2009). The inventory is based on emission reports for North American and European NPPs issued between 1995 and 2005. Generic source terms were calculated for all reported sites and used to estimate the average releases of NPPs for which no data are available. The inventory distinguishes between continuous and batch emissions of all four xenon isotopes with regard to the sum of annual releases. Exact time courses of batch emissions could not be provided due to the lack of available data. Only for some NPPs, frequencies and durations of batch emissions were known averaged over quarter years, but could not be used for purposes of this study since the monitoring results are 24 or 12 h averages. According to the inventory, a total of 0.74 10^{15} Bq of ^{133}Xe is released by all facilities collectively each year, two-thirds of it continuously and about one-third as batch emissions.

The global supply for medical isotopes is covered by a few radioisotope production facilities (RPFs) that work on the basis of uranium irradiation. Experiences with the data from INGE have already shown that these facilities are by far the strongest sources of xenon that exist worldwide. A single extraction plant can emit on the order of 10^{15} Bq of ^{133}Xe per year (RIECHMANN and KALINOWSKI, 2008; SAEY, 2009), which thus exceeds the emissions from all NPPs collectively.

For purposes of this study, the total ^{133}Xe emissions from the NPPs (sum of continuous plus batch releases) were resampled on the same grid as the SRS fields ($1° \times 1°$, 3 h) and assumed to be constant in each 3-h interval during the year. That means the impact of the emission time patterns caused by batch releases from NPPs had to be neglected. Furthermore, constant emissions of 10^{15} Bq/a were assumed at the sites in Fleurus (Belgium), Chalk River (Canada) and Pelindaba (South Africa; see Fig. 2). These facilities are among the four largest producers of medical isotopes worldwide. The fourth facility located in Petten (Netherlands) was not considered here, since it operates long decay lines and thus emits three orders of magnitude less than the other three.

To calculate model-predicted ^{133}Xe concentrations for all RN samples, the calculated sample-specific SRS fields are folded with the assumed emission inventory. For each radionuclide measurement, the concentration c can be computed as follows (see WOTAWA et al., 2006):

$$c = \sum_{i,j,k} M_{ijk} S_{ijk}. \tag{1}$$

M here is the discrete SRS field in grid cell i, j and time step k, while S is the source field available on the same grid. The advantage of this method compared with any explicit forward dispersion modelling approach is that concentration vectors can be predicted with very little demand on computational resources, making the testing of a number of different source scenarios feasible.

In this study, the first emission scenario looked at was the sum of the NPP emissions plus the emissions from the three RPFs. Subsequently, four different emission scenarios were considered separately, namely (i) NPP emissions, (ii) emissions from Chalk River, (iii) from Fleurus and (iv) from Pelindaba. Multiple linear regression analyses were done to determine the possible influence of each of the different sources individually on the temporal as well as spatial variance of the measurement values. The total computation period was one full year (July 2007 to June 2008).

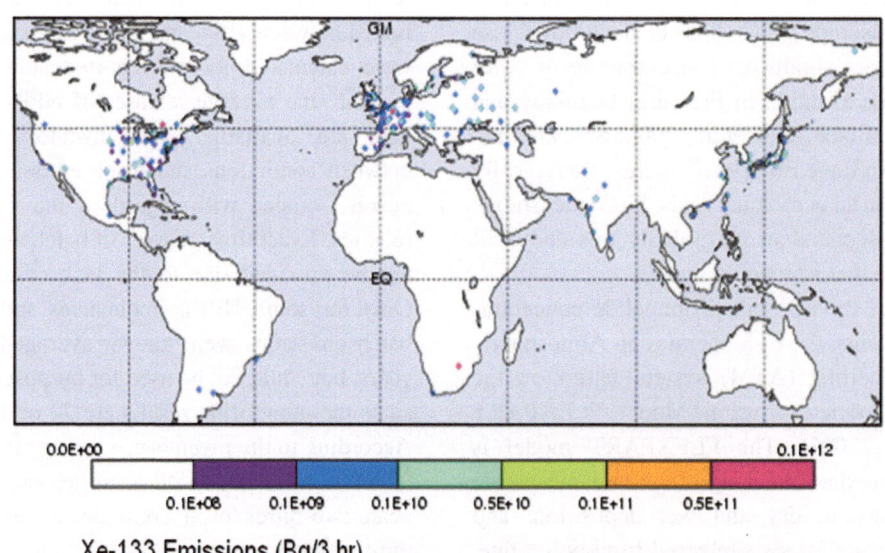

Xe-133 Emissions (Bq/3 hr)

Figure 2
Emission inventory of ^{133}Xe (Bq/3 h) used for this study. The inventory considers emissions from NPPs and four RPFs

3. Calculated ¹³³Xe Concentrations from Civilian Sources at all IMS Sites

First, the model-predicted influence of the known civilian emissions of ¹³³Xe on all 79 IMS sites (see Fig. 1) was investigated. For these purposes, the influence was defined and referred to as "strong" in case that the predicted median value for ¹³³Xe would exceed an assumed typical low-background MDC value for the measurements (in our case 0.1 mBqm⁻³), and as "occasional" in case that the predicted 95% percentile value (but not the median value) would exceed the assumed MDC value. According to this definition, a strong influence would mean that the model indicates 50% or more detections caused by civilian sources at the respective site. No influence would mean that the site would

experience less than 5% detections according to the model.

According to the model, strong influence (and hence a significant ¹³³Xe background) is predicted for 9 IMS sites (11.4% of all); all situated in the mid-latitudes of the Northern Hemisphere (see Fig. 1). Occasional influence is predicted for 19 sites (24.1%), mostly situated in the Northern Hemisphere. At almost two-thirds of all IMS sites, however, the model predicts that there is only a minor, if any, influence from known emissions.

Regarding the influence of the known emissions on the 39 noble gas locations of the IMS, the situation is as follows (see Figs. 3, 4): The most affected station would be DEX33 located in Freiburg, Germany, where even the predicted 5% percentile value is well above the MDC. This station is followed by SEX63

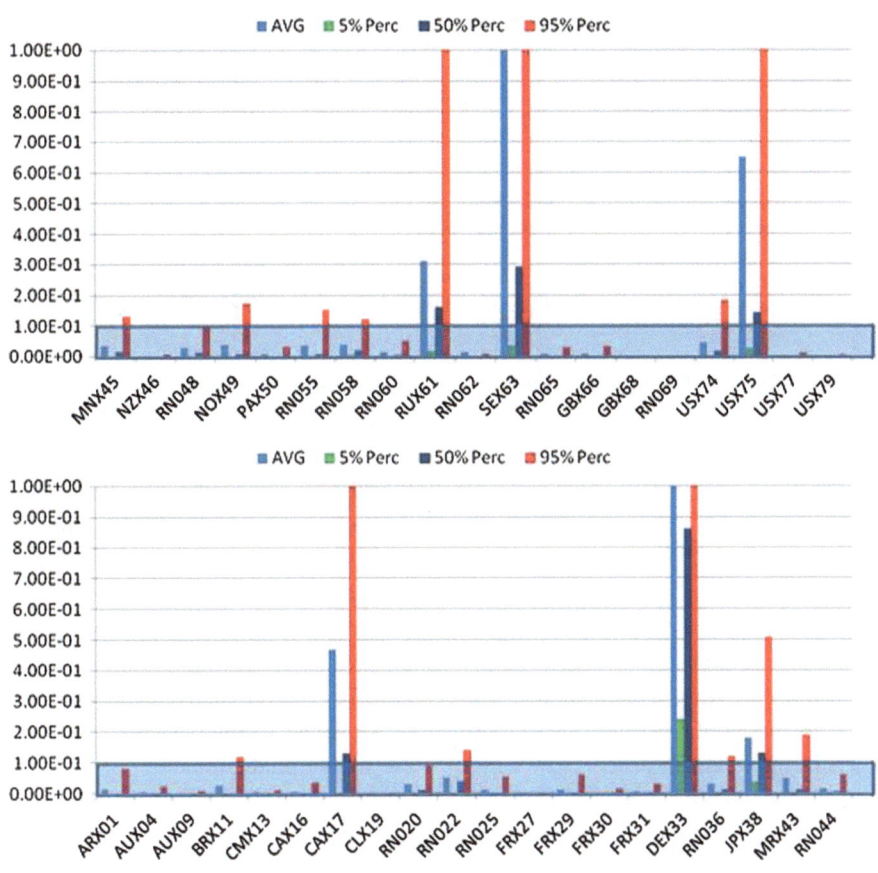

Figure 3
Model predicted ¹³³Xe concentrations at those of the future IMS sites with additional noble gas capabilities, pertaining to the standard source scenario "emissions from NPPs plus RPFs". Diplayed are average concentration and 5%/50%/95% percentile values. The lower image shows sites numbered 1-44, the upper 45-79. The figures are scaled from 0 to 1 mBqm⁻³

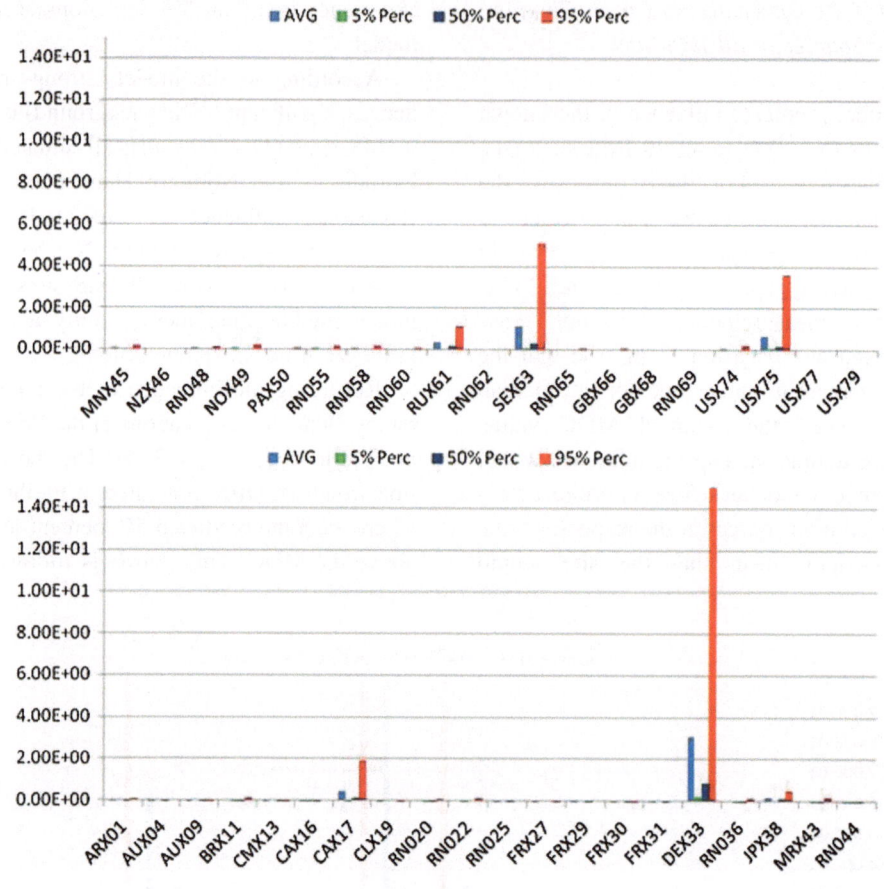

Figure 4
Same as Fig. 3, but figures are scaled from 0 to 15 mBqm^{-3}

(Stockholm, Sweden; Fig. 4), USX75 (Charlottesville, USA), CAX17 (St. Johns, Canada), RUX61 (Dubna, Russia) and JPX38 (Takasaki, Japan).

4. Comparison with Available INGE Measurements

During our calculation period (July 2007 to June 2008), radioxenon concentrations were measured at 12 stations (Table 1, Fig. 5). Most of them are located in the mid-latitudes of the Northern Hemisphere, but data from sites where the model expects a low concentration level are also available. In order to examine the validity of the computations and of the underlying emission assumptions, a comparison of predicted and observed ^{133}Xe concentrations was done.

In general, the model was found to perform reasonably well in describing the global ^{133}Xe

distribution (see Table 2; Figs. 5, 6). The measurements confirm the predicted higher background at stations DEX33, SEX63, and JPX38. At station CAX17, however, the measured background was by one order of magnitude higher than expected from the model. An underprediction of a factor of 5 was also seen at station USX75. A similar result, but on a considerably lower level, was obtained for station CNX22 (Guangzhou, China). Stations AUX09, CAX16, NZX46 and PAX50 were mostly measuring ^{133}Xe values below the MDC, in full accordance with our model predictions.

In order to calculate the total bias of the model (and thus the emission estimate used therein) relative to the available observations, the following linear regression model for predicted versus measured ^{133}Xe median concentrations from all 12 sites during the whole simulation period was set up:

Table 1

List of IMS sites currently equipped with noble gas instruments (longitude/latitude in degrees, sampling interval in hours, station name and country, measurement system)

Station	Lon.	Lat.	Sampl. interval	Name	Country	System
AUX09	130.90	−12.40	12	Darwin	Australia	SAUNA
CAX16	−114.48	62.45	24	Yellowknife	Canada	SPALAX
CAX17	−52.70	47.60	24	St. John's	Canada	SPALAX
CNX20	116.20	39.75	24	Beijing	China	SPALAX
CNX22	113.30	23.00	12	Guangzhou	China	SAUNA
DEX33	7.90	47.90	24	Schauinsland Mountain	Germany	SPALAX
JPX38	139.00	36.31	12	Gunma	Japan	SAUNA
NOX49	15.40	78.20	12	Spitsbergen	Norway	SAUNA
NZX46	−176.50	−43.80	12	Chatham Island	New Zealand	SAUNA
PAX50	−79.53	8.98	24	Panama City	Panama	SPALAX
SEX63	17.57	59.23	12	Stockholm	Sweden	SAUNA
USX75	−78.40	38.00	12	Charlottesville	United States	SAUNA

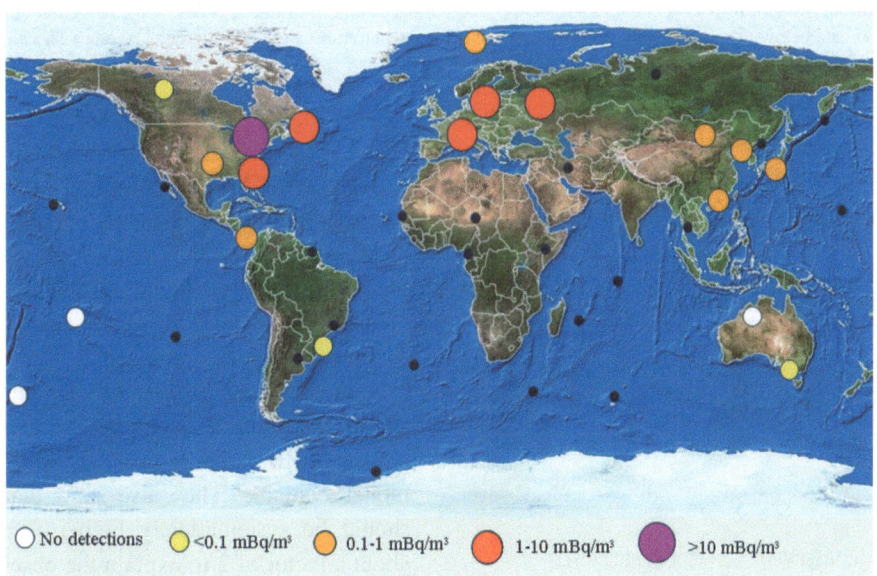

Figure 5

Observed ^{133}Xe concentrations at the 13 INGE sites that are currently sampling noble gas. Please note that station CAX05 is not an IMS but a Canadian national site and was thus discarded from this study

$$c_k = c_0 + ac_{\text{mod},k}. \qquad (2)$$

c_k is the observed median value of ^{133}Xe at station k and $c_{\text{mod},k}$, the respective model-predicted median value ($k = 1, 12$). C_0 and a are the parameters to be specified by the regression analysis and are subsequently referred to as unexplained concentration level and model bias factor, respectively. A scatter-plot for the regression analysis can be seen in Fig. 7. The analysis yielded an unexplained concentration level of 0.13 mBqm^{-3} and a model bias factor of 2.1, with a correlation coefficient of 0.75. The major outlier was station CAX17. Without this site, the unexplained concentration level as calculated with the regression model was 0.03, the model bias factor 2.2 and the correlation coefficient increased to 0.97. These bias factors show that the model systematically underpredicts the measurements, and thus indicate a respective underestimation of the annual emissions driving it. The high correlation indicates that the

175

Table 2

Listing of observed and model predicted concentrations of ^{133}Xe (average observed concentrations, median values of observed/predicted concentrations, 95% percentile values of observed/predicted concentrations) at the 12 IMS sites that are currently sampling noble gas as part of the INGE project

Station	Number of data	Average meas. with L_C	Median meas. without L_C	Median meas. with L_C	Median pred.	95% perc. meas. with L_C	95% perc. pred.
AUX09	458	0.036	0.026	<0.1	0.00	0.12	0.01
CAX16	251	0.011	0.002	<0.1	0.00	<0.1	0.02
CAX17	157	5.00	1.4	1.2	0.07	23.4	1.96
CNX20	121	0.32	0.23	<0.1	0.01	0.53	0.12
CNX22	249	0.35	0.16	0.11	0.04	1.06	0.16
DEX33	141	5.82	1.39	1.2	0.54	26.00	16.76
JPX38	515	0.250	0.14	0.17	0.10	0.75	0.45
NOX49	428	0.156	0.096	0.1	0.00	0.58	0.26
NZX46	191	<0.000	0.006	<0.1	0.00	<0.1	0.00
PAX50	117	0.076	0.045	<0.1	0.00	0.30	0.03
SEX63	249	2.43	0.64	0.63	0.26	15.00	6.79
USX75	247	4.324	0.689	0.47	0.09	16.80	2.78

All values are in mBqm^{-3}. With respect to the measurements, "<0.1" means that the value was below the detection threshold. Data "with L_C" means that also data below the critical limit (minimum detectable concentration) were considered. The one-year calculation period was July 2007 to June 2008

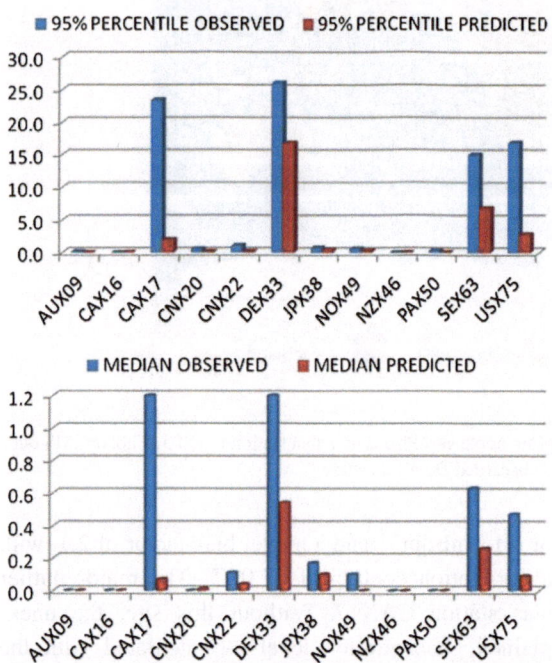

Figure 6
Comparison between predicted and observed ^{133}Xe values (mBqm^{-3}) at the 12 IMS sites that are currently sampling noble gas as part of the INGE project. The upper image shows the median value, the lower image the 95% percentile value. The calculation period was July 2007 to June 2008, and the source scenario was the standard one (emissions from NPPs plus RPFs)

variance of ^{133}Xe concentrations within the network is well represented and thus that the geographical distribution of the emissions is correct. Considering the 95% percentile values for ^{133}Xe instead of the median values yields a similar statistical relationship. The model bias factor, however, decreases to 1.59.

Altogether, these results demonstrate that the global distribution of the ^{133}Xe concentrations is reasonably well understood, based on the emission model available. They also show that the emissions should be systematically higher than predicted by about a factor of 2 to explain the observations, which is well within their inherent range of uncertainty. Regarding the outlier station CAX17, a possible explanation, namely the strong underestimation of emissions from Chalk River, is provided in one of the following sections.

5. Influence of the Known Sources on the Global Network Coverage

One of the topics of interest is the capability of the final 39-station noble gas network to detect ^{133}Xe from a nuclear explosion. For this purpose, the threshold source strength of the network is defined.

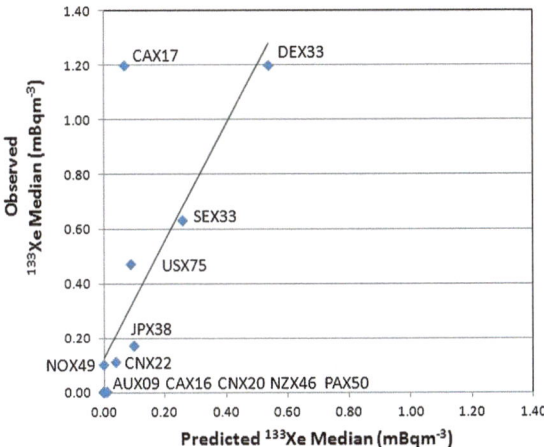

Figure 7

Scatter plot observed versus predicted ^{133}Xe median (mBqm^{-3}) at the 12 IMS sites that are currently sampling noble gas. The calculation period was July 2007 to June 2008, and the source scenario was the standard one (emissions from NPPs plus RPFs)

This is the source strength as a function of space and time that would cause one or more detections (measurement above a defined threshold) within the network. To calculate the global distributions of the threshold source strengths, the SRS fields from the 39 sites were utilized. One ^{133}Xe release per grid cell ($1° \times 1°$) and time step (3 h) was assumed and the respective threshold source strengths were calculated on the same grid (can also be infinite, meaning that even the strongest releases would not be detected within 14 days). Subsequently, a median value of the threshold source strengths for each grid point on the globe was calculated for every month during the one-year calculation period.

For the "clean" case (i.e., without considering any background from known nuclear facilities), the threshold value for a detection is assumed to be 0.1 mBqm^{-3}, which is the typical low-background MDC value for ^{133}Xe. In this case, it can be seen that the mid-latitudes of both hemispheres would be very well covered by the station network, with the threshold source strengths staying orders of magnitude below 10^{15} Bq (see Fig. 8, upper image). This means that underground nuclear explosions with yields substantially below 1-kiloton and 90% containment could still be detected in these areas. Considerably higher threshold source strengths do exist in parts of the subtropics and tropics, where average wind speeds are low and diffusion processes

become more important. There, releases of 10^{15} Bq or more may occasionally go undetected by the network. The possible expansion of the noble gas network to all RN stations would improve the coverage, although not fully remove all non-coverage areas.

Subsequently, investigation is necessary as to how the general coverage situation would be impacted in the case where a possible nuclear explosion signal would coincide with the noise created by continuous emissions from NPPs and RPFs. In such a case, there would be the MDC value for the measurement, a concentration value from known sources, and on top of it a signal from the explosion that should exceed both. Therefore, the threshold value for the detection at a sampling station is set as equal to the maximum of 0.1 mBqm^{-3} and the calculated ^{133}Xe concentrations for the respective sample, assuming the standard emissions scenario from the known sources. As a result, it becomes evident that the noise level from civilian sources mainly impacts the threshold source strengths in the mid-latitudes of the Northern Hemisphere (see Fig. 8, middle image), because the major emitters are located in these regions. The subtropical and tropical areas and the Southern Hemisphere would not be impacted strongly.

To account for (i) the possible general underestimation of some non-nuclear test related emissions (factor 2 to factor 10), and (ii) the possible influence of NPP batch emissions that do not act continuously, an additional sensitivity analysis for the impact of a further increased noise level on the final network coverage related to a possible nuclear explosion signal was conducted. This was achieved by simply multiplying the model-predicted ^{133}Xe background concentrations with a factor of 5. Also in this case, the major impact can be seen in the mid-latitudes (Fig. 8, lower image). Under this scenario, however, the model predicts that the area on the globe where a 10^{15} Bq release could go undetected during more than 50% of the time would slightly enlarge.

In general, the influence of the NPPs and RPFs on the global network coverage can be described as follows: The ^{133}Xe concentration level from civilian sources mainly influences the median value of the threshold source strengths in the mid-latitudes of the Northern Hemisphere where the general detection capability is excellent due to the strong prevailing

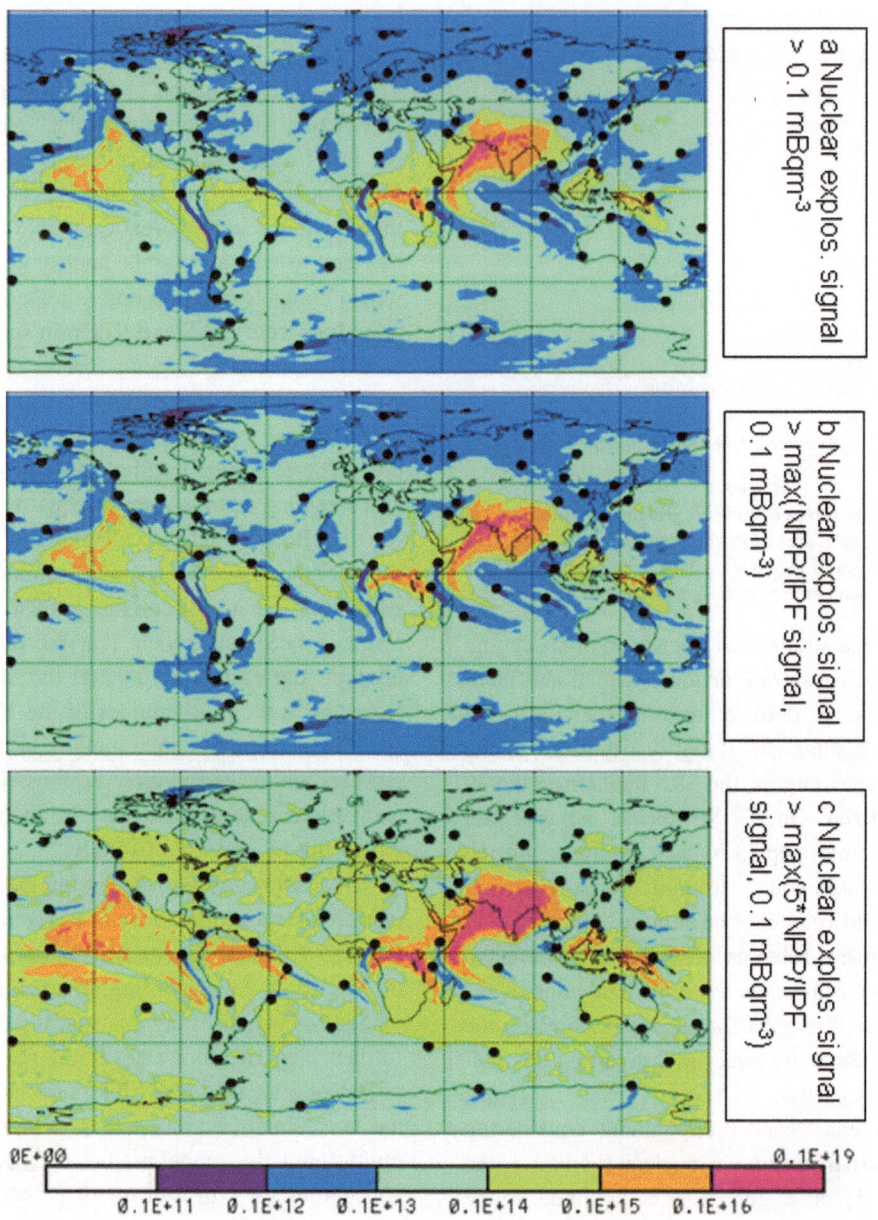

Figure 8
Geographical distribution of the threshold source strength (Bq) for ^{133}Xe based on the following assumptions: **a** detection at least at one site, detected explosion signal >0.1 mBqm^{-3}, **b** like (**a**), but detected explosion signal >0.1 mBqm^{-3} or the simulated background from the known sources, if higher, **c** like (**b**), but the simulated background from the known sources is multiplied with 5; in some tropical areas, a 1-kiloton nuclear explosion (10^{15} Bq release) would, on average, not be detected

winds. This does not substantially harm the capability of the network to detect ^{133}Xe from a 1-kiloton underground nuclear explosion under the assumption that the corresponding source term is 10^{14} to 10^{15} Bq in the first 3 h after the explosion (3 h is the time step of the SRS data utilized).

Regarding the detection capability, one should also note that the calculations utilized in this study do assume maximum transport durations of 14 days. This duration was chosen for computational reasons, taking into account the final network density as well as the dilution and the half-life time of ^{133}Xe. It is, however, still possible that explosion signals are detected after more than 14 days of transport, as seen, for example, in the DPRK case (SAEY et al., 2007).

6. Multiple Linear Regression Analysis

For the 12 IMS sites that are currently measuring xenon, a multiple linear regression analysis of daily observations with different model predictions was performed. The model predictions used for the multiple regression analysis considered the emission scenarios (i) emissions from NPPs only, emissions from Fleurus only and emissions from Chalk River only for all Northern Hemisphere stations, and (ii) emissions from NPPs only and emissions from Pelindaba only for all Southern Hemisphere stations.

Regarding the Southern Hemisphere, the two stations AUX09 and NZX46 do not detect ^{133}Xe regularly, and hence there is also no correlation to known emissions. Regarding the Northern Hemisphere, low ^{133}Xe levels and no correlation to known emitters was also seen for the two stations in CNX20 and CNX22 in China. Stations PAX50, CAX16 and NOX49 also exhibit low ^{133}Xe levels, however they show moderate correlation values to known emitters (regarding CAX16 and NOX49, see also SAEY et al., 2006).

Regarding the five stations with significant background, the situation is as follows (see Table 3): The German station DEX33 is strongly influenced by emissions from the RPF in Fleurus (in terms of the average of the model predictions) and shows a significant correlation towards this predictor ($r = 0.79$). There is also some predicted influence regarding NPP emissions there, but no correlation. The Canadian station CAX17 is, according to the model, predominantly influenced by emissions from the Chalk River facility, and there is also a significant correlation. The correlation towards the NPP emissions also seen at this site is mainly caused by the strong cross correlation between transport from the NPPs and transport from Chalk River. Regarding stations SEX63 (Sweden) and USX75 (USA), the simulations show that the predominant influence there originates from Fleurus and Chalk River, respectively, and there is a correlation regarding the respective predictor. At both stations, the simulations also exhibit impact from the NPPs, but no correlation. Finally, station JPX38 in Japan is the only one that is mainly influenced by the reactors. However, also in this case, there is no correlation whatsoever with the model predictions based on the reactor emissions.

These results can be interpreted as follows. First, the low correlation of observations with model predictions related to reactor emissions indicates that the continuous source hypothesis for NPPs is probably not fully adequate. According to the emission inventory, the total activities emitted through batch releases comprise between one-sixth and one-third of those released continuously over the year. However, the batch releases exhibit irregular and unknown time courses, and the distribution of the released activities per batch event can span

Table 3

Observed ^{133}Xe concentrations (mBqm^{-3}) and comparison with three different model predictions for all INGE stations with significant xenon background (observed median value > 0.2 mBqm^{-3})

Station	Median meas.	Mean NPP	r NPP	Mean CRL	r CRL	Mean Fleur	r Fleur
CAX17	1.2	<0.1	0.41$^+$	0.46	0.63*	<0.1	<0.10
DEX33	1.2	0.61	<0.10	<0.1	<0.10	2.57	0.79*
JPX38	0.17	0.15	<0.10	<0.1	<0.10	<0.1	<0.10
SEX63	0.63	0.2	0.12	<0.1	0.26	1.1	0.38*
USX75	0.47	0.12	<0.10	0.49	0.33	<0.1	<0.10

Average simulated concentrations and correlation coefficients considering emissions from the NPPs (Mean NPP, r NPP), from the Chalk River facility (Mean CLR, r CLR) and from the facility in Fleurus (Mean Fleur, r Fleur). Correlation coefficients marked with (*) are statistically significant on the 95% level. Correlation coefficients marked with (+) are subject to significant cross correlation with a more significant predictor. The basis for these statistics is the daily measurement values

several orders of magnitude, with the largest batch release ever reported totaling 1.1 TBq. Thus, we presume that NPP emissions cannot easily be used as criterion to categorize the noble gas measurements, unless the batch releases are known in near-real-time and can be entered into the model. Second, there is a correlation of most measurements with predictions based on emissions from Fleurus and Chalk River, and this correlation improves better the stronger the impact from these emitters on the respective site is. The continuous source hypothesis seems thus to be a reasonable first-order approximation for RPFs, and potential transport from RPFs can be better used as a criterion to flag a measurement. Third, the generally low correlations at some stations are not only an indication of other sources (e.g., some batch releases not included into the modelling, or emissions from facilities nor included in the emission inventory), but also that even the RPFs do not actually emit totally continuous. As has been reported, the emissions at these sites exhibit a time pattern in accordance with the working hours and production cycles (SAEY, 2009), and there have also been downtimes of facilities over days to weeks within the one-year time period our study refers to, that have also not been systematically considered in the modelling.

Subsequently, a multiple linear regression analysis was performed for all stations together, based on the predicted versus observed median values of ^{133}Xe for the entire one-year period. For such a long period, the daily emissions' fluctuations do not play a major role, therefore the model exhibits the correction factors to be applied to the annual emission values from the NPPs, Chalk River and Fleurus. For this analysis, model predictions based on the emissions of Pelindaba were not considered, since this facility does not affect any of the existing noble gas stations significantly. Thus, the following multiple linear regression model is formulated:

$$c_k = c_0 + a_1 c_{\mathrm{mod,\,NPP}} + a_2 c_{\mathrm{mod,\,CR}} + a_3 c_{\mathrm{mod,\,Fleur}} \quad (3)$$

with c_k being the observed median value of ^{133}Xe at station k, $c_{\mathrm{mod,\,NPP}}$ the median value of the model prediction for station k based on NPP emissions only, and $c_{\mathrm{mod,\,CR}}$, $c_{\mathrm{mod,\,Fleur}}$ the respective predictions based solely on the emissions from Chalk River and

Fleurus, respectively. c_0, a_1, a_2 and a_3 are the parameters to be specified by the regression analysis.

The analysis for the twelve stations yields the following results:

$c_0 = 0 \pm 0.04$ mBqm^{-3} (Unexplained concentration level)
$a_1 = 2.1 \pm 0.5$ mBqm^{-3} (Model bias factor, NPP)
$a_2 = 16.2 \pm 1.8$ mBqm^{-3} (Model bias factor, Chalk River)
$a_3 = 1.7 \pm 2.0$ mBqm^{-3} (Model bias factor, Fleurus)

The correlation coefficient r of the statistical relationship is 0.97, which means that about 95% of the variability of the different station measurements can be explained by the model. The unexplained concentration level (c_0) is low and reflects the minimum detectable concentrations and ^{133}Xe being transported for more than 14 days (and thus not simulated by the model). A comparison between the measured and simulated ^{133}Xe median values based on this statistical model is provided in Fig. 9 and shows very good correspondence. An analogue analysis was also conducted based on the 95% percentile values of the twelve stations rather than the median values. In this case, the model bias factor for Fleurus would be 2.4, and for Chalk River 9.6. Both parameters are significantly different from zero on the 95% confidence level.

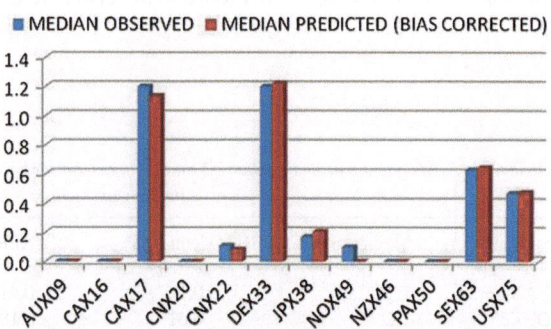

Figure 9
Comparison between predicted and observed ^{133}Xe values (mBqm^{-3}) at the 12 IMS sites that are currently sampling noble gas as part of the INGE project. The predicted concentrations were obtained using the emission bias correction factors based on the multiple linear regression model for the emissions from the NPPs, Fleurus and Chalk River ($r = 0.97$)

These results combined strongly indicate that the annual emissions from Fleurus and from the NPPs could be underestimated by a factor of 2, as also is seen in the simple linear regression analysis. The annual emissions from Chalk River, on the other hand, could be underestimated by a factor of ten to fifteen. This result means that annual ^{133}Xe emissions from all NNPs collectively numbering 1.6×10^{15} Bq, from Fleurus amounting to 2×10^{15} Bq, and from Chalk River totaling 5×10^{16} Bq are needed to best explain the measured global distribution of ^{133}Xe concentrations.

7. Simulated versus Measured Frequency Distributions of Activity Concentrations

Data from measurements and simulations are not directly comparable for two reasons. First, the measured data have a statistical error that becomes relevant for low activity concentrations. Second, the application of a critical limit (CURRIE, 1968) for distinguishing samples with significant xenon activity from zero samples can bias further statistical analysis of the data. If the null hypothesis cannot be rejected, it is current practice not to report any concentration but to give an upper limit instead. However, this reporting practice renders calculation of averages questionable.

In a different approach discussed here, the analysis method has been changed and all measurements including those below the critical limit are considered. The results are given in Table 2. This approach is relevant for stations in the Southern Hemisphere (AUX09, NZX46) and stations remote from source regions (NOX49, PAX50, CAX16). They are of minor relevance for stations CNX20, CNX22 and USX75 which are under some reactor release influence and irrelevant for stations with a high background from major sources (DEX33, SEX63, CAX17).

In Fig. 10, the measured distribution of ^{133}Xe concentrations for systems at three different types of sites is shown. DEX33 is a typical site with a high xenon background and measurements above the detection limit most days. Accordingly, the distribution is typically log-normal (Fig. 10a) and the data

below L_C play a minor role. This role becomes more important when the xenon background approaches the detection limit, as observed at station JPX38 (Fig. 10b). This station is located in a reactor region which is, unlike DEX33, not strongly influenced by an RPF. The corresponding distribution is visibly a convolution of a lognormal distribution of the real measurements with the detector response function, which is a Gaussian distribution. Finally, the station AUX09 in Australia is remote from significant emitters and therefore displays a practically zero background (Fig. 10c). The resulting ^{133}Xe distribution is the pure detector response function. A small asymmetrical shape can be seen and there are indeed few data points that are not explained by pure detector statistics. Thus the analysis of the distribution corroborates that there is indeed some radioxenon present at the site. Stations USX75 (Charlottesville, USA; Fig. 10d) and CAX17 (St. Johns, Canada; Fig. 10e) exhibit a ^{133}Xe distribution similar to DEX33.

Examining the simulated frequency distribution of data from station DEX33, the bias-corrected model results based on the results of multiple linear regression analysis (multiplication of emissions from NPPs and RPFs according to the model bias factors obtained) show a good agreement with the measurements (see Fig. 11a). The log-normal character of the distribution is well represented.

A similar comparison of the measured and simulated distribution for sites with low xenon concentration such as, for example, JPX38 needs a more sophisticated approach: Values below the critical limits have to be included and the uncertainty of the detector system also has to be considered in the simulation. The activity concentrations calculated with an atmospheric transport model were convoluted with the appropriate detector response function, namely a Gaussian distribution with zero average and a sigma $\sigma = 0.118$ mBqm^{-3} derived from the uncertainty of the measurements. The accordingly corrected model data set (Fig. 11c) matches the observed distribution better than the original one (bias-correction only; Fig. 11b). The good match affirms that the total amount of assumed releases in this region is consistent with observations. The simulation could not reproduce, however, observed

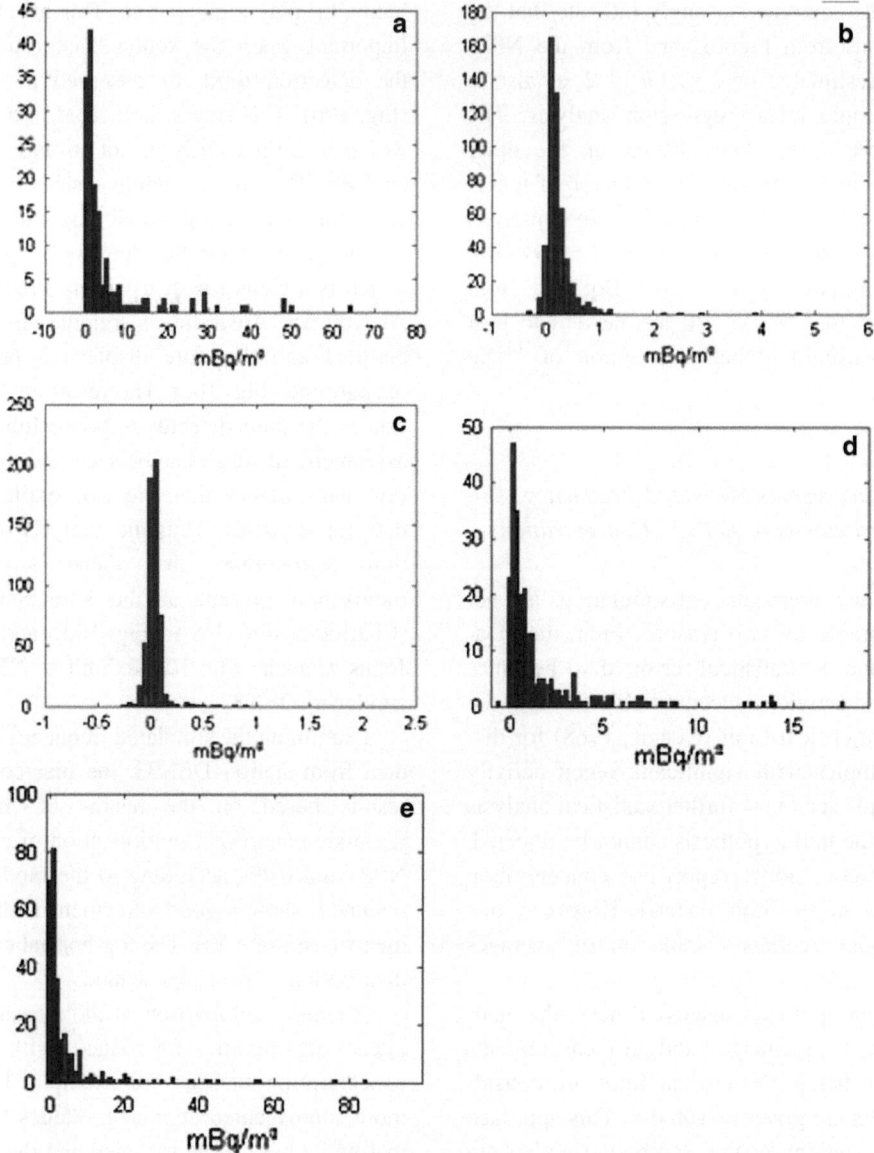

Figure 10
Observed frequency distributions of ^{133}Xe concentrations [mBqm^{-3}] at stations DEX33 (**a**), JPX38 (**b**), AUX09 (**c**), USX75 (**d**) and CAX17 (**e**)

extreme values at this station. This again indicates that the extreme values are caused by individual batch releases from NPPs located in this region, which are currently not reported with their start and stop times and thus only included in the model as part of the annual emission sum.

A similar comparison regarding the two North American stations USX75 (Fig. 11d) and CAX17 (Fig. 11e) also shows that the corrected model results

with Gaussian error function convolution yield good comparisons between measured and simulated concentration distributions. This result further supports the finding that the emissions from the facility in Chalk River are a factor of 15 higher compared with the original emission model.

A comparison of simulated concentrations at site AUX09 was not possible because the simulated data were practically zero. The correct interpretation of

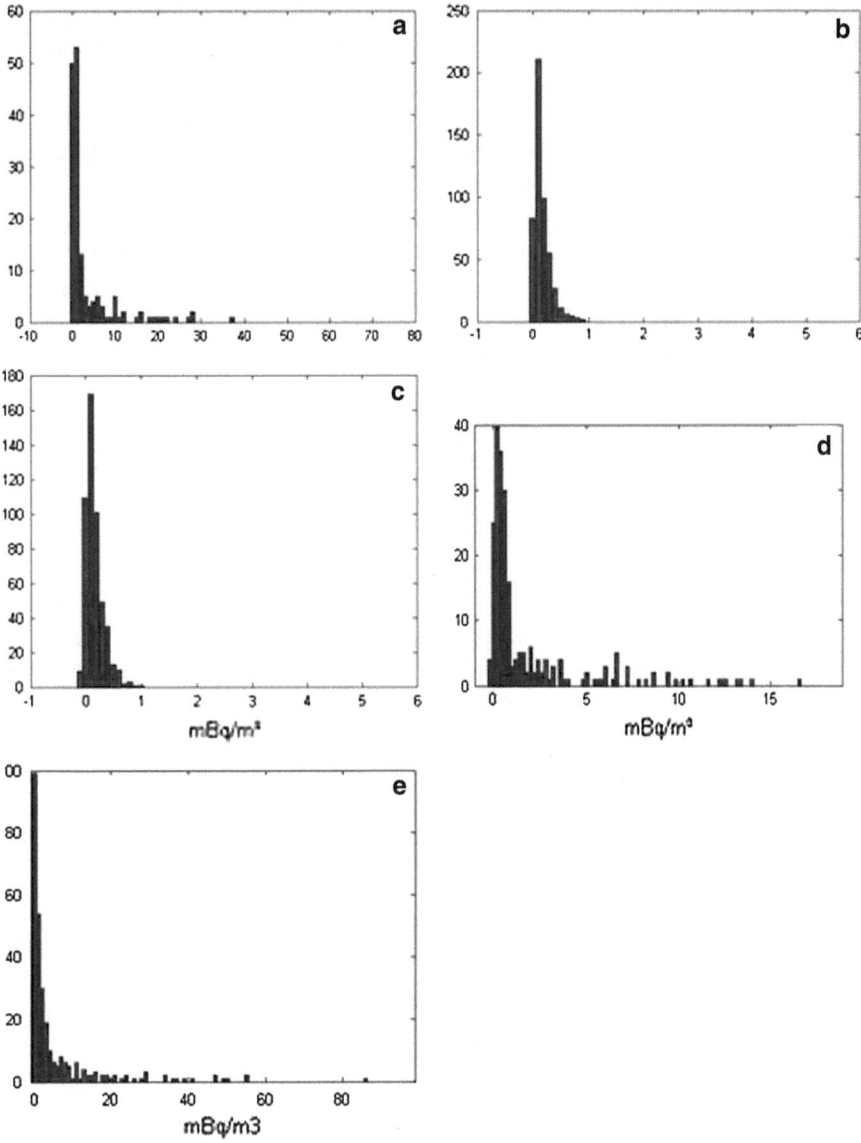

Figure 11

Simulated distribution of ^{133}Xe (mBqm^{-3}) at stations DEX33 (**a** transport only, bias-corrected), JPX38 (**b** transport only, bias-corrected), JPX38 (**c** transport plus detector error), USX75 (**d** transport only, bias-corrected) and CAX17 (**e** transport only, bias-corrected)

the occurrence of small but real ^{133}Xe concentrations needs further investigation. The ^{133}Xe there may come from sources not included in the emission inventory (e.g., medical sources).

8. Summary and Conclusions

Known emissions of radioxenon isotopes, in particular ^{133}Xe, are an important factor of influence for

the noble gas monitoring network constituted as part of the verification system foreseen under the provisions of the Comprehensive Nuclear-Test-Ban Treaty. After a 1 kiloton underground nuclear explosion, up to 10^{15} Bq of ^{133}Xe is available to be injected into the atmosphere. Emissions of the same order of magnitude are estimated from all civilian nuclear power reactors collectively, and also for each of a few radioisotope production facilities. These emissions, however, are spread more or less continuously over a period of 1 year. The understanding of the influence of known emissions on the observed radioxenon level is of crucial importance for treaty verification.

First, with a simplified model based on the folding of global $1° \times 1°$ source receptor sensitivity fields calculated with the atmospheric transport model FLEXPART for all available measurement data with an $1° \times 1°$ emission inventory, which was resampled assuming continuous releases from the above-mentioned nuclear facilities, it was possible to describe the global concentration distribution of ^{133}Xe reasonably well. Also the observed concentration level matched reasonably well with the predictions, with strong indications that the real emissions may be higher than the assumed ones by a factor of 2. For the facility in Chalk River (Canada) the emissions may even be higher by more than one order of magnitude. Second, it was demonstrated that the known emissions do influence the detection capability of the verification network mainly in the mid-latitudes of the Northern Hemisphere, where the capability is significantly better than the one assumed during the network design computations in the phase the treaty was negotiated. In the other regions, the ability of the network to detect xenon from a partly contained 1-kiloton underground nuclear explosion is not further impaired by the noise created by the known sources. It was also demonstrated that about two-thirds of all determined future monitoring sites are unlikely to be regularly influenced by known emissions. Third, it became clear that especially the emissions from the facilities in Fleurus (Belgium) and in Chalk River contribute strongly to the correlation between simulations and measurements at the downwind stations. On the other hand, emissions from civilian power reactors were found to contribute much less to the correlation of model results with observations. This is even true for stations situated within the reactor belts. This means that event categorization schemes for noble gas measurements should make use of transport information related to the isotope production facilities. Regarding the power reactors, however, such an approach may not be generally useful. Continuous emissions from the power plants are clearly important to explain the mean concentration level, nonetheless individual peaks are rather

attributable to irregular batch emissions. The timing of these emissions is not globally known and thus only estimates of the sum of the batch emissions per year could be used in our model. Fourth, predicted distributions of measurement values for all stations under testing were compared with the measured distributions, and an overall good correspondence was found. All the basic features of the distribution observed at sites with a high, medium and low ^{133}Xe level were well resembled. Furthermore, the comparison between model and observations was further improved by correcting the model results with the detector response function; a method that allows providing for measurement values below the detection limit at stations where the ^{133}Xe level is low. The analysis of the distribution also confirmed the possible under-prediction of emissions from certain isotope production facilities, as well as the potential influence of the irregular batch emissions at stations located in areas with a high density of nuclear power reactors.

To further improve understanding on the global xenon levels, future studies could consider temporal variations of emissions and production cycles from the facilities in Chalk River and Fleurus. Furthermore, an attempt to better account for batch releases from reactors statistically, based on the emission studies available, could be tried. Finally, the regression analysis could be further enhanced by including not only daily and annual observation values, but also monthly median values from each station.

REFERENCES

AUER, M. et al. (2004), *Intercomparison experiments of systems for the measurement of xenon radionuclides in the atmosphere*, Appl. Radiat. Isot. *60*(6), 863–877.

BECKER, A. et al. (2007), *Global backtracking of anthropogenic radionuclides by means of a receptor oriented ensemble dispersion modeling system in support of Nuclear-Test-Ban Treaty verification*, Atmos. Environ. *41*, 4520–4534.

BECKER, A., WOTAWA, G., RINGBOM, A. and SAEY, P.J.R. (2010), *Backtracking of noble gas measurements taken in the aftermath of the announced October 2006 event in North Korea by means of PTS methods in nuclear source estimation and reconstruction*, Pure Appl. Geophys. *167*, 415, 2010, in press.

BOWYER, T.W., MCINTYRE, J.I., and REEDER, P. L. (1999), *High-sensitivity detection of Xenon isotopes via beta-gamma coincidence counting*, Abstract for the 21st Seismic Res. Symp.,

Technologies for Monitoring the Comprehensive Nuclear-Test-Ban Treaty, Las Vegas, USA, September 21–24.

CTBT (1996), *Text of the Comprehensive Nuclear-Test-Ban Treaty*. See Web Page of the United Nations Office for Disarmament Affairs (UNODA), Status of Multilateral Arms Regulation and Disarmament Agreements, CTBT (http://disarmament.un.org/TreatyStatus.nsf).

CURRIE, L.A. (1968), *Limits for qualitative detection and quantitative determination. Application to radiochemistry*, Analy. Chem. *40*(3), 586–593.

DE GEER, L.-E. *Atmospheric radionuclide monitoring: A Swedish perspective, in Monitoring a Comprehensive Nuclear-Test-Ban Treaty*, eds. E. S. Huseby and A. M. Dainty, pp. 157– 177 (Kluwer Acad., Dordrecht, Netherlands 1996).

DUBASOV, Y.V., POPOV, Y.S., PRELOVSKII, V.V., DONETS, A.Y., KAZARINOV, N.M., MISHURINSKII, V. V., POPOV, V.Y., RYKOV, Y.M., and SKIRDA, N.V. (2005), *The АРИКС-01 Automatic Facility for Measuring Concentrations of Radioactive Xenon Isotopes in the Atmosphere*, Instruments Experim. Techniques *48*, 3, 373–379.

FONTAINE, J. P., POINTURIER, F., BLANCHARD, X., and TAFFARY, T. (2004), *Atmospheric Xenon radioactive isotope monitoring*, J. Environ. Radiat. *72*, 129–135.

HOFFMANN, W., KEBEASY, R., and FIRBAS, P. (2000), *Introduction to the verification regime of the Comprehensive Nuclear-Test-Ban Treaty*, Phys. Earth Planet. Inter. *113*, 5–9.

KALINOWSKI, M.B., BECKER, A., SAEY, P.J.R., TUMA, M., and WOTAWA, G. (2008), *The complexity of CTBT verification. Taking noble gas monitoring as an example*, Complexity *14*, 1, 89–99.

KALINOWSKI, M. and TUMA, M. (2009), *Global radioxenon emission inventory based on nuclear power reactor reports*, J. Environ. Radioact. *100*, 58–70.

PERKINS, R. W., and CASEY, L. A. (1996), *Radioxenons: Their role in monitoring a Comprehensive Test-Ban Treaty*, Rep. DOE/RL-96-1, Pac. Northwest Natl. Lab., Richland, Wash.

RIECHMANN, B. and KALINOWSKI, M. (2008), *Implications of Xe emissions from medical isotope production*, Geophys. Res. Abstracts, Vol. 10, EGU2008-A-07898.

RINGBOM, A., LARSON, T., AXELSSON, A., ELMGREN, K., and JOHANSSON, C. (2003), *SAUNA: A system for automatic sampling, processing and analysis of radioactive xenon*, Nucl. Instrum. Methods, Phys. Res., Sect. A, *508*, 542–553.

RINGBOM, A., ELMGREN, K. and LINDH, K. (2007), *Analysis of radioxenon in ground level air sampled in the Republic of South Korea on October 11–14, 2006*, FOI Report. FOI-R–2273—SE.

SAEY, P. R. J., WOTAWA, G., DE GEER, L.-E., AXELSSON, A., BEAN, M., d'AMOURS, R., ELMGREN, K., PETERSON, J., RINGBOM, A., STOCKI, T. J., and UNGAR, R. K. (2006), *Radioxenon background at high northern latitudes*, J. Geophys. Res. *111*, D17306, doi: 10.1029/2005JD007038.

SAEY, P. R. J., BEAN, M., BECKER, A., COYNE, J., d'AMOURS, R., DE GEER, L.-E., HOGUE, R., STOCKI, T. J., UNGAR, R. K., and WOTAWA, G. (2007), *A long distance measurement of radioxenon in Yellowknife, Canada, in late October 2006*, Geophys. Res. Lett. *34*, L20802, doi:10.1029/2007GL030611.

SAEY, P.R.J. (2009): *The influence of radiopharmaceutical isotope production on the global radioxenon background*, J. Environm. Radioact. doi:10.1016/j.jenvrad.2009.01.004, in press.

SAEY, P.R.J. and DE GEER, L.-E. (2005), *Notes on radioxenon measurements for CTBT verification purposes*, Appl. Radiat. Isot. *63*, 765–773.

SCHULZE, J., AUER, M., and WERZI, R. (2000), *Low level radioactivity measurement in support of the CTBTO*, Appl. Radiat. Isot. *53*, 23–30.

STOCKI, T.J., BLANCHARD, X., D'AMOURS, R., UNGAR, R.K., FONTAINE, J.P., SOHIER, M., BEAN, M., TAFFARY, T., RACINE, J., TRACY, B.L., BRACHET, G. JEAN, M., and MEYERHOF, D. (2005), *Automated radioxenon monitoring for the Comprehensive Nuclear-Test-Ban Treaty in two distinct locations: Ottawa and Tahiti*, J. Environm. Radioact. *80*, 305–326.

STOHL, A., FORSTER, C., FRANK, A., SEIBERT, P., and WOTAWA, G. (2005), *Technical note: The Lagrangian particle dispersion model FLEXPART version 6.2*, Atmos. Chem. Phys. *5*, 2461–2474.

WERNSBERGER, B. and SCHLOSSER, C. (2004), *Noble gas monitoring within the international monitoring system of the Comprehensive Nuclear Test-Ban Treaty*, Radiation Phys. Chem. *71*, 775–779.

WOTAWA, G. et al. (2003), *Atmospheric transport modelling in support of CTBT verification: Overview and basic concepts*, Atmos. Environ. *37*(18), 2529–2537.

WOTAWA, G., DE GEER, L.-E., BECKER, A., D'AMOURS, R., JEAN, M., SERVRANCKX, R., and UNGAR, K. (2006), *Inter- and intra-continental transport of radioactive cesium released by boreal forest fires*, Geophys. Res. Lett. *33*, L12806, doi:10.1029/2006GL026206.

(Received December 17, 2008, revised March 27, 2009, accepted March 31, 2009, Published online January 19, 2010)

Pure Appl. Geophys. 167 (2010), 559–573
© 2009 Birkhäuser Verlag, Basel/Switzerland
DOI 10.1007/s00024-009-0030-3

Radioxenon Time Series and Meteorological Pattern Analysis for CTBT Event Categorisation

WOLFANGO PLASTINO,[1,2] ROMANO PLENTEDA,[3] GEORGE AZZARI,[1] ANDREAS BECKER,[3] PAUL R. J. SAEY,[3]
and GERHARD WOTAWA[3]

Abstract—Understanding radioxenon time series and being able to distinguish anthropogenic from nuclear explosion signals are fundamental issues for the technical verification of the Comprehensive Nuclear-Test-Ban Treaty. Every radioxenon event categorisation methodology must take into account the background at each monitoring site to uncover anomalies that may be related to nuclear explosions. Feedback induced by local meteorological patterns on the equipment and on the sampling procedures has been included in the analysis to improve a possible event categorisation scheme. The occurrence probability of radioxenon outliers has been estimated with a time series approach characterising and avoiding the influence of local meteorological patterns. A power spectrum estimator for radioxenon and meteorological time series was selected; the randomness of the radioxenon residual time series has been tested for white noise by Kolmogorov–Smirnov and Ljung–Box tests. This methodological approach was applied to radioxenon data collected at two monitoring sites located at St. John's, Canada and Charlottesville, USA, equipped with two different noble gas systems. It shows different feedback with local meteorological patterns and randomness for the radioxenon data recorded at the selected sites of St. John's and Charlottesville as well as a different occurrence probability of the outliers in the normalized radioxenon original and residual time series.

Key words: Radioxenon, Time Series, CTBT event categorisation.

1. Introduction

The International Monitoring System (IMS) is a verification component of the Comprehensive Nuclear-Test-Ban Treaty (CTBT). It includes, beside three-waveform technologies (seismic, hydro-acoustic and infrasound), two radionuclide technologies: global monitoring of radioactive aerosols and of radioactive noble gases. Atmospheric transport modelling is part of the system to establish a source geolocation capability.

The knowledge of the activity concentration and isotopic composition of radioactive noble gases in the atmosphere indicates the nuclear processes governing their formation. Furthermore, by use of atmospheric transport modelling (ATM), knowledge of possible source characteristics, points of origination and potential contamination from other sources in the area of the sampling point can be established.

The noble gas-monitoring technology is a fundamental and highly sensitive technique for the detection of nuclear explosions (PERKINS and CASEY, 1996; BOWYER *et al.*, 2002; SAEY and DE GEER, 2005; SAEY, 2007). It is the only technique, together with radionuclide particulate monitoring, that has the potential to provide unambiguous proof as to whether an explosion was nuclear or not (DE GEER, 1996; SCHULZE *et al.*, 2000).

In this framework, the International Noble Gas Experiment (INGE) of the Provisional Technical Secretariat (PTS) for the CTBT Organization (CTBTO) was established to develop and test suitable radioxenon monitoring systems (AUER *et al.*, 2004; SAEY and DE GEER, 2005; SAEY *et al.*, 2007). In order to ensure the quality and accuracy of the IMS noble gas measurement capabilities, it is of eminent importance to characterise the noble gas background (BOWYER *et al.*, 1997; WEISS *et al.*, 1997; IGARASHI *et al.*, 2000; STOCKI *et al.*, 2005; KALINOWSKI and PISTNER, 2006; SAEY *et al.*, 2006; KALINOWSKI *et al.*, 2008; SAEY, 2009). It is important for the assessment

---
[1] Department of Physics, University of Roma Tre, Rome, Italy. E-mail: plastino@fis.uniroma3.it
[2] National Institute of Nuclear Physics, Section of Roma Tre, Rome, Italy.
[3] Preparatory Commission for the Comprehensive Nuclear-Test-Ban Treaty Organisation, Provisional Technical Secretariat, P.O. Box 1200, 1400 Vienna, Austria.

of measurements also to calculate and analyse the source–receptor relationships (WOTAWA *et al.*, 2003; STOHL *et al.*, 2005; BECKER *et al.*, 2007).

In late October 2006, after the low yield announced underground nuclear explosion on the Korean peninsula, it needed the integrated evaluation of long-distance measurements of radioxenon in Yellowknife, Canada, backward ATM (SAEY *et al.*, 2007), and seismic data (KIM and RICHARDS, 2007) to show their consistency with assumed simple leak scenarios. The closest noble gas stations delivering data during that period were in Ulaanbaatar (Mongolia) in Spitsbergen (Norway) in Stockholm (Sweden) and in Yellowknife (Canada). The relevant atmospheric circulation was generally from west to east disfavouring the first three sites while favouring the last, even though it is more than 7,000 km away (SAEY *et al.*, 2007). Nevertheless, long-range transport of particulate radionuclides from Asia to this site has been observed in the past (WOTAWA *et al.*, 2006) and the reconstruction of the originally simple leak scenario has been confirmed and refined by Becker et al. (2009) by integration of closer mobile noble gas measurements (RINGBOM *et al.*, 2007) taken on the Korean peninsula.

A methodological approach for event screening of radioxenon time series (categorising) detected by the INGE noble gas stations needs to be developed and tested. Such categorisation is fundamental for scenarios different from that which occurred on the Korean peninsula in late October 2006, particularly for undeclared and disguised nuclear tests. Therefore, it is of at most importance to characterise all the components of the background, the trend patterns and outliers of the radioxenon time series, and the physical–chemical processes, which could affect the sampling efficiency at the measurement site.

This paper studies the possible link between meteorological patterns at the sampling site and radioxenon activity concentration data. The methodological approach uses a power-spectrum estimator for the radioxenon and the meteorological time series. The coupled frequencies in both series are characterised to screen them out from the original ones. A white noise process is applied to test the randomness of the radioxenon residual time series by means of Kolmogorov–Smirnov and Ljung–Box

tests, and to estimate the occurrence probability of radioxenon outliers. The meteorological pattern at the sampling site plays a crucial role for radioxenon event categorisation. Its possible links with local known sources should be considered and analysed for a complete characterisation of the radioxenon time series.

2. Data Analysis and Results

The INGE noble gas systems extract xenon out of the air and then measure the activity concentration of four isotopes/isomers, namely ^{131m}Xe, ^{133m}Xe, ^{133}Xe and ^{135}Xe with different measurement techniques. The "Système de Prélèvement d'air Automatique en Ligne avec l'Analyse des radioxenon" (SPALAX) uses a high purity germanium detector for the radioactivity measurement (FONTAINE *et al.*, 2004), while the "Swedish Automatic Unit for Noble Gas Acquisition" (SAUNA) system uses a beta–gamma coincidence detector (RINGBOM *et al.*, 2003). After the measurement, the spectra are sent to the International Data Centre (IDC) in Vienna, Austria, for processing and analysis. The following selection criteria for the radioxenon data used in this study (Fig. 1 and Table 1) were applied to collection period September 2005 to June 2008:

- radioxenon data availability (Fig. 2);
- radioxenon isotopes presence (Figs. 3, 4);
- meteorological data availability at the station: air temperature, air pressure, relative humidity, rainfall, wind direction and wind speed.

Two stations successfully passed all screening criteria: CAX17 (St. John's, Canada), using a SPALAX system and USX75 (Charlottesville, USA), using a SAUNA system. In Fig. 5, the ^{133}Xe time series for both stations are shown for the samples with activity concentrations above the critical limit (Figs. 3, 4). The data of Charlottesville comprise series 1 and 2 (Fig. 2).

The aim of this analysis is to characterise the meteorological patterns at the measurement site and their effects on sampling efficiency and on the time series. Radioxenon concentration could be affected by local sources and meteorological conditions and

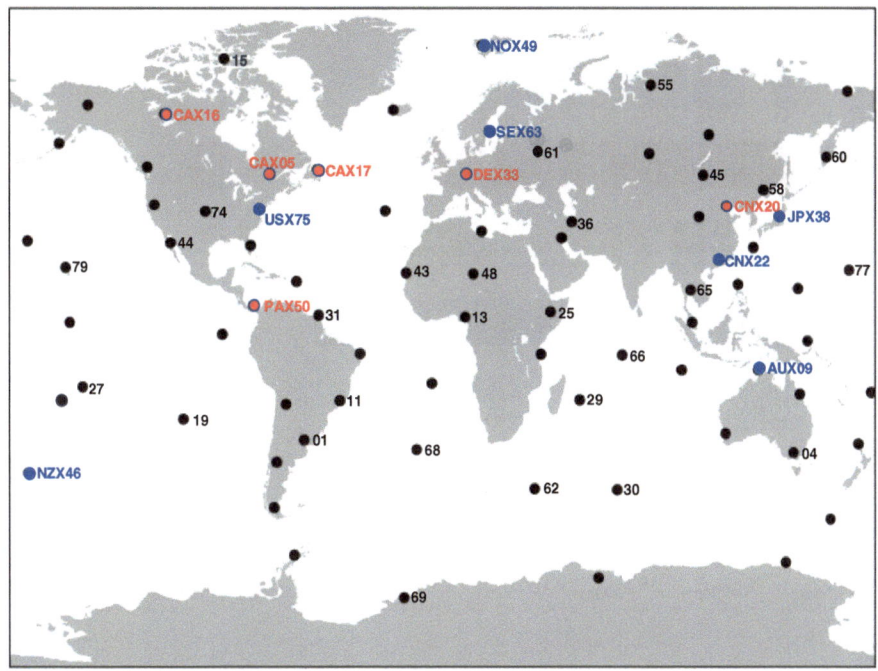

Figure 1

Map showing the location of all the active SPALAX and SAUNA stations in the INGE network in June 2008. The *colour codes* indicate the system type: *red*—SPALAX, *blue*—SAUNA. All other scheduled INGE network locations are indicated by *numbered black dots*. *Unnumbered black dots* indicate IMS RN stations scheduled to have no noble gas measurement capability (colour figure online)

Table 1

Latitude, longitude and elevation of the location for all the SPALAX and SAUNA stations in the INGE network in June 2008

Station	Location	Latitude	Longitude	Elevation (m)
SPALAX Systems				
CAX05	Ottawa, Canada	45.374	−75.686	93
CAX16	Yellowknife, Canada	62.476	−114.469	206
CAX17	St. John's Canada	47.590	−52.740	133
CAX20	Beijing, China	39.950	116.400	40
PAX50	Panama	8.980	−79.530	90
DEX33	Schauinsland, Germany	47.920	7.910	1,208
SAUNA Systems				
AUX09	Darwin, Australia	−12.400	130.700	10
CNX22	Guangzhou, China	23.100	113.300	12
JPX38	Takasaki, Gunma, Japan	36.310	139.080	92
NOX49	Spitsbergen, Norway	78.230	15.390	469
NZX46	Christchurch, NZ	−43.820	176.480	22
SEX63	Stockholm, Sweden	59.380	17.950	54
USX75	Charlottesville, VA	38.000	−78.400	104

catch regional sources, it is very important to discriminate to the best extent possible all the local components at the measurement site that could modify the radioxenon pattern. The occurrence probability of radioxenon outliers has been estimated with a time series approach and has been computed, characterising and avoiding the influence of local meteorological patterns by selecting a power-spectrum estimator for radioxenon and meteorological time series. The coupled frequencies in both series and their possible combinations have been screened out from the original ones. The randomness of the radioxenon residual time series has been tested for their white noise process by means of Kolmogorov–Smirnov and Ljung–Box tests.

The power-spectrum estimator of the radioxenon and meteorological time series has been selected in the Blackman–Tukey method which helps reduce the estimate's variance and bias and attenuate the leakage effects of the periodogram. This method is an indirect estimator, as the power is calculated from the covariance function (CHATFIELD, 1991)

their stochastic coupling, which could generate outliers or background variations in the radioxenon time series. Therefore, to better understand and

189   Reprinted from the journal

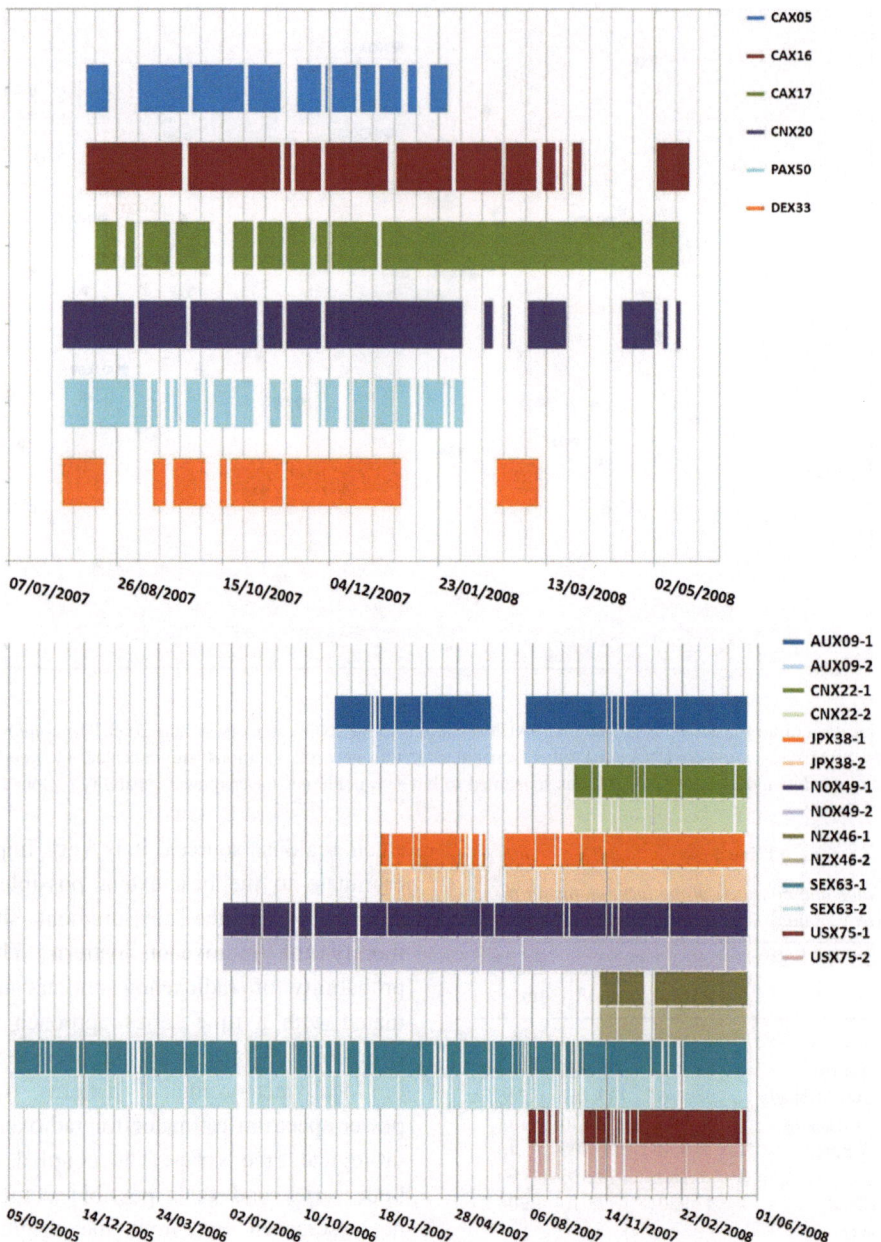

Figure 2

Radioxenon data availability of the INGE noble gas stations SPALAX (*top*): CAX05 (Ottawa, Canada), CAX16 (Yellowknife, Canada), CAX17 (St. John's, Canada), CNX20 (Beijing, China), PAX50 (Panama City, Panama), DEX33 (Schauinsland-Freiburg, Germany); and for SAUNA (*bottom*): AUX09 (Darwin, Australia), CNX22 (Guangzhou, China), JPX38 (Takasaki-Gunma, Japan), NOX49 (Spitsbergen, Norway), NZX46 (Chatham Island, New Zealand), SEX63 (Stockholm, Sweden), USX75 (Charlottesville, USA)

$$\widehat{S}(\omega) = \frac{1}{\pi} \left\{ \lambda(0)\widehat{C}(0) + \sum_{k=1}^{M} \lambda(k)\widehat{C}(k)\cos(\omega k) \right\}, \quad (1)$$

where $\widehat{S}(\omega)$ is the estimated power spectrum for frequency ω, $\widehat{C}(k)$ the estimated covariance function

for the kth lag and $\lambda(k)$ is the weighting function, known as lag-window, which is used to assign less weight to the covariance estimates as the lag increases. It is noticeable that the estimated covariance function is less reliable for large lags. The Tukey

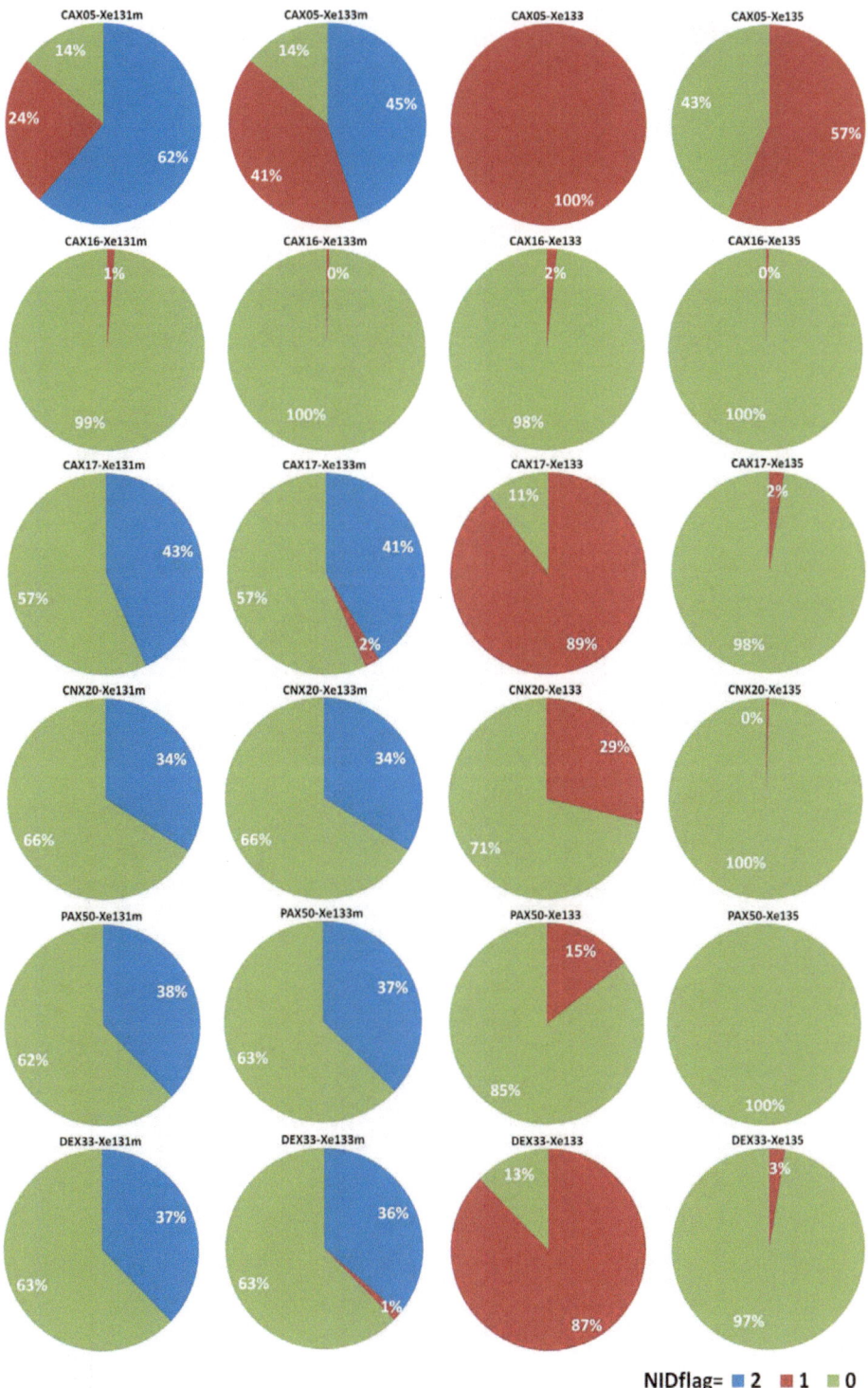

Figure 3

Radioxenon isotopes (^{131m}Xe, ^{133m}Xe, ^{133}Xe, and ^{135}Xe) presence by NIDflag codes (*NID* nuclide identification; *2*—0 but X-rays above critical limit, *1*—nuclide found above critical limit at key line position, *0*—no nuclide found above critical limit at key line position) of the INGE noble gas stations SPALAX: CAX05 (Ottawa, Canada), CAX16 (Yellowknife, Canada), CAX17 (St. John's, Canada), CNX20 (Beijing, China), PAX50 (Panama City, Panama), DEX33 (Schauinsland-Freiburg, Germany)

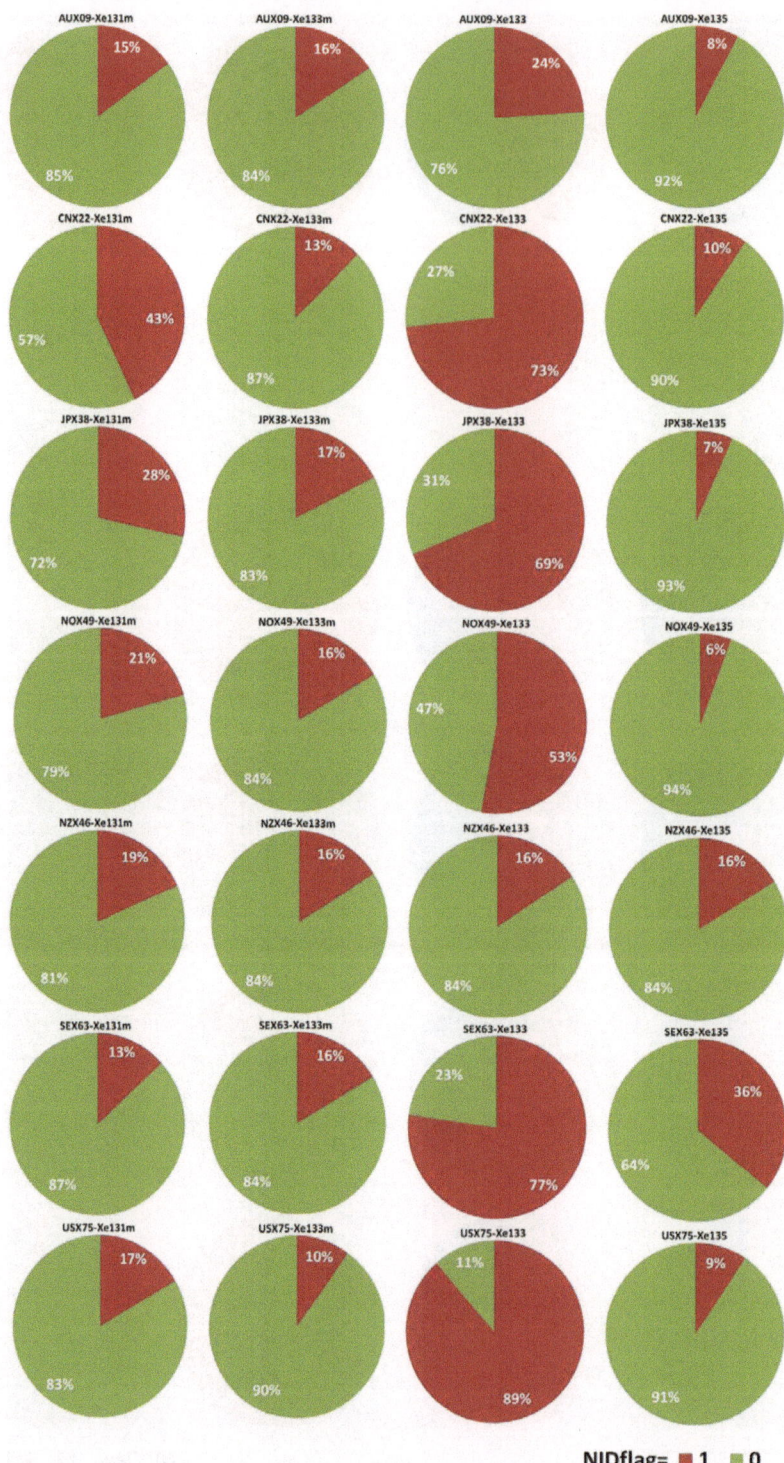

NIDflag= ■ 1 ■ 0

Figure 4
Radioxenon isotopes (^{131m}Xe, ^{133m}Xe, ^{133}Xe, and ^{135}Xe) presence by NIDflag codes (*NID* nuclide identification; *1*—nuclide found above critical limit at key line position, *0*—no nuclide found above critical limit at key line position) of the INGE noble gas stations SAUNA comprising series 1 and 2 (Fig. 2): AUX09 (Darwin, Australia), CNX22 (Guangzhou, China), JPX38 (Takasaki-Gunma, Japan), NOX49 (Spitsbergen, Norway), NZX46 (Chatham Island, New Zealand), SEX63 (Stockholm, Sweden), USX75 (Charlottesville, USA)

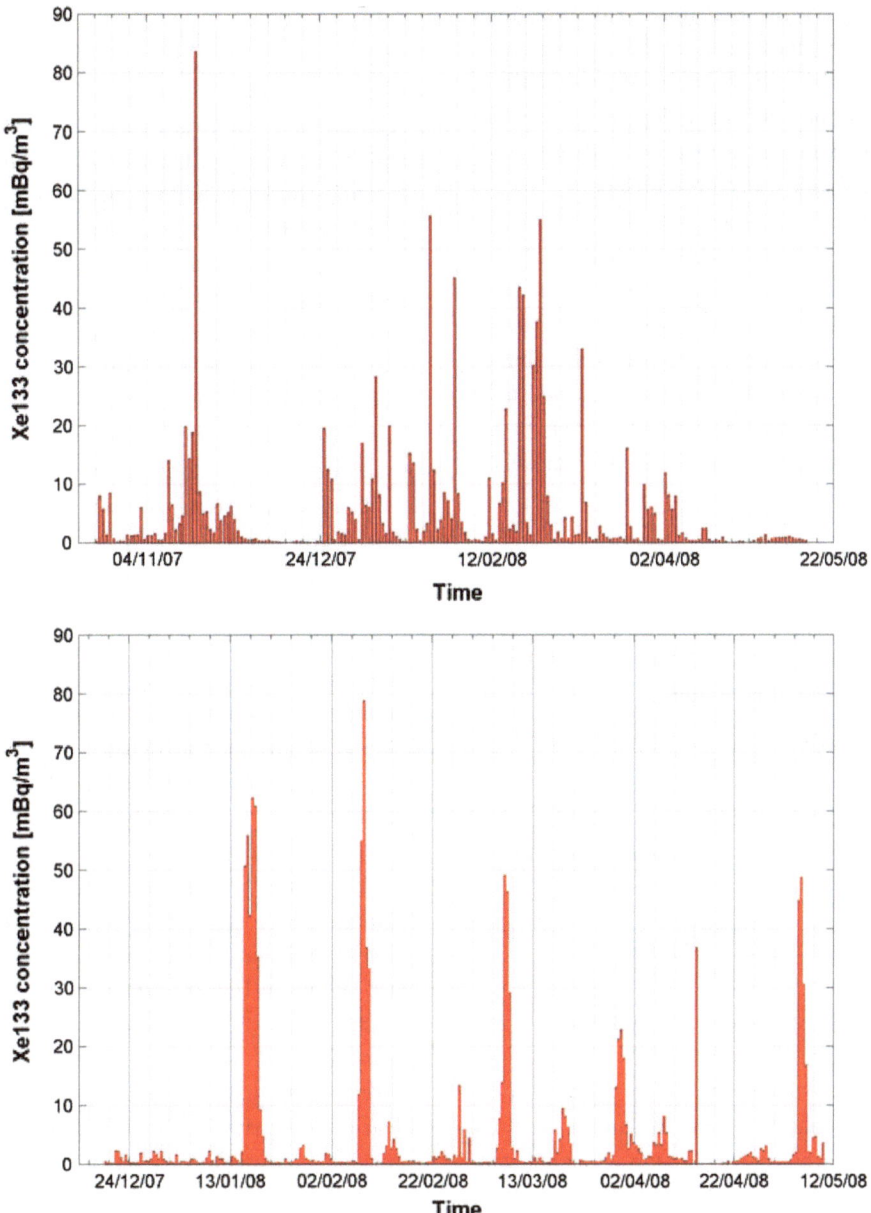

Figure 5
Radioxenon time series for the selected INGE noble gas stations located in St. John's, Canada (*top*), and located in Charlottesville, USA (*bottom*)

window has been selected as lag-window (CHATFIELD, 1991):

$$\lambda(k) = \frac{1}{2}\left\{1 + \cos\left(\frac{\pi k}{M}\right)\right\} \quad 0 \le k \le M. \quad (2)$$

M is the maximum number of lags for the covariance function used in the spectral estimation. The maximum number of lags is $N - 1$, with N being the number of the experimental data; however, with large values of M a great number of peaks will be seen in the estimated power spectrum, most representing spurious cycle. On the other hand, if M is very small, significant cycles would not be seen in the estimated power spectrum. For this reason a value of $M = N/2$ has been used in order to resolve peaks, and a value of $M = N/4$ to find out which were the most significant peaks. For the Blackman–Tukey estimate with a Tukey lag-window, the number of degrees of freedom is 2.67 N/M, with $\chi^2_{v,\alpha}$ being the α quantile of a Chi-square distribution with v degrees of freedom and α being the significance level: confidence levels of 90, 95 and 99% have been established.

The Kolmogorov–Smirnov test (K–S) compares an empirical distribution function with the distribution function of the hypothesized distribution (LAW and KELTON, 1991; VON STORCH and ZWIERS, 1999). Therefore, for the K–S test, an empirical distribution function $F_n(x)$ from the experimental data and a fitting distribution function $\widehat{F}(x)$ are defined: the natural assessment for goodness of fit is the closeness between these functions

$$D_n = \sup_x \left\{ \left| F_n(x) - \widehat{F}(x) \right| \right\}. \qquad (3)$$

To estimate μ and σ^2 of the hypothesized distribution $N(\mu, \sigma^2)$ by $\bar{X}(n)$ and $S^2(n)$ the distribution function $\widehat{F}(x)$ is defined as follows:

$$\widehat{F}(x) = \Phi \left\{ \left[x - \bar{X}(n) \middle/ \sqrt{S^2(n)} \right] \right\}, \qquad (4)$$

where Φ is the distribution function of the standard normal distribution. The hypothesized distribution is rejected if

$$\left(\sqrt{n} - 0.01 + \frac{0.85}{\sqrt{n}} \right) \; D_n > c'_{1-\alpha}, \qquad (5)$$

where α has been selected 0.05 and thus $c'_{1-\alpha} = 0.8950$ (LAW and KELTON, 1991).

The Ljung–Box statistic has been used to investigate model adequacy, i.e., to ascertain that the residuals are white noise: zero mean, constant variance, uncorrelated process and normally distributed (BROCKWELL and DAVIS, 2002). The Ljung–Box test is based on the autocorrelation function

$$Q_{LB} = n(n + 2) \sum_{j=1}^{h} \frac{\rho^2(j)}{n - j}, \qquad (6)$$

where n is the sample size, $\rho(j)$ is the autocorrelation at lag j, and h is the number of lags tested.

The Ljung–Box statistics reports the correlogram of the model residuals. The null hypothesis of white noise residuals is rejected if

$$Q_{LB} > \chi^2_{1-\alpha,h}, \qquad (7)$$

where χ^2 is the percent point function of the Chi-square distribution for degree of freedom h (the lag), and α is the significance level. Furthermore, the residual autocorrelations are also used to test for goodness of fit in the frequency domain.

The power spectra of the ^{133}Xe time series and their coupling with the frequencies of the local meteorological parameters are shown in Fig. 6 for the St. John's and Charlottesville stations, respectively. A different spectral density characterises two monitoring sites above the white noise threshold spectrum calculated by Monte Carlo simulation as well as a different coupling in the frequencies bandwidth from 0.06 to 0.15 cycles per day. Although a common meteorological pattern due to prevalent air temperature and wind direction for both stations was identified, only Charlottesville emphasised a specific feedback with relative humidity in the above-mentioned bandwidth as well as a residual spectral density decoupled by meteorological parameters above the white noise threshold in the frequencies bandwidth from 0.20 to 0.28 cycles per day.

Therefore, the radioxenon residual time series, detrended by local meteorological parameters using a filtering technique with cutoff coincidence frequencies shown in Fig. 6 (red lines), have been analysed to characterise the randomness and the occurrence probability of the outliers. These are defined as observations that lie outside the overall pattern of a distribution (MOORE and MCCABE, 2006), and from this data analysis the outlier on the white noise series obtained was identified when the modulus was greater than three times the estimated standard deviation, i.e., confidence level of 99.7%.

In Figs. 7 and 8 the distribution functions of the radioxenon original and residual time series, the fit of the residual one with a normal distribution by the K–S test, and the L–B test for the original and residual series are shown for the St. John's and Charlottesville stations, respectively. Particularly, the residual time series with calculated outliers for the selected systems are shown in Fig. 9.

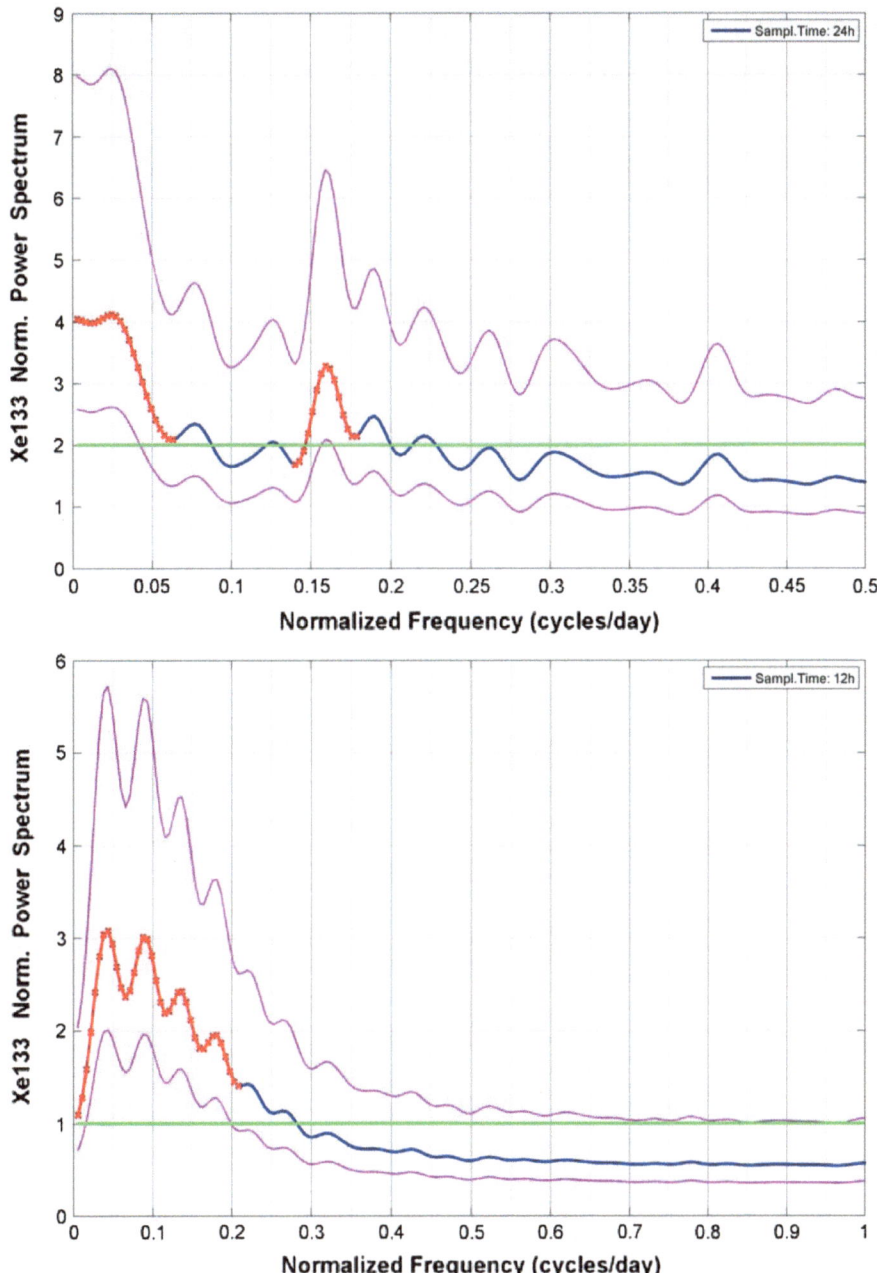

Figure 6
Power spectra for the selected INGE noble gas stations in St. John's, Canada (*top*), and in Charlottesville, USA (*bottom*): the power spectrum of the radioxenon time series (*blue line*); the χ^2 confidence level of 90% (*pink lines*), the threshold spectrum calculated for a white noise by Monte Carlo simulation (*green line*), the coupled frequencies of the local meteorological parameters with the radioxenon time series (*red line*) (colour figure online)

3. Discussion

The radioxenon time series analysis for St. John's, Canada and Charlottesville, USA detrended by the local meteorological patterns emphasised two different behaviours: the residual time series for the former system is characterised by a randomness with a white noise structure because it passed the K–S and L–B tests (Fig. 7); on the other hand, the latter system failed both tests (Fig. 8), indicating the existence of a random

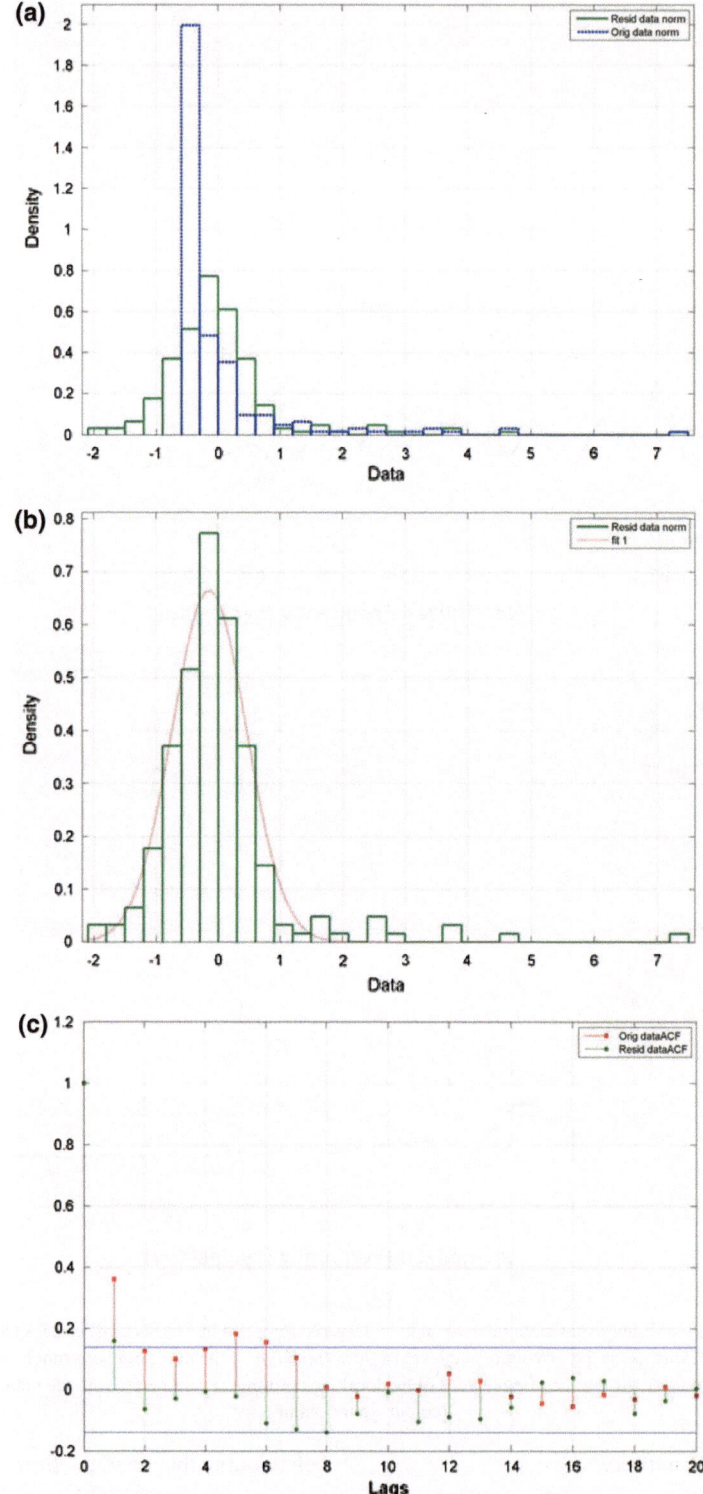

Figure 7
The INGE noble gas station located in St. John's, Canada: **a** The distribution function of the original and residual radioxenon time series; **b** the Kolmogorov–Smirnov test for the normal distribution on the residual time series; **c** the Ljung–Box test for the original and residual time series

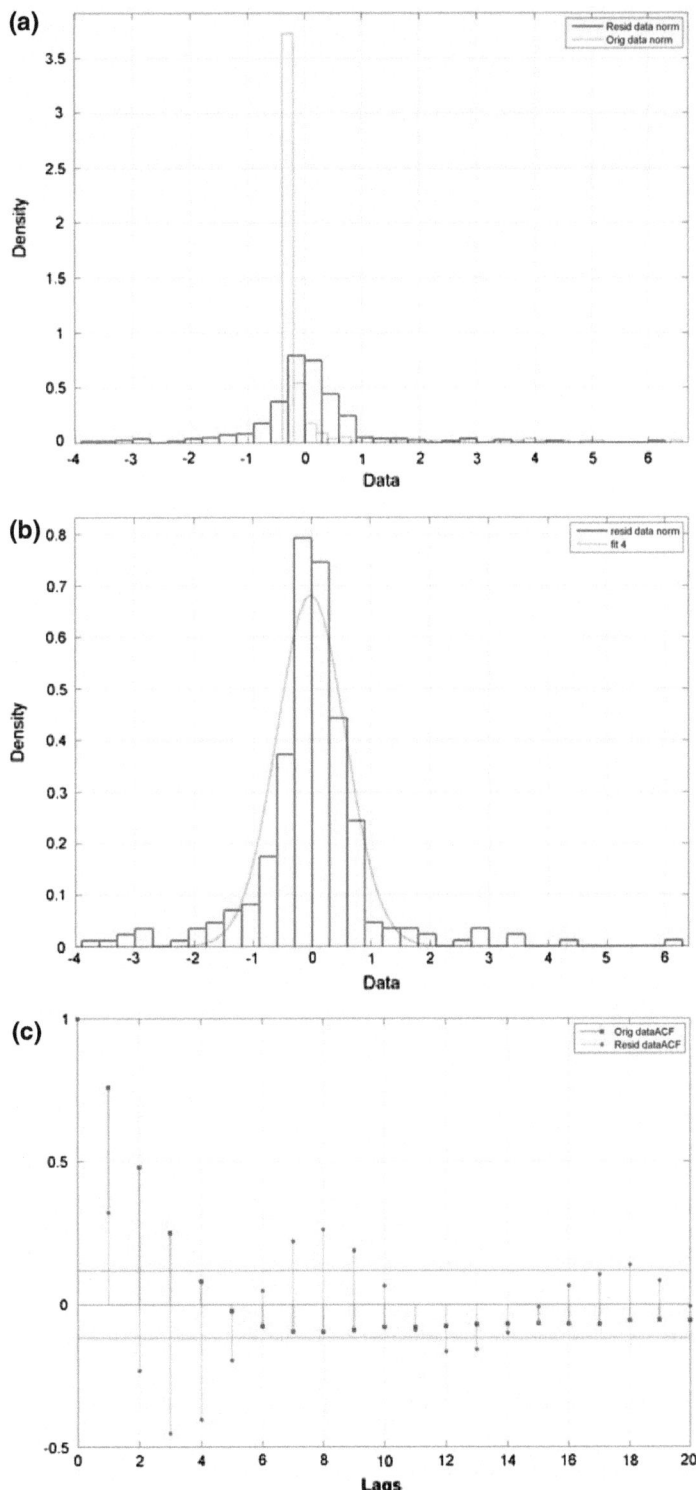

Figure 8
The INGE noble gas station located in Charlottesville, USA: **a** the distribution function of the original and residual radioxenon time series; **b** the Kolmogorov–Smirnov test for the normal distribution on the residual time series; **c** the Ljung–Box test for the original and residual time series

197

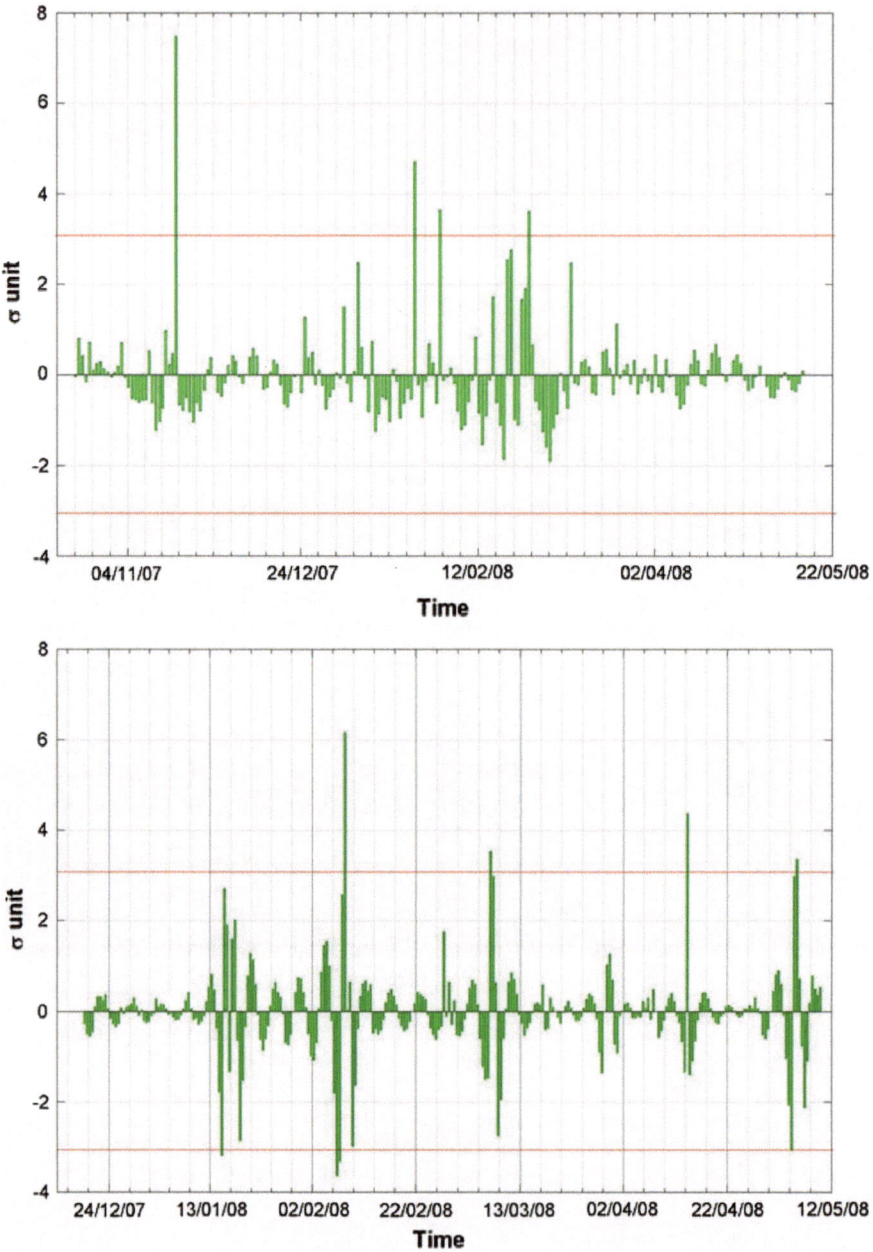

Figure 9
The residual time series normalised to null mean value and unit standard deviation for the INGE noble gas stations located in St. John's, Canada (*top*) and in Charlottesville, USA (*bottom*). The estimated 3σ, i.e. the thresholds used to identify the outliers, are also shown

component in the residual time series that is different from white noise.

Although the aim of this analysis was to characterise the occurrence probability of the outliers at St. John's and Charlottesville, the different randomness identified at the two sites needs to be taken into account for a correct outlier's interpretation. The radioxenon residual time series in Fig. 9 emphasizes that a comprehensive outliers analysis for the Charlottesville station needs a further randomness characterisation of its time series after detrending from its power spectrum to account for some identified but not characterised frequencies not linked with the local meteorological parameters (Fig. 6). In fact,

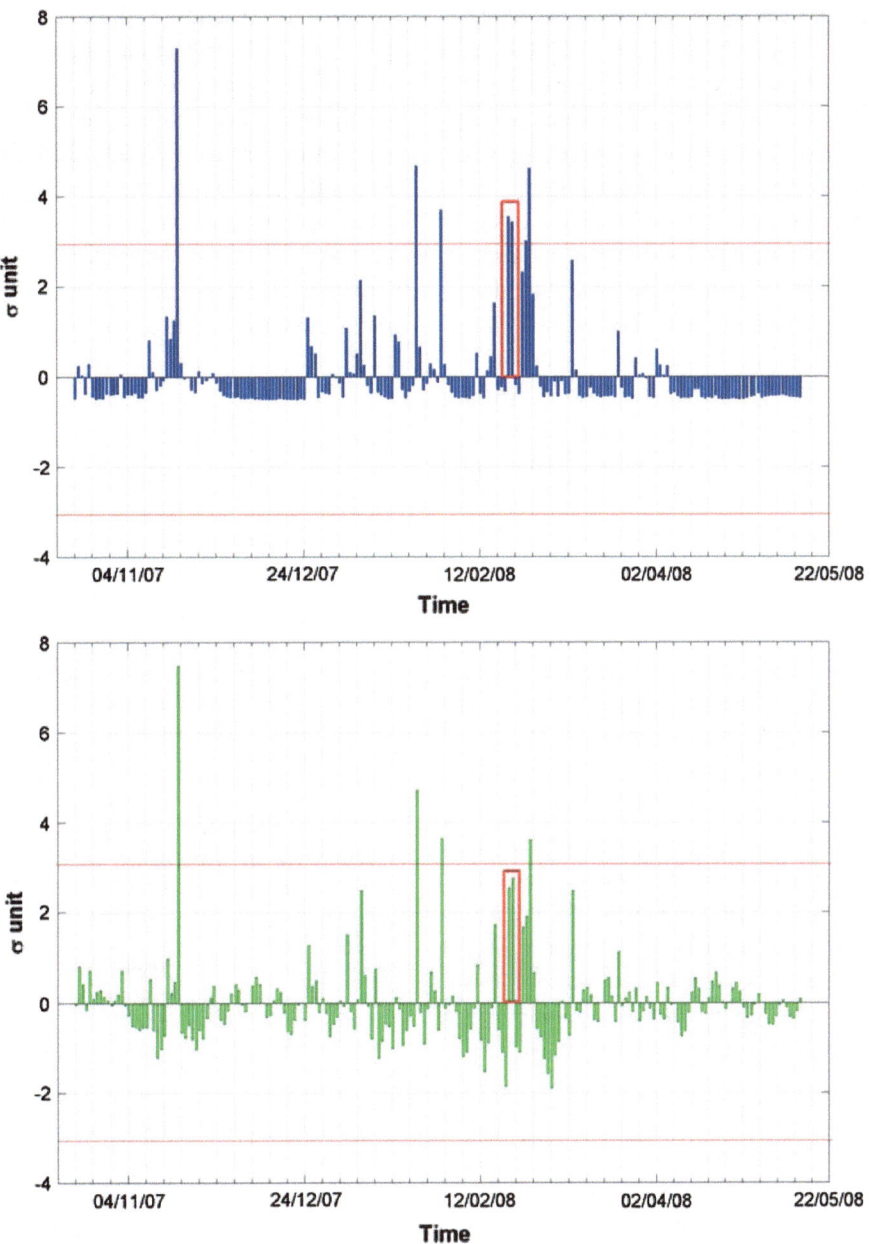

Figure 10
The original (*top*) and residual (*bottom*) time series normalised to null mean value and unit standard deviation for the INGE noble gas station located in St. John's, Canada. The estimated 3σ, i.e., the thresholds used to identify the outliers, are also shown. The *red rectangles* show the different occurrence probability for the same outlier in the original and residual time series (colour figure online)

such identified components in the radioxenon time series are possible reasons for the failure of the K–S and L–B tests at the Charlottesville station. Conversely, the occurrence probability of the outliers at the St. John's station allows to perform an event categorisation and to try locating regional sources.

These may be nuclear power plants as well as a large radioisotope production facility (STOCKI *et al.*, 2005).

A main difference between the two sites might be that St. John's is situated on an isolated island, east of Canada and receives regular diluted signals from the continent. The closest nuclear power plant is

1,000 km away and the large radiopharmaceutical facility in Chalk River is at a distance of 2,000 km. The station in Charlottesville, however, is located in an industrial area with more than 20 nuclear power plants within a 500-km radius and a hospital nearby that uses radioxenon for medical treatments.

Figure 10 presents another important issue of this analysis related to the possible false outliers in the raw data. In fact, the different occurrence probability of an outlier in the normalised original and residual time series highlights the necessity for event categorisation to characterise all the components of the series to identify statistical anomalies. This was evidence of the local meteorological pattern influence on sampling and time series analysis, because the difference between normalised original and residual data was solely from the meteorological parameters. Hence the original data included all meteorological parameters that were detrended in the residual one. Therefore, it can happen that a false outlier is screened out or overestimated in the raw data (Fig. 10).

In 2000, a comparison exercise between the different types of noble gas systems was held in Freiburg, Germany (AUER *et al.*, 2004). They ran in parallel for almost 1 year. No systematic differences were found that might have been traced back to memory effects or other technical differences between the different equipment. Therefore, the revealed differences between the two sites of this study were not related to the different equipment (SPALAX vs. SAUNA) utilised.

4. Conclusions

The radioxenon time series analysis detrended by the local meteorological patterns for measurements taken with a SPALAX noble gas system in St. John's, Canada and a SAUNA system in Charlottesville, USA is a useful method to better understand and characterise the occurrence probability of outliers. The method is capable of checking different randomness for the time series examined: white noise for St. John's and a more complex and not yet fully characterised noise at Charlottesville. Furthermore, this approach identifies possible sources of false alarms triggered by identification of wrong outliers in

the raw data. Therefore, the future event categorisation for CTBT verification purposes needs to include a methodological approach of this kind, although this is only one of the possible scenarios. Further work is required to consider not only the outlier but also the temporal variation of the randomness in the time series as possible anomalous signals. Besides the local meteorological characteristics, such a scheme must also consider regional transport patterns that stem from known radioxenon sources such as nuclear facilities.

REFERENCES

AUER, M., AXELSSON, A., BLANCHARD, X., BOWYER, T.W., BRACHET, G., BULOWSKI, I., DUBASOV, Y., ELMGREN, K., FONTAINE, J.P., HARMS, W., HAYES, J.C., HEIMBIGNER, T.R., McINTYRE, J.I., PANISKO, M.E., POPOV, Y., RINGBOM, A., SARTORIUS, H., SCHMID, S., SCHULZE, J., SCHLOSSER, C., TAFFARY, T., WEISS, W., and WERNSPERGER, B. (2004), *Intercomparison experiments of systems for the measurement of xenon radionuclides in the atmosphere*, Appl. Radiat. Isot. *60(6)*, 863–877.

BECKER, A., WOTAWA, G., DE GEER, L.E., SEIBERT, P., DRAXLER, R.R., SLOAN, C., D'AMOURS, R., HORT, M., GLAAB, H., HEINRICH, P., GRILLON, Y., SHERSHAKOV, V., KATAYAMA, K., ZHANG, Y., STEWART, P., HIRTL, M., JEAN, M., and CHEN, P. (2007), *Global backtracking of anthropogenic radionuclides by means of a receptor-oriented ensemble dispersion modeling system in support of Nuclear-Test-Ban Treaty verification*, Atmos. Environ. *41*, 4520–4534.

BECKER, A., WOTAWA, G., RINGBOM, A. and SAEY, P.R.J. (2010), *Backtracking of noble gas measurements taken in the aftermath of the announced October 2006 event in North Korea by means of PTS methods in nuclear source estimation and reconstruction.* J. Appl. Geophys. *167*, 415, in press.

BOWYER, T.W., ABEL, K.H., HENSLEY, W.K., PANISKO, M.E., and PERKINS, R.W. (1997), *Ambient Xe-333 levels in the Northeast US*, J. Environ. Rad. *37 (2)*, 143–153.

BOWYER, T.W., SCHLOSSER, C., ABEL, K.H., AUER, M., HAYES, J.C., HEIMBIGNER, T.R., McINTYRE, J.I., PANISKO, M.E., REEDER, P.L., SARTORIUS, H., SCHULZE, J., and WEISS, W. (2002), *Detection and analysis of xenon isotopes for the Comprehensive Nuclear-Test-Ban Treaty international monitoring system*, J. Environ. Rad. *59 (2)*, 139–151.

BROCKWELL, P.J. and DAVIS, R.A., *Introduction to Time Series and Forecasting, 2nd Ed.*, (Springer-Verlag, New York 2002) 36 pp.

CHATFIELD, C., *The Analysis of Time Series* (Chapman and Hall London 1991) 241 pp.

DE GEER, L.E., *Atmospheric radionuclide monitoring: A Swedish perspective*, In *Monitoring a Comprehensive Nuclear-Test-Ban Treaty*, eds. E. S. Huseby and A. M. Dainty (Kluwer Acad., Dordrecht, Netherlands 1996) pp. 157–177.

FONTAINE, J.P., POINTURIER, F., BLANCHARD, X., and TAFFARY, T. (2004), *Atmospheric xenon radioactive isotope monitoring*, J. Environ. Rad. *72*, 129–135.

IGARASHI, Y., SARTORIUS, H., MIYAO, T., WEISS, W., FUSHIMI, K., AOYAMA, M., HIROSE, K., and INOUE, H. (2000), ^{85}Kr and ^{133}Xe monitoring at MRI, Tsukuba and its importance, J. Environ. Rad. 48 (2), 191–220.

KALINOWSKI, M.B. and PISTNER, C. (2006), Isotopic signature of atmospheric xenon released from light water reactors, J. Environ. Rad. 88(3), 215–235.

KALINOWSKI, M.B., BECKER, A., SAEY, P.R.J., TUMA, M.P., and WOTAWA, G. (2008), The complexity of CTBT verification. Taking noble gas monitoring as an example, Complexity 14(1), 89–99.

KIM, W.Y. and RICHARDS, P. (2007), North Korean nuclear test: Seismic discrimination at low yield, EOS Trans. AGU, 88(14), 158.

LAW, A.M. and KELTON, W.D., Simulation Modeling and Analysis (McGraw-Hill, New York 1991) pp. 387–390.

MOORE, D.S. and McCABE, G.P., Introduction to the practice of statistics (5th ed.) (W.H. Freeman & Company, New York 2006).

PERKINS, R.W. and CASEY, L.A. (1996), Radioxenons: Their role in monitoring a Comprehensive Test-Ban Treaty, Rep. DOE/RL-96-1, Pac. Northwest Natl. Lab., Richland, Washington.

RINGBOM, A., LARSON, T., AXELSON, A., ELMGREN, K., and JOHANSSON, C. (2003), SAUNA: A system for automatic sampling, processing and analysis of radioactive xenon, Nucl. Instr. Methods Phys. Res., Sect. A 508, 542–553.

RINGBOM, A., ELMGREN, K., and LIND, K. (2007), Analysis of radioxenon in ground level air sampled in the Republic of South Korea on October 11-14, 2006, FOI Report. FOI-R–2273—SE.

SAEY, P.R.J. (2007), Ultra-low-level measurements of Argon, Krypton and Radioxenon for Treaty verification purposes, ESARDA Bull. 36, 42–55.

SAEY, P.R.J. (2009), The influence of radiopharmaceutical isotope production on the global radioxenon background, J. Environ. Rad., doi:10.1016/j.jenvrad.2009.01.004.

SAEY, P.R.J. and DE GEER, L.E. (2005), Notes on radioxenon measurements for CTBT verification purposes, Appl. Radiat. Isot. 63, 765–773.

SAEY, P.R.J., WOTAWA, G., DE GEER, L.E., AXELSSON, A., BEAN, M., D'AMOURS, R., ELMGREN, K., PETERSON, J., RINGBOM, A., STOCKI, J.T., and UNGAR, R.K. (2006), Radioxenon background at high northern latitudes, J. Geophys. Res. 111, D17306, doi:10.1029/2005JD007038.

SAEY, P.R.J., BEAN, M., BECKER, A., COYNE, J., D'AMOURS, R., DE GEER, L.E., HOGUE, R., STOCKI, T.J., UNGAR, R.K., and WOTAWA, W. (2007), A long distance measurement of radioxenon in Yellowknife, Canada, in late October 2006, Geophys. Res. Lett. 34, L20802, doi:10.1029/2007GL030611.

SCHULZE, J., AUER, M., and WERZI, R. (2000), Low-level radioactivity measurement in support of the CTBTO, Appl. Radiat. Isot. 53 (1–2), 23–30.

STOCKI, T.J., BLANCHARD, X., D'AMOURS, R., UNGAR, R.K., FONTAINE, J.P., SOHIER, M., BEAN, M., TAFFARY, T., RACINE, J., TRACY, B.L., BRACHET, G., JEAN, M., and MEYERHOF, D. (2005), Automated radioxenon monitoring for the comprehensive nuclear Test-Ban Treaty in two distinctive locations: Ottawa and Tahiti, J. Environ. Rad. 80 (3), 305–326.

STOHL, A., FORSTER, C., FRANK, A., SEIBERT, P., and WOTAWA, G. (2005), Technical note: The Lagrangian particle dispersion model FLEXPART version 6.2, Atmos. Chem. Phys. 5, 2461–2474.

VON STORCH, H. and ZWIERS, F.W., Statistical Analysis in Climate Research (Cambridge University Press, Cambridge 1999) 484 pp.

WEISS, W., SARTORIUS, H., and SCHLOSSER, C. (1997), The background levels of radionuclides of the noble gas xenon in the environment and the existing source detection capability, Informal Radionuclide Workshop on Radionuclide IMS Network Specifications, US DOE Environmental Measurements Laboratory, New York.

WOTAWA, G., DE GEER, L.E., DENIER, P., KALINOWSKI, M., TOIVONEN, H., D'AMOURS, R., DESIATO, F., ISSARTEL, J.P., LANGER, M., SEIBERT, P., FRANK, A., SLOAN, C., and YAMAZAWAJ, H. (2003), Atmospheric transport modelling in support of CTBT verification: Overview and basic concepts, Atmos. Environ. 37(18), 2529–2537.

WOTAWA, G., DE GEER, L.E., BECKER, A., D'AMOURS, R., JEAN, M., SERVRANCKX, R., and UNGAR, R.K. (2006), Inter- and intracontinental transport of radioactive cesium released by boreal forest fires, Geophys. Res. Lett. 33, L12806, doi:10.1029/2006GL026206.

(Received October 10, 2008, revised March 7, 2009, accepted April 7, 2009, Published online December 11, 2009)

Reprinted from the journal

Pure Appl. Geophys. 167 (2010), 575–580
© 2009 The Author(s)
This article is published with open access at Springerlink.com
DOI 10.1007/s00024-009-0029-9

❙ Pure and Applied Geophysics

Estimation of Explosion Energy Yield at Chernobyl NPP Accident

SERGEY A. PAKHOMOV[1] and YURI V. DUBASOV[1]

Abstract—The value of the ^{133}Xe/^{133m}Xe isometric activity ratio for the stationary regime of reactor work is about 35, and that for an instant fission (explosion) is about 11, which allowed estimation of the nuclear component of the instant (explosion) energy release during the NPP accident. Atmospheric xenon samples were taken at the trajectory of accident product transfers (in the Cherepovetz area); these samples were measured by a gamma spectrometer, and the ^{133}Xe/^{133m}Xe ratio was determined as an average value of 22.4. For estimations a mathematic model was elaborated considering both the value of instant released energy and the schedule of reactor power change before the accident, as well as different fractionation conditions on the isobaric chain. Comparison of estimated results with the experimental data showed the value of the instant specific energy release in the Chernobyl NPP accident to be $2 \cdot 10^5$–$2 \cdot 10^6$ J/Wt or $6 \cdot 10^{14}$–$6 \cdot 10^{15}$ J (100–1,000 kt). This result is matched up to a total reactor power of 3,200 MWt. However this estimate is not comparable with the actual explosion scale estimated as 10t TNT. This suggests a local character of the instant nuclear energy release and makes it possible to estimate the mass of fuel involved in this explosion process to be from 0.01 to 0.1% of total quantity.

Key words: Xenon-133, isomeric ratio, Chernobyl, estimation, energy, explosion.

To date there is no general idea regarding the physical nature of the Chernobyl NPP accident. According to the main version, it was an explosion of chemical character, that is, the explosion of hydrogen formed in the reactor at high temperature as a result of water reaction with zirconium and other elements.

The alternative version is based on the assumption of a large instant energy release of nuclear energy. Convincing evidence in favor of this version was for the first time obtained by Radium Institute employees on the basis of an analysis of atmospheric xenon radionuclide samples collected in the area of Cherepovetz and of the analysis of the value of ^{133}Xe/^{133m}Xe isomers activity ratios. For long-term reactor work at constant power this ratio obtains a value close to 35. In the extreme case of instant fission of nuclear fuel (an explosion) this ratio value recalculated for zero time becomes close to 11. The relatively long half-life of these radionuclides (2.19 days for ^{133m}Xe and 5.24 days for ^{133}Xe) makes it possible to carry out their monitoring at a significant distance from the source; that is the case in the town of Cherepovetz located 1,000 km from Chernobyl (PAKHOMOV et al., 1991), 2,000 km downwind.

As a result of the fourth reactor block explosion at the Chernobyl NPP on the night of 26, April 1986, most xenon and krypton isotopes in the reactor were ejected into the atmosphere, as well as a significant part of other products accumulated in the reactor to the moment of its destruction. The emitted products were dispersed over a significant part of the European continent under the influence of atmospheric processes (UNSCEAR, 1988, p. 7). The early phase of the transfer is shown in Fig. 1.

Large-scale monitoring of radioactive contamination of atmospheric air began in the USSR at once after the accident. Atmospheric noble gases (RNG), that is, krypton and xenon, radioactivity measurement and determination of their radionuclide structure were included in the monitoring program.

Samples of these gases for measurement were manufactured from an industrial krypton–xenon mixture (KXM). This mixture was a byproduct from large air-separating plants (ASP) manufacturing oxygen for the metallurgical industry. Atmospheric xenon samples separated from other gaseous and

---
[1] V.G. Khlopin Radium Institute, 28, 2nd Murinskiy av., 194021 St.-Petersburg, Russia. E-mail: pakhomov@khlopin.ru; yuri@dyuv.spb.su

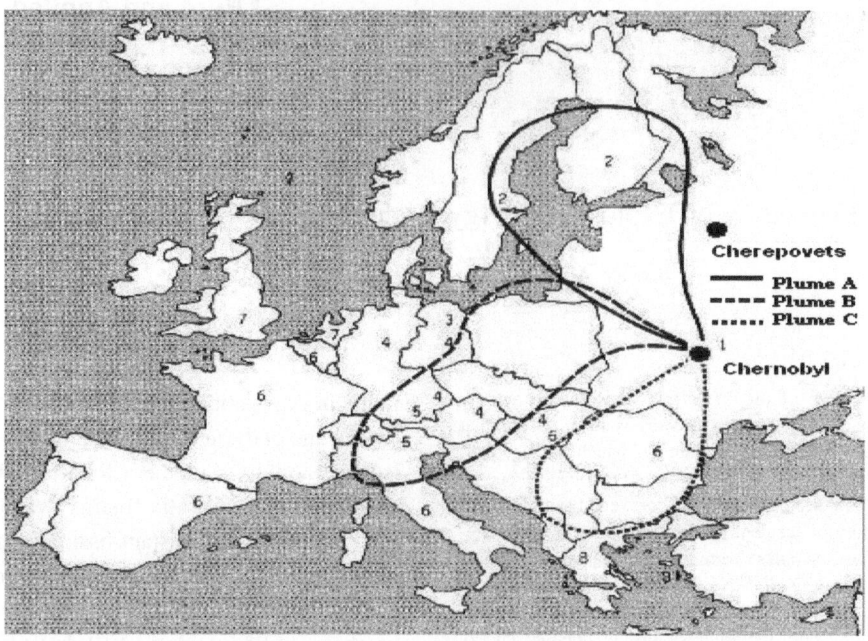

Figure 1
Trajectories of transfer of the Chernobyl NPP accident ejection and the days of their first arrival (after the accident) (UNSCEAR, 1988, p. 7). A, B, C correspond to the transfer trajectories on 27–28 and 29–30 of April

radon were measured by a gamma-spectrometer with a Ge(Li) detector.

The filling of cylinders at the air-separating plants takes on average 2–3 days. A new bottle is produced every day. Filled cylinders are stored for a certain time, and only after that are they sent to a consumer. Thus, the use of air-separating plants for atmospheric RNG sampling provided the unique possibility to obtain both background samples (taken before the accident) as well as samples containing noble gases from the accident.

The volume of pure xenon samples separated from the industrial krypton–xenon mixture was within the range of 1–1.5 l. These samples were adsorbed by charcoal put in the spectrometric samplers, formed as a cylinder of 5 cm diameter and of 2.5 cm height. The quantity of xenon was measured both by gas-chromatographic and gravimetric manner. The typical xenon weight in the sampler was 5–7 g. At the maximum of contaminated clouds passage the activity of ^{133}Xe in the sampler exceeded thousands of Bq; the activity of ^{133m}Xe reached hundreds of Bq.

The measurements were carried out with a low-background spectrometer equipped with the Ge (Li) detector. The detector with 80 sm^3 volume was placed inside of passive shield (lead of 80 mm thickness), and active anti-coincidence shield (NaI detector of large volume with a well). The analyzer on 4,096 channels, the "VECTOR" made in the standard, by being analogue of system "CAMAC" was used. Since the spectrometer had not been equipped with the computer the information output was only carried out on a printer. Processing of peaks was accomplished manually, a method of channel summation was used, in strict accordance with the methodical instructions accepted in that time.

The most important information was obtained from analysis of krypton–xenon mixture samples done in Cherepovetz from 22, April, to 6, May, 1986. Values of ^{133}Xe/^{133m}Xe volumetric activity in the atmospheric air of Cherepovetz obtained as a result of the analysis of these samples are given in Table 1. (The middle of KXM cylinders filling up period was taken as a conventional sampling date). Values of activity ratios of these isomers obtained both for the sampling date and decay corrected to the date of the accident are also given in Table 1.

Table 1

Values of $^{133}Xe/^{133m}Xe$ 'specific activities in atmospheric air of Cherepovetz and their ratios

Cylinder filling-up period	Estimated time of sampling	^{133}Xe, Bq/m^3 (for sampling date)	^{133m}Xe, Bq/m^3 (for sampling date)	$^{133}Xe/^{133m}Xe$ (for sampling date)	$^{133}Xe/^{133m}Xe$ (for the accident date)
22.04–25.04	23.04	$(7.4 \pm 1.0)\ 10^{-3}$	Less than detecting limit	–	–
27.04–29.04	28.04	$(7.41 \pm 0.74)\ 10^{-1}$	$(2.5 \pm 0.6)\ 0.10^{-2}$	29.6 ± 8.9	20.5 ± 6.2
27.04–30.04	29.04	1.62 ± 0.13	$(4.1 \pm 1.0)\ 0.10^{-2}$	39.5 ± 11.9	22.9 ± 6.9
29.04–02.05	30.04	1.54 ± 0.13	$(3.1 \pm 0.8)\ 0.10^{-2}$	49.7 ± 14.9	23.9 ± 7.2
30.04–03.05	02.05	$(2.71 \pm 0.29)\ 10^{-1}$	$(4.1 \pm 1.0)\ 0.10^{-3}$	66.1 ± 19.8	22.2 ± 6.7
02.05–06.05	04.05	$(8.0 \pm 1.1)\ 10^{-3}$	Less than detecting limit	–	–

As is seen from Table 1, the arrival of air masses contaminated by the Chernobyl accident outburst to the Cherepovetz area was observed from 28, April, to 2, May. The obtained values of ^{133}Xe and ^{133m}Xe activity ratios are homogeneous enough, which is probably due to characteristic properties of their transfer from the area of the accident to the Cherepovetz area. This indicates the arrival of only "fresh" products and not the arrival of products from the later outburst. In Fig. 2 values of the average $^{133}Xe/^{133m}Xe$ activity ratio, as well as corresponding "reactor" and "explosion" values, are shown.

As can be seen from Fig. 2, the obtained values of $^{133}Xe/^{133m}Xe$ activity ratios and their mean value (22.4 ± 3.4) substantially differ from a "reactor" value of 35, which enhances the validity of the nuclear mechanism hypothesis assuming a huge instant energy release in the Chernobyl accident.

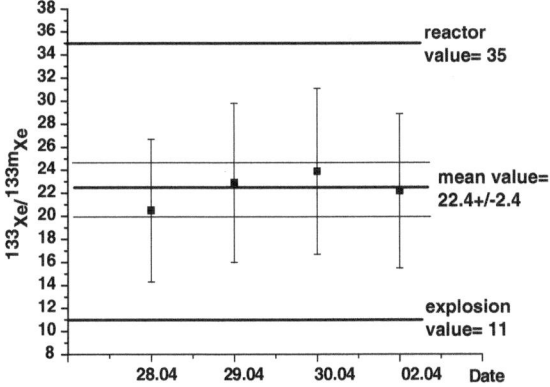

Figure 2
Values of $^{133}Xe/^{133m}Xe$ activity values shown at the accident date, and an estimation of their mean value

A special computer program was developed for numerical estimation of the specific (normalized for the reactor power before the accident) instant energy yield. This program simulates $^{133}Xe/^{133m}Xe$ ratio as a superposition of the "reactor" and "explosion" values taking into account both different contributions of instant energy yield and changes of the reactor power before the accident.

The following data are necessary to carry out model calculations of $^{133}Xe/^{133m}Xe$ ratio:

1. a scheme of radioactive transformations in mass chain 133 and the values of half-lives;
2. values of radionuclides fission yield;
3. data on fuel composition;
4. estimations of possible ^{133}Xe and ^{133m}Xe and their precursors fractionation values;
5. a graph of reactor power changes before the accident.

A scheme of radioactive transformations in mass chain 133 taken from DE GEER (2007) is shown in Fig. 3. In Table 2 values of fissionable components content in the fuel of IV-reactor block at the moment of the accident are given, as well as cross sections of fission by thermal neutrons and their contribution to total energy yield (UNSCEAR, 1988, p. 5; JEFF 3.1., 2008). Half-lives and independent radionuclides yield of mass chain 133 for ^{235}U, ^{239}Pu, ^{241}Pu and their composition by thermal neutrons are given in Table 3 (ENGLAND and RIDER, 1994).

The data of work (WINKELMANN, et al., 1987) in which the authors were investigating chemical forms of ^{131}I presence in products of the Chernobyl accident yield taken within, the territory of Germany, could be used to estimate the value of probable ^{133}Xe and

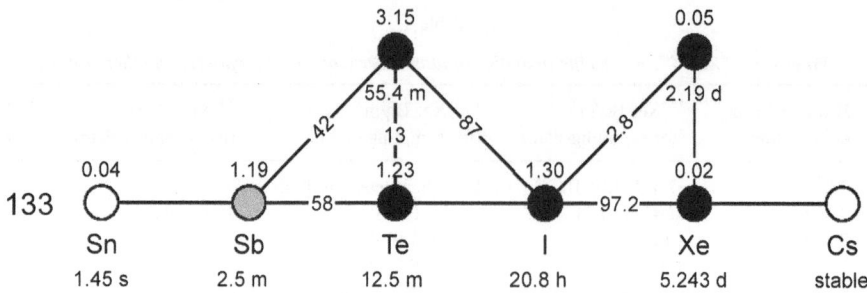

Figure 3
Radioactive transformations scheme in mass chain 133 (DE GEER, 2007)

Table 2

Fissionable components content in reactor fuel for the accident moment, cross-sections of fission by thermal neutrons and its contribution into total energy yield

Fissionable component	Content, (UNSCEAR, 1988, p. 5)	Cross sections of fission by thermal neutrons (JEFF 3.1., 2008)	Contribution into energy yield (%)
^{235}U	4.5 kg/ton	584 barn	51.7
^{236}U	2.4 kg/ton	0.06 barn	–
^{239}Pu	2.6 kg/ton	748 barn	38.3
^{240}Pu	1.8 kg/ton	0.06 barn	–
^{241}Pu	0.5 kg/ton	1012 barn	10.0

Table 3

Half lives and independent yields of mass chain 133 radionuclides in ^{235}U, ^{239}Pu, ^{241}Pu decay and their composition by thermal neutrons (ENGLAND and RIDER, 1994)

Nuclide	Half life	^{235}U (*), thermal	^{239}Pu (*), thermal	^{241}Pu (*), thermal	Fuel composition
$^{133}In_{49}$	0.18 s	0.000171	3.81e-05	0.000335	0.000123
$^{133}Sn_{50}$	1.44 s	0.137730	0.034296	0.159381	0.10028
$^{133}Sb_{51}$	2.5 month	2.256714	1.173643	2.506718	1.866898
$^{133m}Te_{52}$	55.4 month	2.986257	2.891422	2.859248	2.937234
$^{133}Te_{52}$	12.4 month	1.148216	1.765602	1.167859	1.386639
$^{133}I_{53}$	20.8 h	0.165036	1.107495	0.031506	0.512645
$^{133m}Xe_{54}$	2.19 day	0.001886	0.033799	0.000614	0.013981
$^{133}Xe_{54}$	5.243 day	0.000666	0.009448	0.000251	0.003988

^{133m}Xe fractionation from its immediate precursor in a chain of ^{133}I transformations. The ^{131}I yield into the atmosphere was estimated at 20%. 40% of iodine was in the form of aerosol, with 35% of it in elementary gaseous form. The aerosol form of iodine is subjected to gravitational sedimentation and washed out by sediments. Gaseous iodine is also subjected to washing out. Only organically-bound iodine is relatively stable and can accompany xenon radioisotopes in their long-distance transfer without any noticeable fractionation. Thus, assuming that ^{131}I and ^{133}I from the accident are of identical form, the ^{133}I fraction keeping its genetic connection with the daughters ^{133}Xe and ^{133m}Xe can be estimated to range from 5% (far transfer, washing out) to 20% (near transfer without sedimentation).

A graph of thermal power changing of the reactor IVth block Chernobyl NPP in the pre-accidental period according to the data of the report (ABAGYAN et al., 1996) is shown in Fig. 4.

To calculate the values of radionuclides activity in an isolated isobar chain, a linear differential

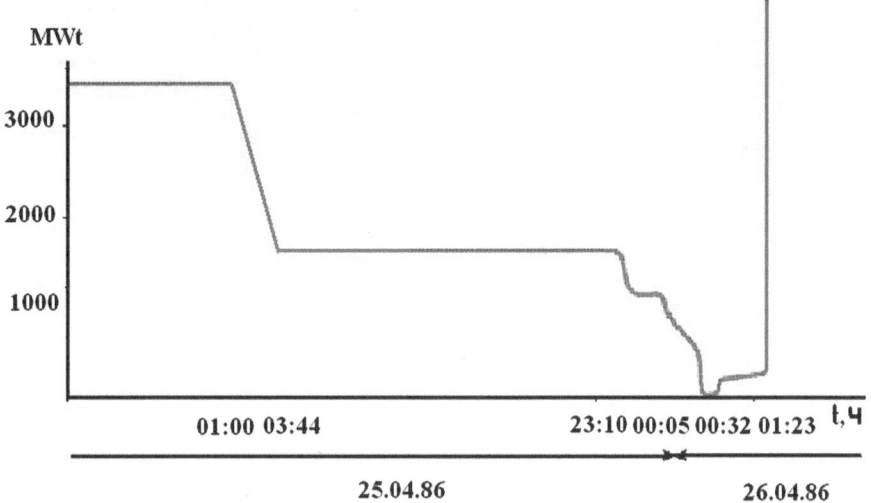

Figure 4
Graph of thermal power changing of the reactor IVth block Chernobyl NPP in pre-accidental period (ABAGYAN et al., 1996)

equations system (1a) with starting conditions (1b) should be solved.

$$\frac{dN_j}{dt} = -\lambda_j N_j + \sum_{k=1}^{j-1} \lambda_k N_k B_{kj} \qquad (1a)$$

$$N_j(0) = Y_j \qquad (1b)$$

The solution of system (1) can be presented as a series (2).

$$N_j(t) = \sum_{k=1}^{j} C_{kj} e^{-\mu_j t}. \qquad (2)$$

The expansion coefficients C in (2) could be determined using recurrent ratios (3a, 3b, 3c).

$$C_{11} = Y_1 \qquad (3a)$$

$$C_{jk} = \sum_{i-j}^{k-1} \frac{C_{ji} \lambda_i B_{jk}}{\lambda_k - \lambda_j} \qquad (3b)$$

$$C_{jj} = Y_j - \sum_{k=1}^{j-1} C_{kj} \qquad (3c)$$

Estimation of fission products accumulation in the reactor could also be made using ratios (3). With this aim additional "virtual" radionuclide simulating continuous replenishment of isobar chain with products formed during the reactor's work could be inserted into the isobar chain.

Calculation results are given in Fig. 5. as a collection of curves showing the $^{133}Xe/^{133m}Xe$ ratio dependence on the value of a specific (normalized for a reactor power value before the accident) instant energy release. Each curve corresponds to the specific value of the ^{133}I fraction keeping its connection with daughter radinuclides ^{133}Xe and ^{133m}Xe (fractionation effects in the course of distant atmospheric transfer are considered in this parameter value). A comparison of these curves with the mean value of $^{133}Xe/^{133m}Xe$ ratio experimentally determined is also shown in this Figure.

Comparison of calculation curves and the experimentally found value of $^{133}Xe/^{133m}Xe$ ratio results in an estimate of the instant specific energy release in the Chernobyl accident of $2 \cdot 10^5$–$2 \cdot 10^6$ J/Wt. This result being formally referred to a nominal reactor power of 3,200 MWt, the estimation of absolute value of the energy released during the accident comprises a value of the order of $6 \cdot 10^{14}$–$6 \cdot 10^{15}$ J (100–1,000 kt). This is not commensurable both with a true reactor destruction scale and seismic estimations (STRAHOV et al., 1997) defining the value of the explosion energy as about 10t TNT. Collectively it points to a local character of instant nuclear outburst stipulated by extremely non homogeneous distribution of the neutron flux in an active reactor zone at the moment of the accident. Fuel mass drawn into this

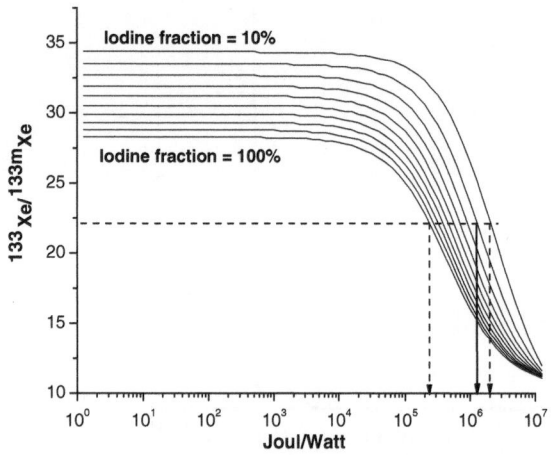

Figure 5

Calculated ^{133}Xe/^{133m}Xe ratio values given at the accident moment depending on the specific value of instant energy release for different fractionation conditions

explosion process could be estimated as from 0.01 to 0.1% of the total mass.

Summarizing given estimations it should be acknowledged that the hypothesis of a nuclear mechanism of enormous instant energy yield in the Chernobyl accident seems quite convincing, as is supported by experimental data; these data are in good agreement with the calculated results.

REFERENCES

ABAGYAN, A.A., ADAMOV, E.O. BURLAKOV, E.V. et al. (1996), *IAEA-J4-TC972*, Vienna, April 1–3, 46–65.
DE GEER, L. E. (2007), *The Xenon NCC method*, FOI-R-2350—SE, October 2007.
ENGLAND, T. R. and RIDER, B. F. (1994), LA-UR-94-3106 (ENDF-349), October, 1994.
JEFF 3.1. NUCLEAR DATES LIBRARY (2008), *OECD Nuclear Energy Agency*
PAKHOMOV, S.A., KRIVOKHVATSTY, K.S., and SOKOLOV, I.A. (1991), *Assessment of the prompt energy release by the accident at the reactor of the Chernobyl NPP, based on estimation of the ratio of Activities of Xenon-133 and Xenon-133 m in the Air*, Radiokhimiya *33*, 6, 125–132 (in Russian).
STRAHOV, V.N. et al. (1997), *Seismic phenomena's in region of Chernobyl NPP*. Geophys. J., (Ukraine) *19, 3*.
UNSCEAR REPORT (1988), Annex D: *Exposures from the Chernobyl accident*, pp. 5 and 7.
WINKELMANN, I. et al. (1987), *Radioactivity measurements in the Federal Republic of Germany after the Chernobyl accident*, ISH-116.

(Received September 30, 2008, revised March 23, 2009, accepted June 3, 2009, onlinedate December 16, 2009)

Pure Appl. Geophys. 167 (2010), 581–599
© 2009 Birkhäuser Verlag, Basel/Switzerland
DOI 10.1007/s00024-009-0025-0

Backtracking of Noble Gas Measurements Taken in the Aftermath of the Announced October 2006 Event in North Korea by Means of PTS Methods in Nuclear Source Estimation and Reconstruction

ANDREAS BECKER,[1] GERHARD WOTAWA,[1] ANDERS RINGBOM,[2] and PAUL R.J. SAEY[1]

Abstract—The announced October 2006 nuclear test explosion in the Democratic People's Republic of Korea (DPRK) has been the first real test regarding the technical capabilities of the verification system built up by the Vienna-based Provisional Technical Secretariat (PTS) of the Comprehensive Nuclear-Test-Ban Treaty Organization (CTBTO) to detect and locate a nuclear test event. This paper enhances the resolution of the DPRK events' xenon source reconstruction published by SAEY et al. (2007, "A long distance measurement of radioxenon in Yellowknife, Canada, in late October 2006", GRL, Vol. 34, L20802) that was based solely on radio-xenon measurements taken at the remote radionuclide station in Yellowknife, Canada by involving additional measurements taken by a mobile noble gas system deployed quite close to the event location in the Republic of Korea (ROK). Moreover the horizontal resolution of the forward and backward atmospheric transport modelling methods applied for the source scenario reconstruction has been enhanced appropriately to reflect the considerably shorter source-receptor distances examined in comparison to the previously published source reconstruction. It is shown that the ^{133}Xe measurements in Yellowknife could register ^{133}Xe traces from the nuclear explosion during the first 3 days after the event, while the mobile measurements were rather sensitive to releases during days 2–4 after the explosion. According to the analysis, the most likely source scenario would consist of an initial (possibly up to 21 h delayed) venting of 1×10^{-15} Bq ^{133}Xe during the first 24 h, followed by a two orders of magnitude weaker seepage during the following 3 days. Both measurements corroborate the scenario of a rather rapid venting and soil diffusion of the ^{133}Xe yielded during the explosion. While the Swedish mobile measurements were crucial to enhancement of the reconstruction of the source scenario, given the installation status of the IMS xenon network at the time of the event, a sensitivity analysis revealed that the fully developed network would have been able to detect ^{133}Xe traces from the Korean explosion at a number of stations and allowed for an even better constraint on the release function. The station Ussuriysk, Russia, being in operation in 2006, would have registered ^{133}Xe within 1 day and with a three orders of magnitudes stronger signal compared to the detection at Yellowknife.

Key words: October 2006 nuclear test of North Korea, case study, atmospheric backtracking, radioxenon monitoring (network based and mobile), CTBT verification.

1. Introduction

As part of its regular operation the International Monitoring System (IMS) being built under the auspices of the Provisional Technical Secretariat (PTS) of the Preparatory Commission for the Comprehensive Nuclear-Test-Ban Treaty Organization (CTBTO) recorded a seismic event with the characteristics of an underground explosion located at 41.3119°N, 129.0189°E in the Democratic People's Republic of Korea (DPRK). The date and time of the event was 9 October, 2006, 01:35:28 UTC (LE BRAS et al., 2007). The US Geological Survey independently diagnosed a body-wave magnitude of 4.2 of the event and located it slightly south-eastward at 41.294°N, 129.094°E (KIM and RICHARDS, 2007), but still within the 880-km^2 error ellipse which the CTBTO assigned to this event as part of the so-called Reviewed Event Bulleting (REB) product disseminated to its member states. It is worth noting in this context that the Democratic Peoples Republic of Korea already televised at 3 October, 2006 an announcement that it would conduct a nuclear explosion. For further detail see SAEY et al. (2007).

As reviewed by SCHLITTENHARDT et al., (2009) the yield estimations of the explosion ranged from 0.4 to 2.2 kilo ton (kt) TNT equivalent. The lower end of

[1] CTBTO Preparatory Commission, Vienna International Centre, P.O. Box 1200, 1400 Vienna, Austria. E-mail: andreas.becker@ctbto.org
[2] Swedish Defence Research Agency (FOI), Defence and Security, System and Technology Division, 172 90 Stockholm, Sweden.

this range gave rise to speculations on whether the event was indeed a nuclear explosion or not. As demonstrated below, the IMS radionuclide (RN) network in full scale operation would provide a robust answer to this question. In its final build-up, the IMS will have 80 highly sensitive particulate RN stations (SCHULZE et al., 2000) with 40 of these stations scheduled to also measure the CTBT relevant radio-xenon isotopes ^{131m}Xe, ^{133m}Xe, ^{133}Xe, ^{135}Xe (Fig. 1; AUER et al., 2004; KALINOWSKI et al., 2008) that are known to be created in substantial amounts after a nuclear explosion. These four isotopes are most interesting as their respective half-life times of 11.94, 2.19, 5.24 days and 9.14 h (ENSDF, 2009) are long enough to allow for detection and also clocking of their time of production by evaluation of the isotopic ratios. Moreover, such isotopes may be the only ones still leaking out from a well-contained underground explosion (CARRIGAN et al., 1996; PERKINS and CASEY, 1996; DE GEER, 1996) which does not diffuse

the many non-gaseous (particulate) fission products through the soil into the atmosphere. Finally the isotopic ratios of the four radio-xenon fission products are characteristic if yielded from a nuclear explosion allowing for their discrimination against radio-xenon released from known civil sources like nuclear power plants or medical isotope production facilities (KALINOWSKI et al. 2009).

For ^{133}Xe a rather conservative estimation of the total fission yield of 1 kt TNT equivalent nuclear explosion is 15.4×10^{-7} mol corresponding to 13.4×10^{15} Bq (CARRIGAN et al., 1996). Release scenarios of different levels of complexity are possible in principle starting from a singular release resulting from immediate venting of the noble gases up to a complex leakage scenario from rock cracks triggered for example by the passage of a low-pressure system. Due to the enormous pressures generated in the explosion cavity the gaseous ^{133}Xe can be pushed quickly via cracks and fissures in the bedrock

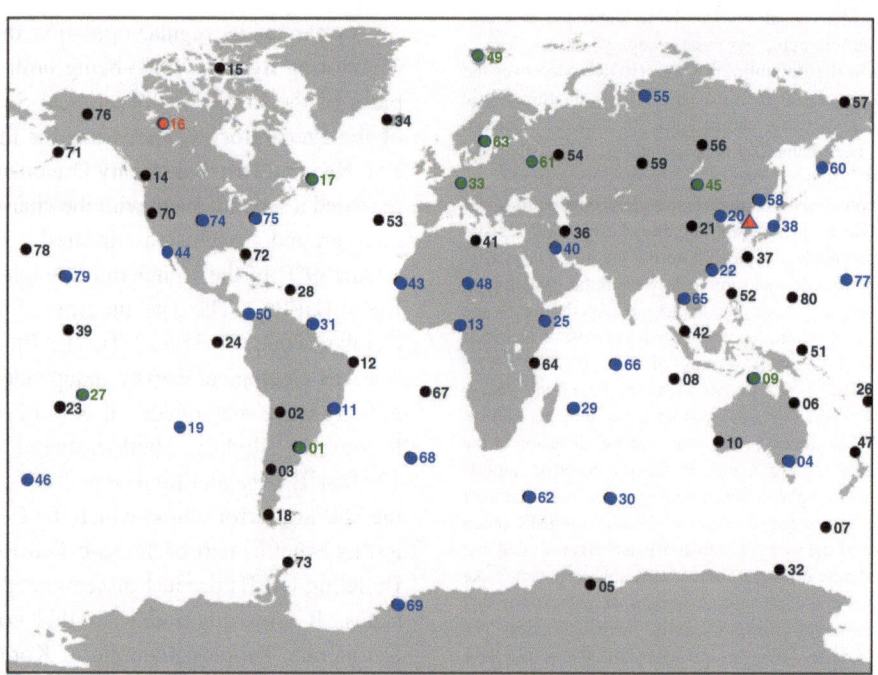

Figure 1

Radionuclide compartment of the International Monitoring System (IMS) consisting of a network of 80 stations (*all numbered dots*) with each of them equipped with a particulate filter sampler supplemented by highly sensitive gamma ray spectroscopy. Forty of these stations (*blue, green and red dots*) also are scheduled to measure radio-xenon isotopes. Station 35, reserved for the Indian subcontinent, is among those but not yet exactly located and thus missing on the map. At the time of the DPRK event only 10 (*green and red dots*) of these stations were already operational. The *red objects* (*dot and triangle*) determine locations where detection of traces of ^{133}Xe originating from the events' release plume was encountered into the atmosphere, yielding an immediate venting or rapid seepage source scenario. On the other hand very well contained underground explosions might not allow for soil diffusion for days or weeks, and the seepage finally occurring is barometrically driven (atmospheric pumping). Resolving a source scenario strongly depends on how well the event location has been covered by radio-xenon monitoring. The current radio-xenon systems reach critical limits (LCs) that correspond to minimum detectable activity concentrations (MDCs) of the stations to 1×10^{-4} Bq/m^3 for ^{133}Xe that enables them to still detect releases of 0.10–1×10^{12} Bq in the DPRK region, assuming a full scale network (WOTAWA et al., 2009). This would be three to four orders of magnitudes more sensitive than required to detect the approximately 10^{15} Bq release of an immediate venting of 10% of the ^{133}Xe fission yield mentioned above.

In a previous study published after the event (SAEY et al., 2007), a data analysis was made based on measurements from the IMS network. It was determined that the measurements would be consistent with the assumption that 10^{15} Bq of ^{133}Xe was released at the DPRK event location within the first days. In the following, the existing study is enhanced by involving additional measurements taken by a mobile noble gas system deployed quite close to the event location in the Republic of Korea (ROK). In doing so, the horizontal resolution of the forward and backward atmospheric transport modelling methods applied for the source scenario reconstruction have been adapted to the much shorter source-receptor distances examined in comparison to the previously published source reconstruction. Given the fact that the exact release location was known already, a multiple regression method introduced by WOTAWA et al. (2006) was applicable to the two xenon readings available to better constrain and reconstruct the most likely ^{133}Xe source scenario associated with the DPRK event. Both forward and backward modelling methods play relevant roles in doing so. Backward modelling is needed to access the xenon readings at Yellowknife and ROK for their potential to contain releases from the DPRK event. Forward modelling plays a role as the seismic event data offer a precise location and a temporal constraint for the earliest possible onset of the release scenario to be resolved.

With the estimate on the course of the release scenario, that must not necessarily start at the time of the event but could also be delayed, at hand, forward modelling can subsequently confirm the release scenario reconstruction and prospect what role other IMS stations could have played in a full scale network. In this context we will also demonstrate that the external mobile measurements were indeed useful as a substitute for IMS stations not yet sending data in October 2006, but that the network after its completion would have been well capable of detecting the event.

2. Methods and Model Set-ups Applied for Nuclear Source Estimation and Reconstruction

In order to simulate the atmospheric dispersion of airborne material and to compute RN sample-specific source-receptor sensitivity (SRS) fields, the PTS has set-up a modular system that utilizes the Lagrangian Particle Diffusion Model (LPDM) FLEXPART (STOHL et al., 1998, 2005) for the dispersion modelling part. FLEXPART can be operated in both, forward (Runs 1–4, Table 1) and backward (Run 5, 6; Table 1) modes on the basis of analysis wind fields provided by the European Centre for Medium-Range Weather Forecasts (ECMWF) or from the US National Centers for Environmental Prediction (NCEP). For the latter provider PTS can also utilize the publicly available forecast wind fields (Run 1, Table 1). Starting from 10 October, 2006 when the seismic event location was resolved with a high level of certainty (LE BRAS et al., 2007), the plume dispersion for the most simple release scenario of immediate surface level venting of 1×10^{15} Bq ^{133}Xe from that location was predicted daily, based on the two wind fields analysis at 1.0° horizontal resolution (Runs 2 and 4, Table 1).

For source estimation of the ^{133}Xe readings at Yellowknife, Canada and Kansong (ROK), the backward modelling-based SRS fields at 1.0° (Run 5) and 0.2° (Run 6) resolution are used for an inversion method that tests the consistency between the singular source assumptions folded with the relevant SRS data and the actual activity concentrations monitored in the network (WOTAWA et al.,

Table 1

List of forward and backward modelling runs performed with the LPDM FLEXPART 5.0 (STOHL et al., 1998, 2005) to reconstruct the DPRK event scenario

Run no. and purpose	Release/sample information			Meteorological input data utilized by LPDM					
	Location and time	Duration	Thousands of particles released per source or sample	Numerical weather prediction System	Resolution and architecture of LPDM input grid				
					$\Delta\lambda \times \Delta\varphi$	Δt (h)	Vertical coordinate	# of levels	"Height" of lowest 3 levels
1. Forecast of activity concentrations (ACs) at IMS stations caused by DPRK xenon release	Source: DPRK event on 9 October, 2006, 1:35 UTC	Puff	2,000	NCEP GFS	1.0° × 1.0°	3	p	26	1,000; 975; 950 hPa if p_{surf} = 1,015 hPa
2. Alternate forward analysis of DPRK xenon release and related IMS stations' ACs			2,000	NCEP GDAS	1.0° × 1.0°	6	p	26	1,000; 975; 950 hPa
3. High resolution forward dispersion modelling of early plume stages	and 9 October, 2006, 0–3 UTC		5,000	ECMWF 4DVAR	0.2° × 0.2°	3	eta	91	1,012; 1,009; 1,005 hPa if p_{surf} = 1,013.25 hPa
4. Standard forward analysis of DPRK xenon release and related IMS stations' ACs (incl. sensitivity studies)	and 10 October, 2006, 12–15 UTC and 11 October, 2006, 09–12 UTC	3 h	2,000	ECMWF 4DVAR	1.0° × 1.0°	3	eta	91	1,012; 1,009; 1,005 hPa if p_{surf} = 1,013.25 hPa
5. PTS standard backward analysis of Yellowknife and mobile xenon detections	Yellowknife, Canada, all samples 21–27 October, 2006 Kansong, ROK, all 5 samples 11–14 October, 2006 (Table 3)	24 h 12 h	240	ECMWF 4DVAR	1.0° × 1.0°	3	eta	91	1,012; 1,009; 1,005 hPa if p_{surf} = 1,013.25 hPa
6. PTS high resolution backward analysis of mobile xenon detections	Kansong, ROK, all 5 samples 12–14 October, 2006 (Table 3)	12 h	2,400	ECMWF 4DVAR	0.2° × 0.2°	3	eta	91	1,012; 1,009; 1,005 hPa if p_{surf} = 1,013.25 hPa

2003; BECKER *et al.*, 2007). In addition, taking the source location information provided by the seismic part of the IMS, the same SRS fields were also utilized for a source reconstruction conducted by means of a multiple regression analysis method introduced and elaborated by WOTAWA *et al.* (2006, 2009) and KALINOWSKI *et al.* (2008). Finally, forward dispersion modelling was applied to predict the activity concentrations for the location and sampling times of the ^{133}Xe measurements with the parabolic kernel method of ULIASZ (1994) serving modelled activity concentrations with a method complementary to the SRS folding method. Moreover the forward modelling method was also applied based on forecast wind fields in order to predict which of the already operational IMS RN stations should be relevant on those days (10–21 October, 2006) when the location of the announced nuclear explosion was known, but no RN detection was yet encountered. All modes of operations and model set-ups utilized and their specific purpose for this study are summarized in Table 1.

3. A Special Situation in CTBT Context: an Announced Nuclear Explosion

The announcement by the DPRK on the 3rd of October, 2006 has provided the PTS an excellent opportunity to test its capabilities regarding monitoring function in the relevant technologies utilized for verification (wave-form for "listening" and RN for "sniffing"). Moreover, the DPRK case has challenged the PTS to provide a prediction capability in support of the RN monitoring effort particularly during the period when a CTBT relevant seismic event was already detected at 9 October, 2006, but an RN detection clarifying the nuclear character of the explosion detected had not yet been encountered. Due to the announcement with 6 days lead time from a country with a relatively small territory, it was possible and worthwhile to utilize forward atmospheric transport modelling (ATM), as a reasonably accurate a priori assumption on the source location was possible. Normally (in the watchdog mode of the PTS) this information is not available, making forward ATM an inefficient approach. It should thus be noted

that the DPRK test does not exactly match the archetype scenario the PTS is in charge of preparing, namely to detect an unannounced nuclear test explosion pursued in an evasive manner and that the forward ATM methods discussed here would only be relevant if also in the archetype scenario the seismic technology can capably provide a quick and reliable source location hypothesis.

4. The Early Stages of the Plume Dispersion

The DPRK event plume dispersion was rather complex, as the event location is in a topographically structured terrain, leading to a heterogeneous wind field in all three spatial dimensions. Moreover, with respect to the seasonal behaviour of the prevailing wind and surface pressure patterns in East Asia, the event (see earthquake epicentre and error ellipse in Fig. 2) happened to take place during the ending of the "summer type" of prevailing southwesterly winds (see Fig. 3, top), due to the establishment of a high pressure system resulting from irradiative cooling of the East Asian land surface in contrast to the thermal inert Sea of Japan that was still warm from the recent summer season. This differential cooling during the East Asian fall season and the transition to the "winter type" of prevailing winds (see Fig. 3, bottom) poses a meteorological condition that gives rise to cold air outflows from the Asian continent into the Sea of Japan. Consequently there are frequently non-stationary wind conditions in October where the overall still south-westerly wind regime is occasionally interrupted. To better understand the flow and resulting dispersion conditions on the event day, a higher-resolution forward model simulation based on ECMWF wind field analysis was conducted (Run 3, Table 1). Compared with the standard simulation (Run 4) already discussed by SAEY *et al.* (2007), the horizontal resolution was increased five times. At this resolution the forward modelling and three-dimensional wind field analysis show that a continental cold air outflow took place after 12:00 UTC on 10 October (Fig. 4, bottom row, left) and persisted until the early hours (UTC time) of 11 October (Fig. 4, bottom row, right). As also depicted in Fig. 5 (corresponding to Run 4 of Table 1) the meso-scale low-level wind

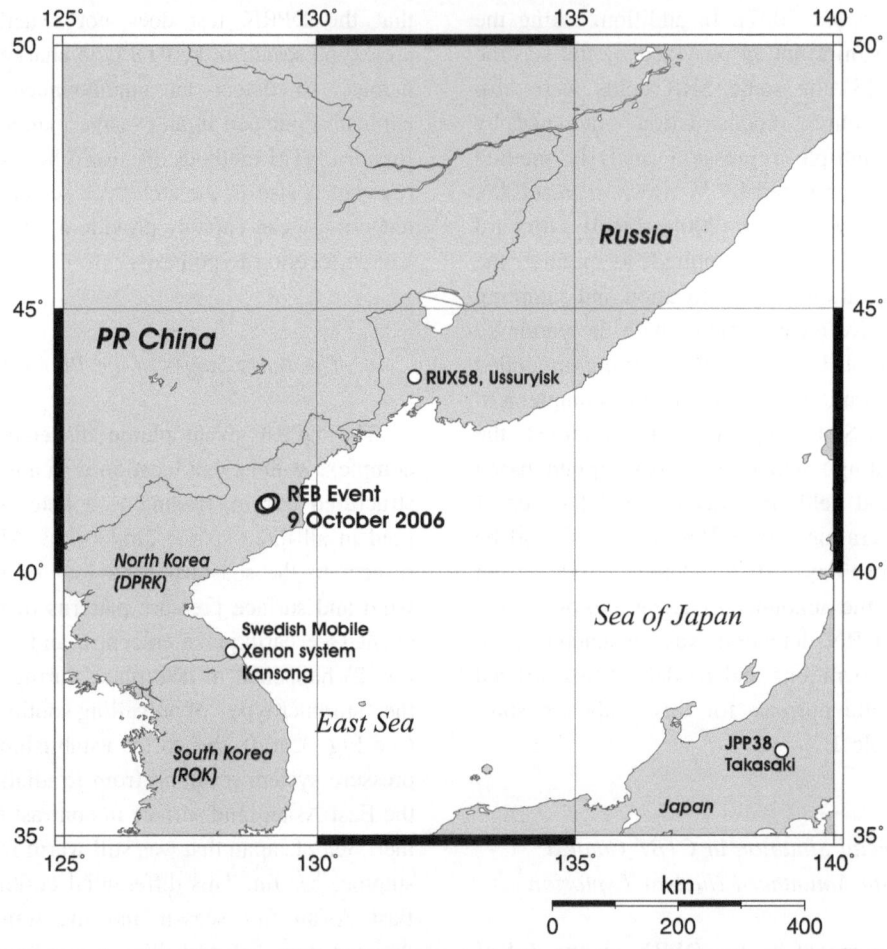

Figure 2
Map of the REB event area and the relevant regions across and around the Sea of Japan where xenon measurements were taken in the first week after the event. The REB event of 9 October, 2006, 1:35 AM with the corresponding error ellipse covering 880 km^2 and the location of the mobile xenon system are indicated. Finally the scheduled IMS radio-xenon station Ussuriysk, Russian Federation (RUX58) that would have been directly exposed to the plume is indicated

conditions across DPRK and the East Asian region remained highly variable throughout days 2–6 after the event so that

a. there should be a high dependency of the plume prediction on the assumed release scenario,
b. the plume dispersion calculation could be less representative and be highly sensitive to the meteorological analysis field utilized.

Indeed, the non-stationary wind field conditions lead to a strong sensitivity of the DPRK event plume dispersion during the first days to the assumed release time (see Fig. 6, where three different release times are examined). However, the sensitivity to the

analysis wind field utilized for the dispersion modelling was comparatively small (see Fig. 5, right column). One can see a big overlap of results from the two different model configurations, one based on wind fields from the US NCEP Global Data Analysis System (GDAS; Run 2 in Table 1) and one on fields from the European Centre for Medium Range Weather Forecasts (ECMWF; Run 4 in Table 1). The good overlap is another indication of the congruence of both wind field representations which typically appear in rather dynamic meteorological conditions under which the impact of local effects, as for example different representations of the topography, is reduced.

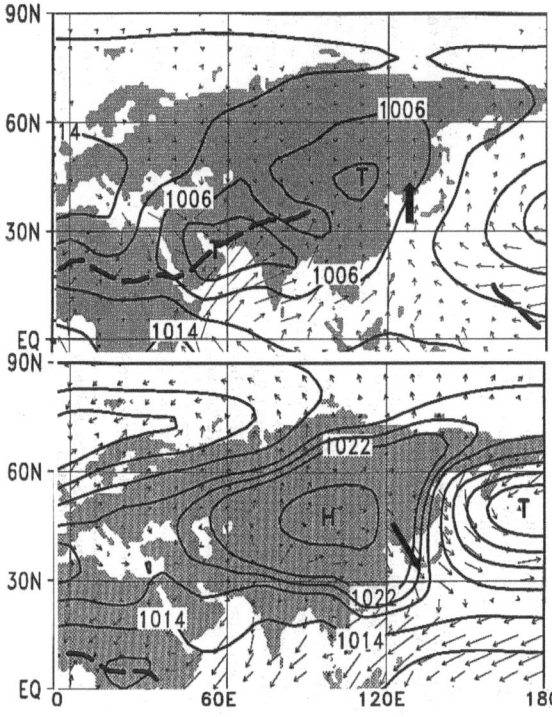

Figure 3

Climatology of wind and surface pressure across the Asian continent during July (*top*) and January (*bottom*) according to KRAUS (2001). The change from the summertime low (denoted by a '*T*') to the wintertime high pressure system regime is obvious, yielding mainly southerly wind directions during summer and mainly north-westerly ones during winter for the area of interest and IMS RN stations around North Korea (see oversized wind barbs). The DPRK event took place during the transition from the summer to the winter regime

5. Identification of the Potentially Relevant IMS RN Stations

Both model configurations (Runs 2 and 4, Table 1) correspondingly predicted that in the first week the plume resulting from an immediate venting at the time and location of the DPRK event, would not reach any of the two closest existing downwind RN particulate stations in the East Asian region, namely Okinawa (JPP37) and Gunma (JPP38) in Japan (Table 2). The first particulate station downwind intercepted by the simulated plume was already very remote at Sand Point, Alaska (USP71, Table 2). However, none of the particulate samples in the weeks after the DPRK event featured the occurrence of treaty-relevant radionuclides, a fact that does not falsify the modelling but confirms that the underground test explosion featured a containment of all particulate fission products as can be expected. In the following the results based on the standard forward model simulation (Run 4, Table 1) are taken to discuss the later stages of the plume dispersion. A quite complex structure of the surface level (0–30 m above model terrain) plume activity concentration has already evolved 8 days after the event at 17 October, when parts of it reach the North American continent (Fig. 7a). The various mountain ranges along the Pacific coastline in Canada and the USA strongly govern the further evolution of the plume that is finally calculated to reach the IMS RN station in Yellowknife, Canada (Table 2) twice (Fig. 7b,c) with concentrations still above the minimum detectable activity concentration (MDC) for ^{133}Xe. As Yellowknife is a 'clear air' station, the critical limit and the MDC is 1×10^{-4} Bq/m^3 or even slightly lower, exceeding the PTS requirement by one order of magnitude. It is also the first RN station predicted to be reached by the plume that features both an RN particulate and a radio-xenon station (Fig. 1, Table 2).

As thoroughly discussed by SAEY et al. (2007) the activity concentration predicted by the PTS forward ATM method predicts a double peak ^{133}Xe detection on the 22nd and 27th of October, 2006, which very well coincides with true ^{133}Xe detection in excess of background on the 21st and 25th of October (Fig. 8) in view of the geo-temporal range of more than 7,000 km and almost 2 weeks examined. Later on the plume is predicted to disperse and decay below the aforementioned typical MDC for ^{133}Xe which corresponds to the fact that no other xenon station further downwind (e.g., Spitsbergen, Norway, and St. John, Canada, Table 2) has encountered anomalous ^{133}Xe detection.

6. Backward Atmospheric Transport Modelling at the PTS upon First RN (Xenon) Detection

With the first ^{133}Xe detections at Yellowknife, the results of the PTS operational daily backward atmospheric transport modelling (see Run no. 5 of Table 1), were utilized for

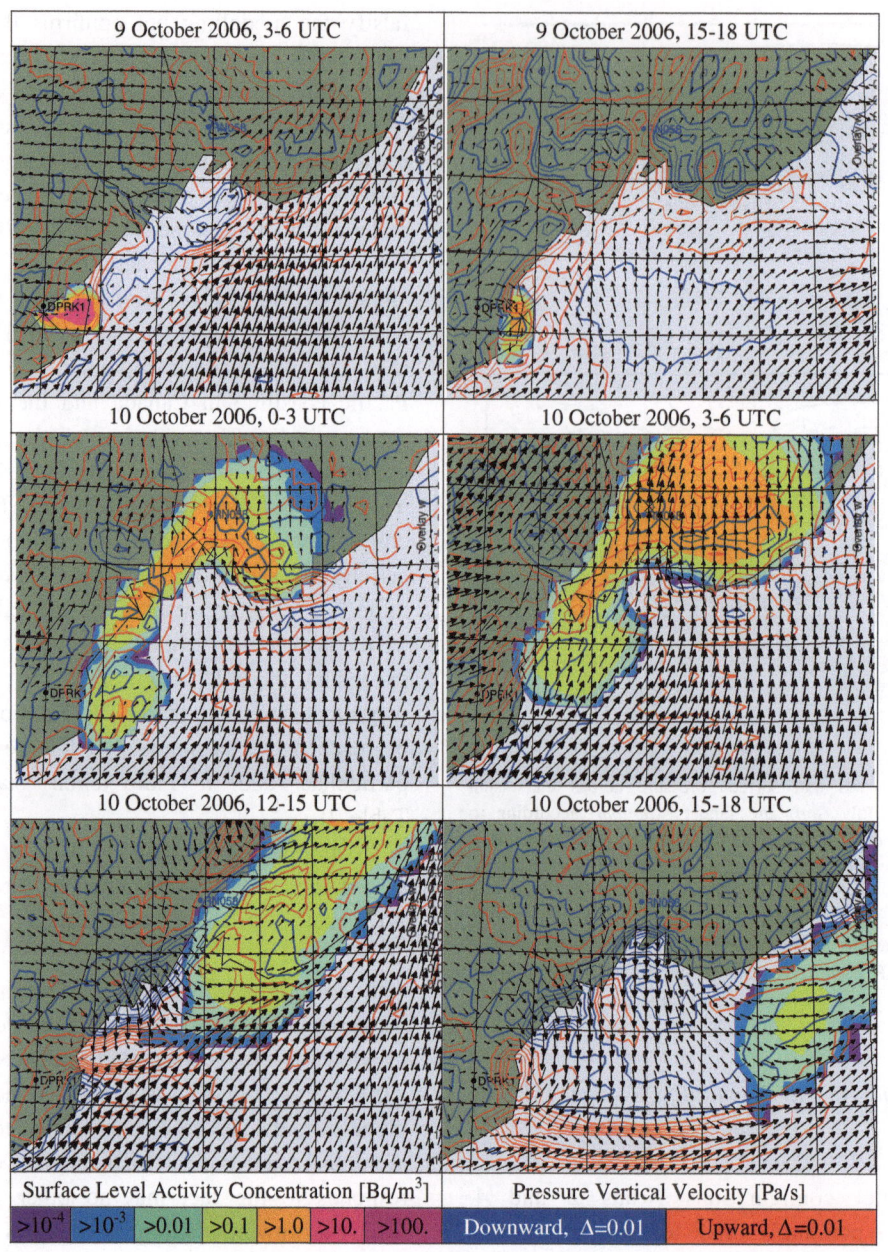

Figure 4
First 42 h of plume dispersion in case of immediate venting from the event location (denoted as 'DPRK1') as calculated with forward ATM at 0.2° × 0.2° resolution (Run no. 3 in Table 1). Besides the colour-coded surface level plume activity concentration in Bq/m^3 (lower colour code) the 3D wind in the second lowest layer ($\sim$20–45 m above ground) of the ECMWF wind field analysis is plotted in terms of *arrows* for the horizontal wind and colour-coded contours for the vertical wind (contour interval, 0.01 Pa/s, *red* denotes lifting)

- Folding of all SRS fields for Yellowknife during October 2006 with a worst case release scenario for a known xenon emitter, the isotopic production facility Chalk River Laboratories (CRL) in Canada, to explore its maximum potential influence on the background activity concentration at Yellowknife (see Fig. 8, bottom row).

- Comparison of forward ATM based ^{133}Xe prediction at Yellowknife with observation (see 'ForwardATM' and 'Observed' labelled results in

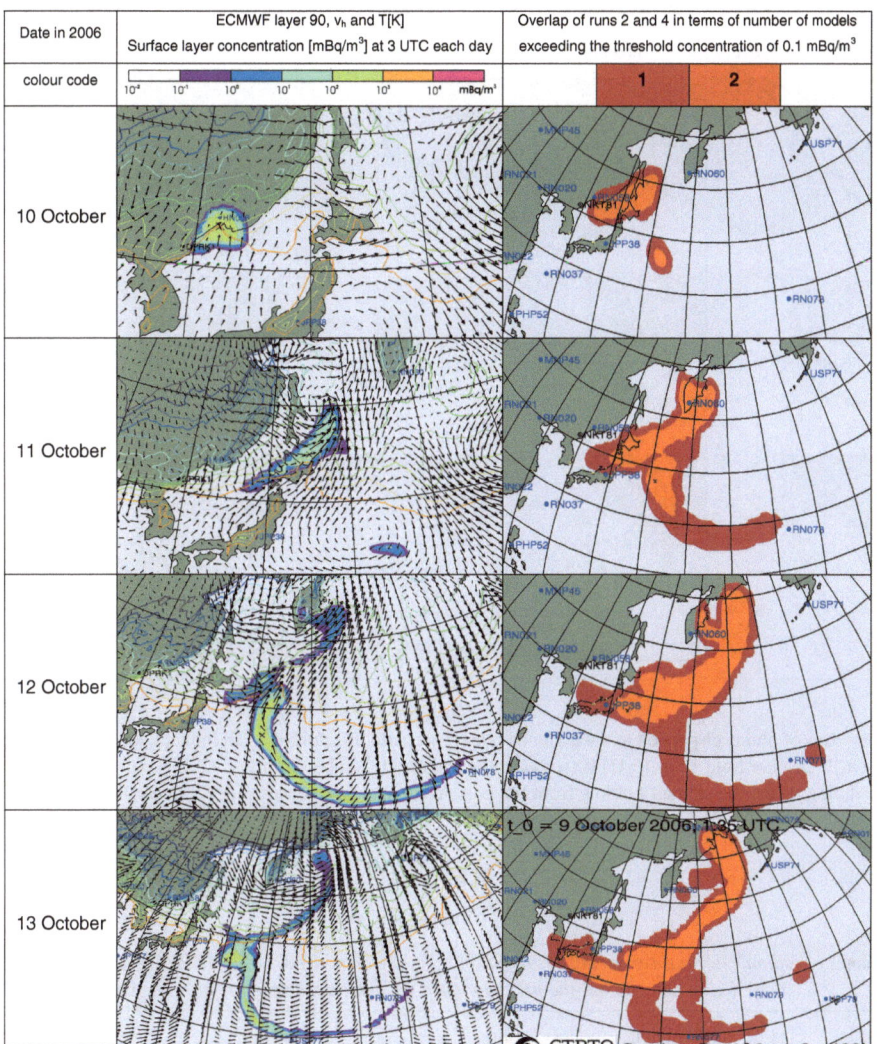

Figure 5

Days 2–6 of xenon plume dispersion in case of immediate venting. *Left column* Horizontal wind (barbs) and temperature (colour-coded from 268 to 300 K, $\Delta T = 2$ K) as retrieved from the second lowest model level 90 of the ECMWF 4DVAR fields used for the 3D forward simulation at $1° \times 1°$ resolution (Run no. 4 in Table 1). The continental cold air outflow starting at 10 October, 2006, 12 UTC (Fig. 4) is also visible on 11 October, 3 AM. Finally the colour-coded surface layer activity concentration of the PTS forward simulation for ^{133}Xe is also shown. *Right column* Overlap between two simulations of the same dispersion model (FLEXPART_5.1) although with different wind field analysis assimilated from global observations by the European Centre for Medium-Range Weather Forecasts (Run 4 in Table 1) and the US National Centers for Environment Prediction (NCEP) Global Data Analysis System (Run 2 in Table 1), respectively

Fig. 8) and the predictions based on the folding of the relevant SRS fields with the various release scenario assumptions (see 'DPRK' labelled results in Fig. 8). A good forward to backward ATM consistency adds to the credibility of the ATM modelling results.

- Backtracking of the Yellowknife detection scenario by calculation and evaluation of the belonging possible source region (Fig. 9).

Due to the long geo-temporal range between the DPRK event (source) and the measurement at CAX16 (receptor), the possible source region (PSR, Fig. 9) related to the first detection at Yellowknife covers a vast area that also contains the DPRK event region but is too large to be suitable to independently identify the DPRK event as the only reasonable source. Nevertheless, the result at least reveals that the isotopic production facility at Chalk River (CLR),

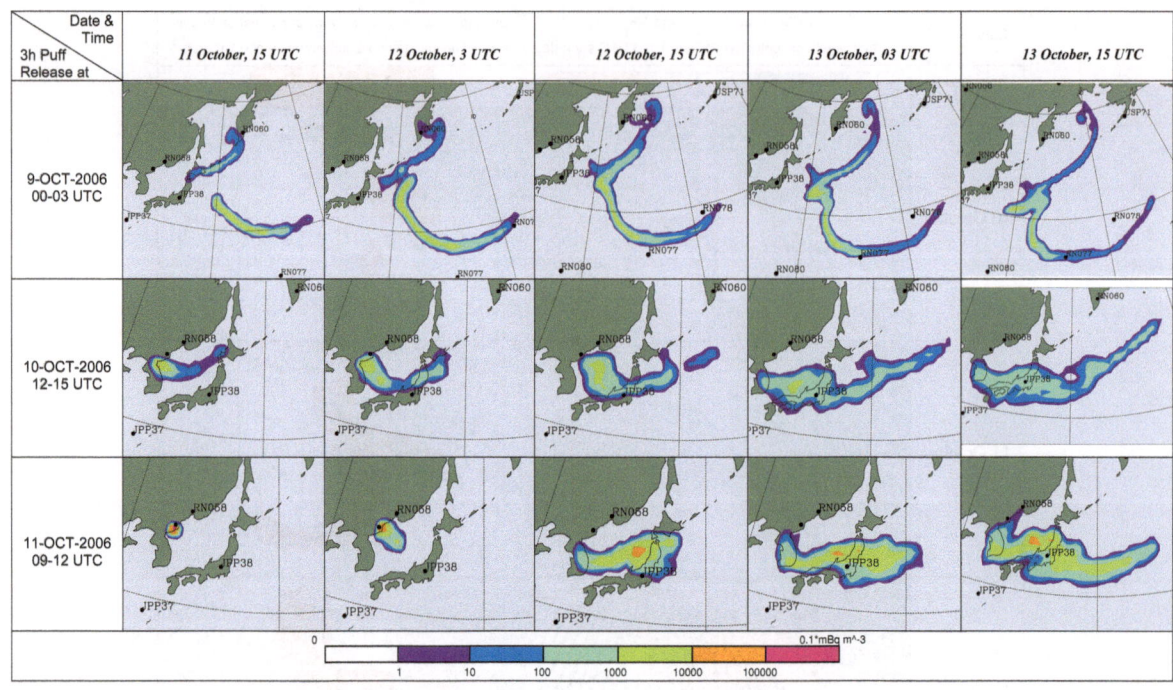

Figure 6

Sensitivity of the first days of plume dispersion to the assumed release scenario. In addition to an immediate venting after the REB event (*first row*) later releases on 10 October from 12 to 15 UTC (*second row*) and on 11 October from 9 to 12 UTC (*bottom row*) are considered. These rows cover periods during which the cold air outflow from the East Asian landmass towards the Japanese Sea interrupted the overall south-westerly wind field regime across DPRK

Table 2

Names and coordinates of radionuclide measurement sites that are in the regional vicinity of the DPRK event. There was no particulate RN detection at any site but the first two listed encountered detections of radio-xenon

Name (ID in maps)	IMS station no.	Host country	Location (long, lat)	Kind of radionuclide isotopes measured in October 2006
Yellowknife (CAX16)	16	NWT, Canada	114.48°W, 62.45°N	Particulate and xenon
Kansong (RNX91-95)	Mobile	Republic of Korea (ROK)	128.45°E, 38.38°N	Xenon
St. John's (CAX17)	17	NL, Canada	52.70°W, 47.60°N	Particulate and xenon
Okinawa (JPP37)	37	Japan	126.50°E, 27.91°N	Particulate
Takasaki, Gunma (JPP38)	38	Japan	139.00°E, 36.61°N	Particulate
Ulaanbaatar (MNX45)	45	Mongolia	106.33°E, 47.89°N	Particulate and xenon
Spitsbergen (NOX49)	49	Norway	15.40°E, 78.20°N	Particulate and xenon
Sand Point (USP71)	71	Alaska, USA	160.49°W, 55.34°N.	Particulate
Ussuriysk (RUN058, RUX58)	58	Russian Federation	131.90°E, 43.70°N	None (not operational)

Canada, does not belong to the PSR, indicating that it most likely could not generate the first xenon signal at CAX16. However, application of further constraints, such as the distribution of known xenon emitters, the land–sea mass distribution below the PSR pattern and scale analysis for the required

release strength reveals that a release of 1×10^{15} Bq at the time and location of the DPRK event is the most likely explanation for the measurements encountered at Yellowknife. As already discussed in the study by SAEY *et al*. (2007), this release corresponds well to the release to be expected from a 1 kt

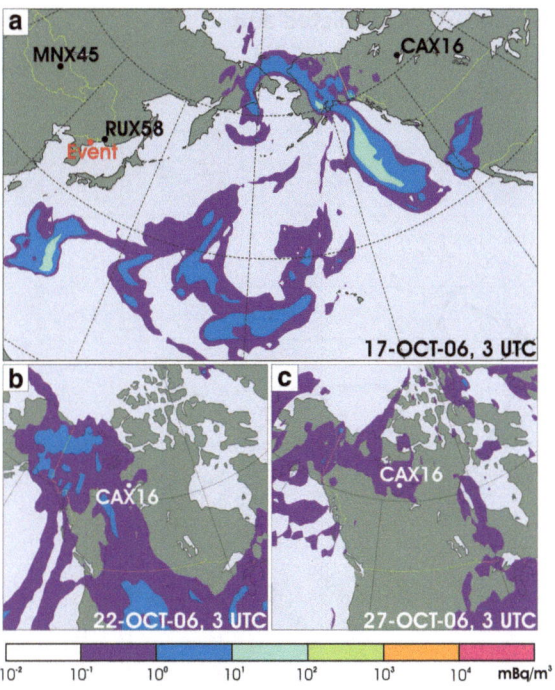

| 10^2 | 10^{-1} | 10^0 | 10^1 | 10^2 | 10^3 | 10^4 | mBq/m³ |

Figure 7

Later interesting stages of the DPRK events' plume evolution as calculated by the Lagrangian Particle Trajectory model FLEX-PART 5.1 (Run no. 4 in Table 1). **a** On 17 October, 2006 when arriving at the Pacific coastline of the Northern American Continent and **b**, **c** on 22 and 27 October, respectively, the days the plume is predicted to arrive at CAX16 and causing the double-peak structure (adopted from SAEY *et al.*, 2007)

TNT equivalent underground nuclear explosion that generates immediately 1×10^{16} Bq ^{133}Xe. A maximum of 10% of it may be vented to the surface within the initial hours. The folding of such a release assumption with the SRS fields yields an activity concentration prediction (see 'DPRK01' in Fig. 8) that is very similar to the forward ATM prediction but also to an activity concentration prediction that applies a permanent 3-day release of the same amount of 1×10^{15} Bq ^{133}Xe (see 'DPRK_Days 1–3' in Fig. 8). Across the first 3 days after the event time there is an insensitivity of the Yellowknife receptor (measurements) to the actual release time for the first 3 days after the DPRK event (see also SAEY *et al.*, 2007). This also shows a very good self-consistency of the ATM forward and backward modelling method. However, the backtracking results do not allow for a further constraint on the release rates that are always assumed to total 1×10^{15} Bq.

Finally all methods applied reproduce the double-peak pattern also encountered in the observations, but all of them correspondingly delay the arrival by 1–2 days, which still seems acceptable in view of the large geo-temporal range (>7,000 km, >12 days) encountered between the release at DPRK and the detection at Yellowknife, Canada. We associate this delay with a weaker down mixing of the high level fast track source-receptor transport of ^{133}Xe (also modelled by D'AMOURS *et al.*, 2007) in the dispersion model in comparison to the reality, as the Eulerian wind field model features only a smoothened representation of the steep mountain ranges located upflow of Yellowknife (grid discretization error).

7. Mobile Measurements Close by the DPRK Event and their Backtracking to DPRK

At the time of the announcement of DPRK a mobile xenon sampling system (Swedish Automatic Unit for Noble gas Acquisition, SAUNA-II), was packed and ready for transport. The team arrived and commenced sampling in Kansong, ROK (Table 2) near the demarcation line to the DPRK by 10 October, 2006 (Fig. 2). This was the closest possible approach to the DPRK event location. Five samples were taken until 14 October, 2008 and sent back to the radioxenon laboratory at FOI in Stockholm, Sweden for the accurate analysis for the four xenon isotopes, ^{133}Xe, ^{133m}Xe, ^{135}Xe and ^{131m}Xe. The results are shown in Table 3 and Fig. 10. For further details see RINGBOM *et al.* (2007). It is worth noting that the meta-stable isotopes were detected rather in the later samples, whereas one would a priori assume them to appear rather in the earlier ones when a potentially fresher part of the xenon release from DPRK is fetched. Moreover backtracking results specific for each sample, the so-called Field-of-Regards (Fig. 11), reveal that the first sample was likely not sensitive across the DPRK event region at any time prior to the measurements. However, the other four samples were found to have monitored the target area but rather 2–4 days after the DPRK event day. Taking uncertainties of the models into account, there is a slight possibility that xenon detections in the late samples could have originated from the

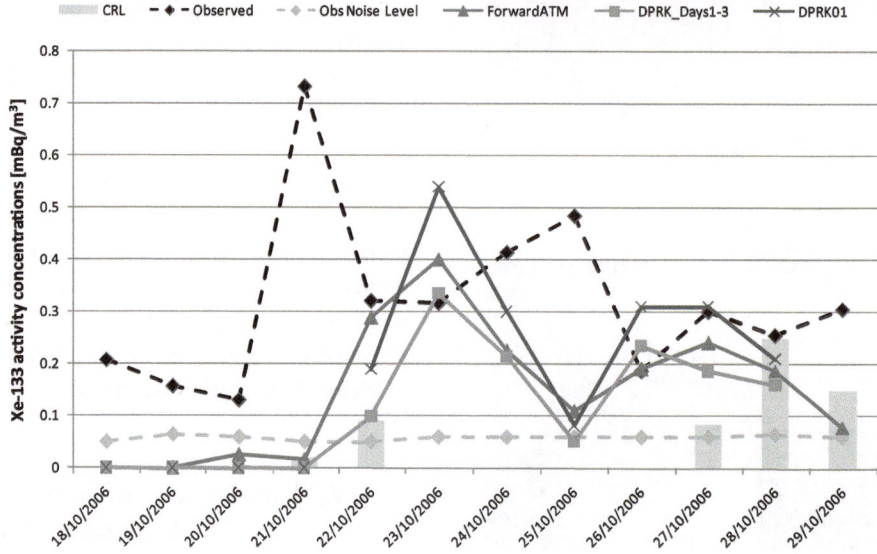

Figure 8

Comparison of observed (*dashed lines* for measurements and corresponding noise levels) and predicted ^{133}Xe signals at the station Yellowknife (CAX16). The predictions stem from three different ATM configurations: 'Forward ATM' assuming a release at the time and place of the October event, and two backward modelling results assuming a 3 hourly release at the event time ('DPRK01') and alternatively a release of the same amount for the first 3 days since the event time ('DPRK_Days 1–3'). The similarity of the results shows the consistency of the forward and backward ATM approach and the insensitivity of the Yellowknife double ^{133}Xe peak prediction to the assumed release time, regarding the first 3 days. Finally the maximum influence of a known Canadian xenon source (Chalk River Laboratories, CRL) is examined (*grey bars*)

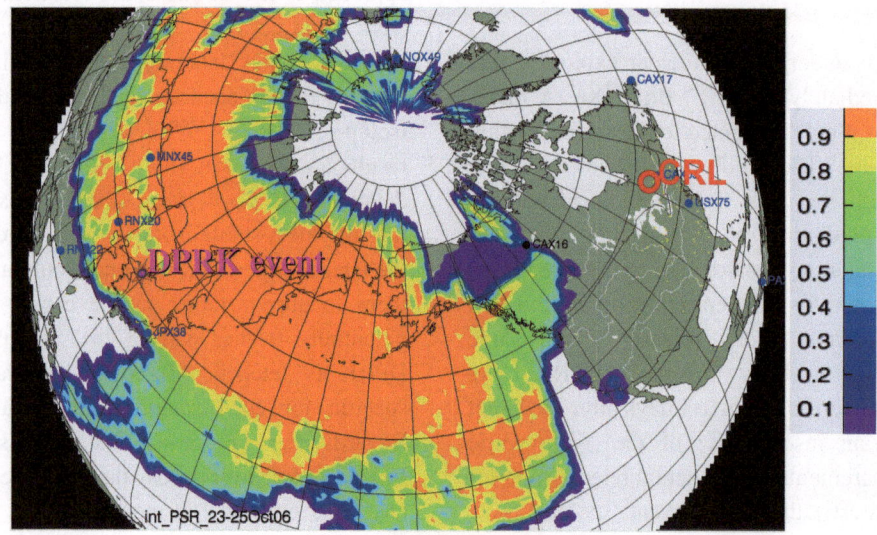

Figure 9

Backtracking of the first ^{133}Xe detection at station CAX16 in Yellowknife, Canada in terms of the possible source region (PSR) map showing regions where a point source surface emission at some time between 11 and 23 October would generate the Yellowknife observation between 23 and 25 October. The regions are ten colour-coded against the best correlation coefficient (range 0.1–1.0) achieved in each region during the time under consideration

Table 3

Results of October 2006 mobile measurements at Kansong (ROK) according to RINGBOM et al. (2007)

Sample	Collection start date and time	Acquisition time (s)	Collection stop date and time	Air volume (l)	CS date and time for ATM run 6	^{133}Xe (mBq/m^3)	^{131m}Xe (mBq/m^3)	^{133m}Xe (mBq/m^3)
RNX91	11 Oct 21:02	43,226	12 Oct 09:02:26	8,730	20061012 09:00:00	1.48 ± 0.17	–	–
RNX92	12 Oct 14:40	43,200	13 Oct 02:40:00	12,770	20061013 03:00:00	7.16 ± 0.51	–	–
RNX93	13 Oct 02:40	43,680	13 Oct 14:48:00	12,890	20061013 15:00:00	2.00 ± 0.24	–	0.58 ± 0.38
RNX94	13 Oct 14:49	43,106	14 Oct 02:47:26	12,940	20061014 03:00:00	1.58 ± 0.18	–	–
RNX95	14 Oct 02:49	43,260	14 Oct 14:50:00	13,710	20061014 15:00:00	–	0.22 ± 0.08	0.9 ± 0.54

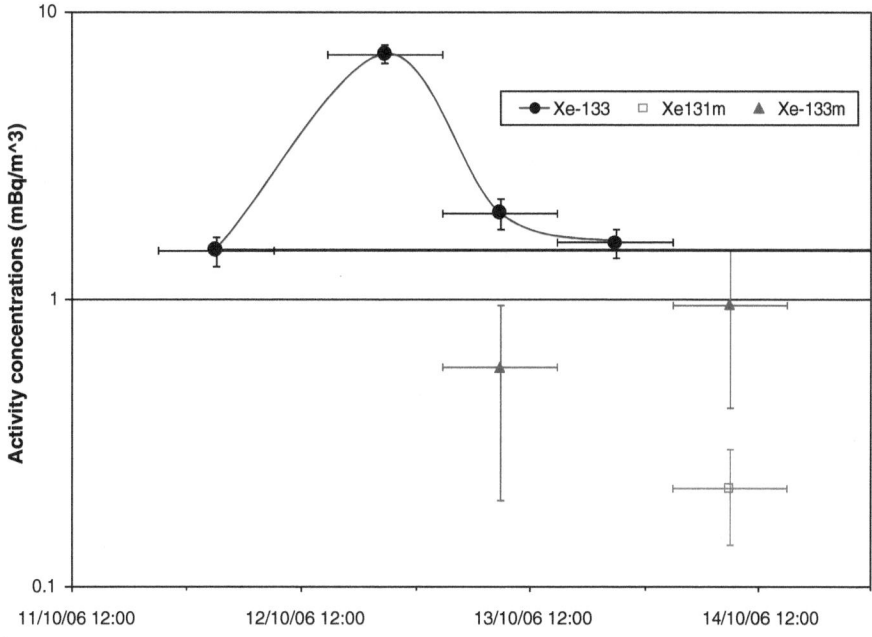

Figure 10

Illustration of Swedish measurements of RINGBOM *et al.* (2007) nearby Kansong, ROK from 11 to 14 October, 2006 also listed in Table 3. The *vertical error bars* denote the measurement uncertainties (at the 1σ level) and the *horizontal error bars* the air collection time

immediate venting. According to the plume dispersion calculations shown in Fig. 5, a part of the cloud is transported back in the direction of the Korean peninsula, and a model run (no. 2a, Table 1) indicates that it reaches the southern part of the peninsula on October 13. All models, however, indicate that the plume did not reach Kansong, ROK, but this can also not be totally ruled out. Regardless, the absence of ^{133m}Xe in the earlier samples indicates that the

detection in those samples originates from later seepage. In this paper we will focus on analysis of the ^{133}Xe results, since this was the only isotope also detected in the later Yellowknife measurements.

Backtracking was performed for all five samples based on ATM results calculated at 1.0° (Run 5) and 0.2° horizontal resolutions (Run 6, Table 1). The respective backtracking results were evaluated by means of a multiple regression analysis introduced

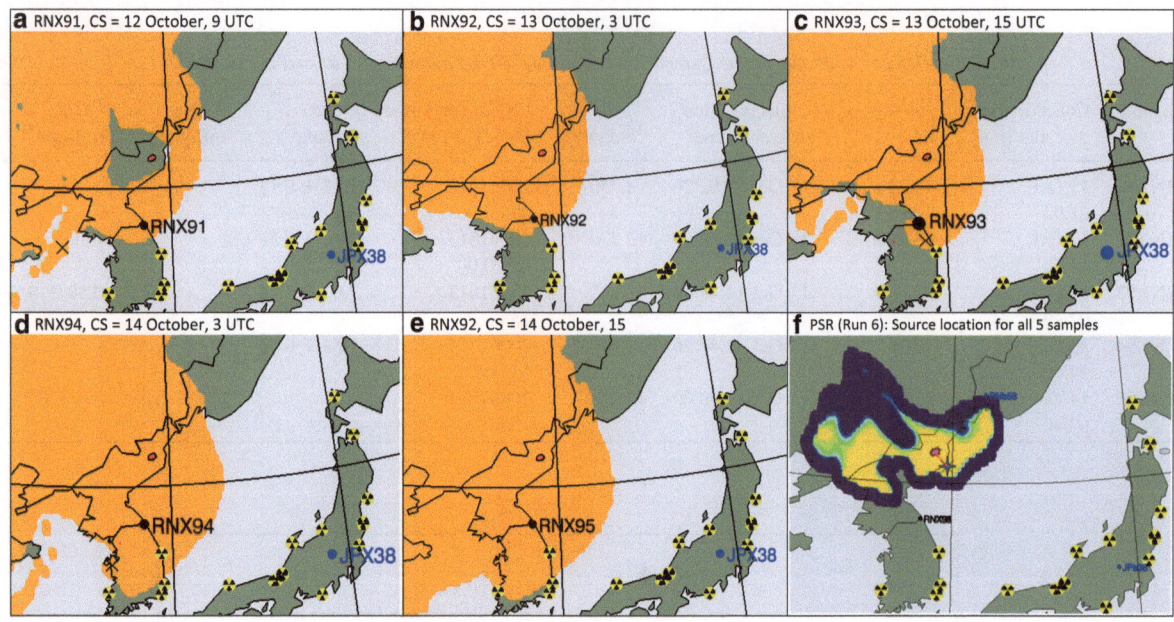

Figure 11

a–e Integral Field-of-Regards (FORs = areas exceeding a threshold sensitivity of $1 \times 10^{-2\circ}$ m^{-3} at least once during the backtracking period regarded) for the five xenon samples (RNX91-95) taken at Kansong (ROK), valid for the periods backward from their respective collections stop dates and times indicated to the day of the announced event at DPRK. The error ellipse of the belonging seismic event is also indicated as a *red object*. **f** Possible Source Region for the 3-h release time period when the maximum correlation of 0.938 (*blue diamond object*) is located closest to the actual seismic event location at $(\lambda, \varphi) = (129.00°E, 41.30°N)$. This gives the best source location estimate at $(\lambda, \varphi) = (129.7°E, 40.7°N)$ and $t = 11$ October (day 3), 21–24 UTC with a geo-temporal offset to the true location (*pink object*) of $\Delta r = 89$ km and $\Delta t = 69$ h. The radioactivity symbols indicate locations of nuclear power plants being potential sources of radio-xenon that would affect the five samples if covered by the FOR

and elaborated by WOTAWA *et al.* (2006, 2009) and KALINOWSKI *et al.* (2008) in order to identify the most consistent singular source hypothesis that would best explain the activity concentrations encountered in those five (four) ^{133}Xe samples in ROK that covered at least once the DPRK event location. Interestingly, a 3-day persistent release of 15×10^{12} Bq (Run 5) or 60×10^{12} Bq (Run 6) from 10 to 13 October, 0 UTC was shown to be most consistent with the measurements taken from the many possible release scenarios examined (Fig. 12). Moreover the correlation coefficient of the regression yielded on the basis of ATM results at 1.0° resolution was substantially higher (0.87) compared to the one on the basis of 0.2° horizontal resolution (0.65). Here the high resolution ATM tends to falsify a simple singular source assumption. In any event, the best estimates of the backtracking results spot the DPRK event location with a horizontal deviation of 89 km and a time delay of 69 h (Fig. 11f). Moreover the PSR covers a

comparatively confined region that concentrates mainly on the northern part of DPRK. This is a by far better confined PSR than the one encountered for the Yellowknife detection (Fig. 9) that covered almost half of the northern hemisphere. The difference, however, is that the Yellowknife PSR showed correspondence with the exact event time, while the PSR based on the mobile measurements corresponded to significantly delayed releases.

8. Reconstruction of the Release Scenario Based on Both Xenon Detections

With both backtracking results at hand, we can now significantly refine the source scenario reconstruction. Assuming a release of 1×10^{15} Bq to be distributed across a certain release time, the Yellowknife detection alone can only be used to state that any release time distribution across the first

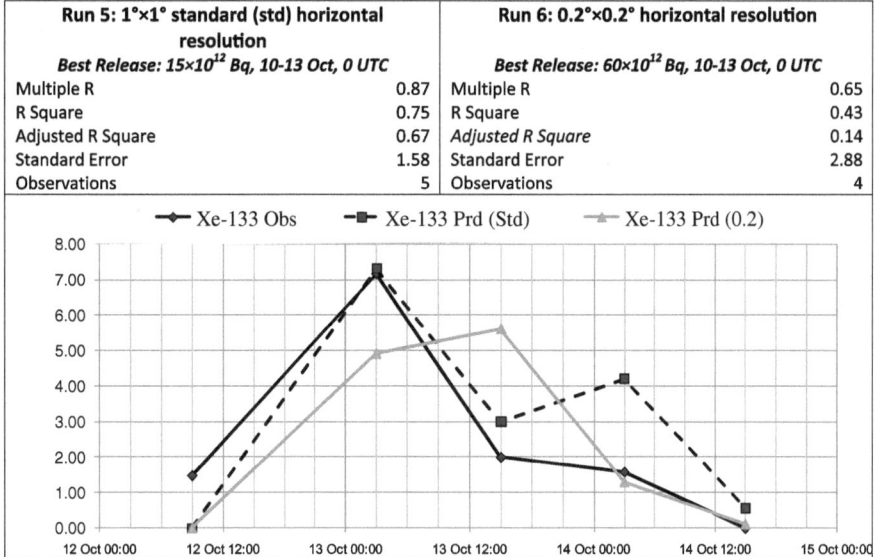

Run 5: 1°×1° standard (std) horizontal resolution		Run 6: 0.2°×0.2° horizontal resolution	
Best Release: 15×10¹² Bq, 10-13 Oct, 0 UTC		*Best Release: 60×10¹² Bq, 10-13 Oct, 0 UTC*	
Multiple R	0.87	Multiple R	0.65
R Square	0.75	R Square	0.43
Adjusted R Square	0.67	*Adjusted R Square*	0.14
Standard Error	1.58	Standard Error	2.88
Observations	5	Observations	4

Figure 12

Multiple regression analysis based on backtracking modelling of ROK measurements at $1.0°$ (Run 5, *left*) and $0.2°$ (Run 6, *right*) horizontal resolution. For both model setups the best fit to the observations is yielded when assuming a permanent release from 10 to 13 October, 0 UTC

3 days would be consistent with the observations. We have illustrated the earliest and latest possible case in Fig. 13 (Run 5 denoted by violet contours). All other possibilities between these extreme cases are equally likely if only the Yellowknife measurements are considered. The mobile measurements at ROK, however, indicate a much weaker and later release across the region (see blue contours of Runs 4 and 6 in Fig. 13). Combining both measurements to reconstruct the source in time (see shaded area in Fig. 13) the range of possible release scenarios is constrained, as the ROK measurements would falsify any assumption of a strong release taking place later than 24 h after the event. Consequently, looking at both measurements simultaneously, one would conclude that there was a strong initial release (within the first 24 h) at the event location, which was subsequently measured in Canada. During the following days, ^{133}Xe was still emitted, however the rate was two orders of magnitude weaker. This source was measured with the mobile instrument. Therefore, the consistent scenario would be an immediate or shortly (12–21 h) delayed venting followed by a later seepage (see shaded areas in Fig. 13). There still remains a range of possibilities, especially for day 1 after the event that has not been monitored by the close-by

mobile measurement at ROK (see green shaded areas and contour lines for Run 5 in Fig. 13). It would need further ^{133}Xe monitoring data taken in the regional vicinity of DPRK to gain a higher resolution of the immediate venting at day 1 of the release scenario.

9. Expected ^{133}Xe Detections during the DPRK Event in Case of a Completed PTS Xenon Network

Forward modelling-based (Run 4, Table 1) predictions of ^{133}Xe detections in a full-scale operating network show that there would have been a more than 500 times stronger detection at the IMS RN station in Ussuriysk, Russian Federation (RUX58) in comparison to the Yellowknife detection. However, none of the other IMS stations scheduled to measure ^{133}Xe in the East Asian full scale operation of the network is predicted to encounter xenon detection above 0.14×10^{-3} Bq/m³. The high ^{133}Xe detection at Ussuriysk is confirmed by the HR forward run no. 3 (Table 1) where the detailed series of plume snapshots (Fig. 4) shows that the plume is predicted to reach Ussuriysk within 1 day, whereby up- and down-mixing of the radionuclide debris obviously plays an

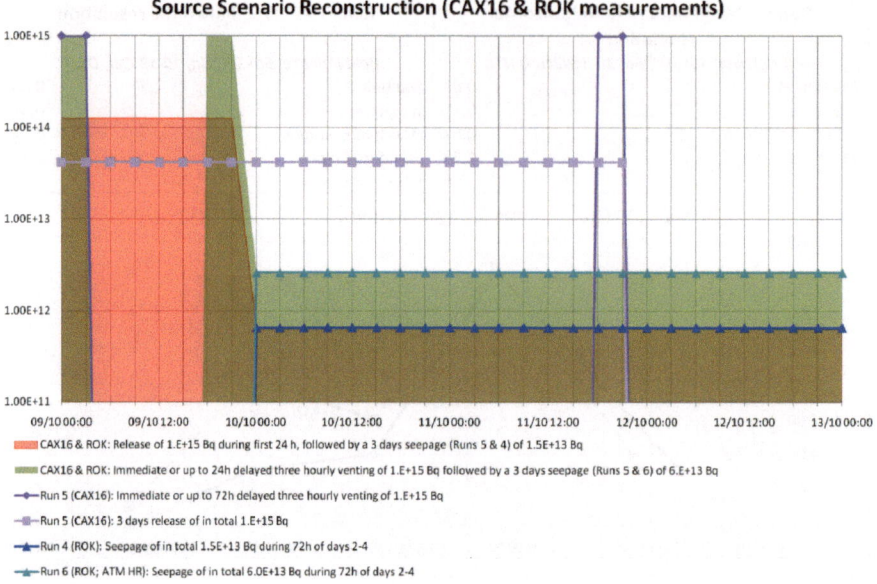

Figure 13
Release scenario reconstruction taking the measurements in Yellowknife (CAX16), Kansong (ROK) separately (*contour lines*) and jointly (*shaded areas*) into consideration. Obviously the ROK measurements help to constrain the range of times when the immediate venting could have taken place. The most likely scenario consists of both a (possibly delayed) strong venting on the first day and a later seepage during the following 3 days. The remaining ambiguity is reflected in the existence of the *greenish* and *red* that cover the range of possible release scenarios consistent with both measurements at CAX16 and ROK

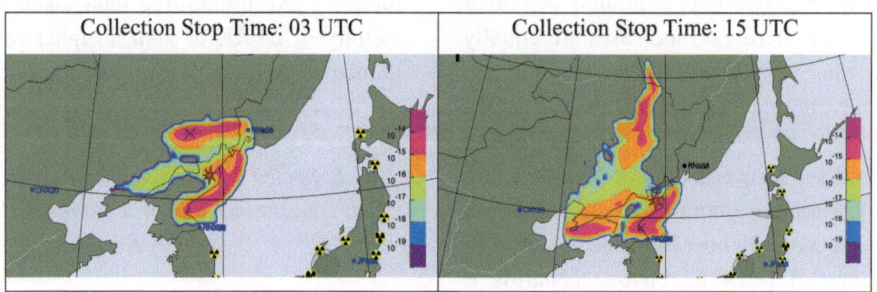

Figure 14
Quantitative FORs for two fictitious 12 h samples at station RUX58 (Ussuriysk, Russian Federation) in case it would have been in operation on 10 October, 2006 just 1 day after the event in DPRK (see *red star object*) occurred. The quantitative FORs are calculated based on Run 5 type (Table 1) of backtracking. The signal estimates derived from the sensitivities across the DPRK event location (*red star*) are 0.1–1 Bq/m^3 (early sample) and 1–10 Bq/m^3 (later sample) and correspond well with the results yielded for RUX58 by forward modelling (Runs 3 and 4, Figs. 4 and 5)

important role in view of the sudden appearance of the northern plume pattern across RN058 on 10 October, 2008, 0–3 UTC (see left middle row plot in Fig. 4). Moreover it is noticeable that the higher resolution run predicts four times higher peak activity concentration at Ussuriysk compared to run 4. Both forward modelling results also are confirmed by the respective sample specific backtracking results (Fields-of-Regard for RUX58, based on Run no. 6, Table 1) for

hypothetical 12 h samples with assumed collection stops at 10 October, 3 UTC and 15 UTC, that indicate a high sensitivity towards the release region that would result in a detection of about 1 Bq/m^3 if folded with a release of 1×10^{15} Bq. Moreover, this detection already would have taken place 1 day after the event, thus 11–16 days earlier compared to Yellowknife and therefore would have contained the shorter-lived xenon isotopes ^{135}Xe and ^{133m}Xe. Finally, the source location result based on such a strong and close detection would have been sufficiently confined to spot the seismic event belonging to the announced nuclear explosion at DPRK as the only possible source (Fig. 14). Nonetheless, even with just 25% of all scheduled noble gas stations (Fig. 1) the DPRK event has been detected and the source region estimation result based on the Yellowknife measurement only would not have screened out the associated geo-temporal location.

Given the fact that the geo-temporal source-receptor range to be examined is crucial for the accuracy of the backtracking result, it was quite useful to have the mobile measurements performed at Kansong, ROK at the time of the DPRK event. Nevertheless the station RUX58, if it would have been operational at the time of the DPRK event, would have provided all required information to clarify the nuclear character of the DPRK event.

10. Conclusions

A comprehensive study of the reconstruction of the most likely ^{133}Xe release scenario related to the announced October 2006 nuclear test event in the DPRK is presented. The study constitutes an enhancement of a prior study by SAEY et al. (2007) which was based only on the ^{133}Xe detections made at the remote IMS radio-xenon station Yellowknife, Canada, by adding the radio-xenon measurements raised by RINGBOM et al. (2007) with a mobile system deployed in ROK rather close to the event location 3–5 days after the event day on 9 October, 2006, to the analysis.

Forward and backward ATM methods have shown that both ^{133}Xe measurements were sensitive within the DPRK event region, however only the IMS station at Yellowknife also monitored the area at the exact time of the explosion. In contrast to this, four of the five measurements at ROK were only monitoring releases 2–4 days later than the event time. In the absence of any further detection, the source reconstruction presented is limited to the first four days, which excludes processes of barometrically driven later seepages from the possible scenarios resolvable.

The meteorological condition that governed the dispersion of potential ^{133}Xe releases from the DPRK event across the region of interest showed a high variability of the wind fields in space and time, and consequently a high sensitivity of the early stage of the plume dispersion to the assumed release scenario, according to our ATM based analysis. Forward ATM and folding of source-receptor sensitivity fields resulting from backward ATM with different release assumptions show that this high sensitivity to the release function is important for the close by measurements at ROK however completely blurred away during the long transport from DPRK to Yellowknife, Canada.

With regard to the source reconstruction of the first 4 days after the DPRK event of 9 October, 2006, the ^{133}Xe measurements in combination with our ATM backtracking methods result in the following findings:

- The most consistent ^{133}Xe source scenario consists of two phases: a strong release of 10^{15} Bq during 9 October, 2006 (day 1) followed by a substantially weaker release of 1.5–6×10^{13} Bq during the following 3 days.
- The scale of the strong release is derived from the Yellowknife detections that have been sensitive within the DPRK event region during 9, 10 and 11 October, 2006, for the first three days after the explosion.
- The sensitivity of the ROK measurements is three orders of magnitude stronger than for the Yellowknife measurements. However, the activity concentrations measured are only one order of magnitude higher, although just 300 km away from the DPRK event, instead of >7,000 km as is true for Yellowknife.
- Consequently the source calibration reveals source strengths that are two orders of magnitudes weaker for days 2, 3 and 4 compared with Yellowknife-based estimates, and the ROK measurements thus serve the temporal constraint for the end of the strong initial release detected at Yellowknife. Canada.
- The ROK observations, however, cannot provide any additional information to improve the source scenario reconstruction for day 1, as they showed no sensitivity for this day towards the DPRK event.

- Therefore, the day 1 release scenarios cannot be higher resolved as they are solely determined by the remote Yellowknife measurements. Consequently the ambiguity exists that the amount could have been released in an immediate venting as a puff at any time during the 24 h of 9 October, or as constant 24 h release with an accordingly smaller release rate.
- The ROK measurements exhibit a stronger sensitivity to the DPRK event location for 11, 12 October (days 3 and 4) than for 10 October (day 2), giving less indication on how exactly the transition form the stronger to the weaker release period took place on that day.
- Nevertheless, the ^{133}Xe source reconstruction of the DPRK event as depicted in Fig. 13 is the most detailed measurement constraint ever published.

Finally we have accessed the role of the mobile measurements to reconstruct the source scenario, and compared this to potential contributions of IMS radio-xenon stations scheduled to be operated in the 40-stations network operating at full scale. In doing so, we can state that the mobile measurements (RINGBOM et al., 2007) have been crucial to constrain the release scenario reconstruction. Moreover, they filled the gaps in the xenon measurement system in 2006 that featured only 25% of the scheduled stations available at the time of the event (Fig. 1).

However, we have shown that in a full-scale xenon network, the two 12 hourly samples taken on 10 October, 2006 (day 1 after the event) at the station in Ussuriysk, Russian Federation (RUX58) would have resolved most of the remaining uncertainties of the emission scenario for that day. Moreover, the xenon station in Takasaki, Japan (JPX38) would have monitored day 2 after the event if in operation at that time.

In any case, for days 3 and 4 the mobile measurements would have made unique contributions, even in a full-scale xenon network. Despite their strong dependence on the notice provided to be duly in the region of interest, mobile measurements can play an important role in supplementing the IMS network with a further short distance monitoring capability. This is truer for the more the event specific xenon release scenario features a seepage that can persist for many days after the date and time of a clandestine nuclear explosion.

REFERENCES

AUER, M., AXELSSON, A., BLANCHARD, X., BOWYER, T.W., BRACHET, G., BULOWSKI, I., DUBASOV, Y., ELMGREN, K., FONTAINE, J.P., HARMS, W., HAYES, J.C., HEIMBIGNER, T.R., MCINTYRE, J.I., PANISKO, M.E., POPOV, Y., RINGBOM, A., SARTORIUS, H., SCHMID, S., SCHULZE, J., SCHLOSSER, C., TAFFARY, T., WEISS, W., and WERNSPERGER, B. (2004), *Intercomparison experiments of systems for the measurement of xenon radionuclides in the atmosphere*, Appl. Radiat. Isot. *60*, 6, 863–877.

BECKER, A., WOTAWA, G., DE GEER, L.-E., SEIBERT, P., DRAXLER, R.R., SLOAN, C., D'AMOURS, R., HORT, M., GLAAB, H., HEINRICH, P., GRILLON, Y., SHERSHAKOV, V., KATAYAMA, K., ZHANG, Y., STEWART, P., HIRTL, M., JEAN, M., and CHEN, P. (2007), *Global backtracking of anthropogenic radionuclides by means of a receptor oriented ensemble dispersion modelling system in support of Nuclear-Test-Ban Treaty verification*, Atmos. Environm. *41*, 4520–4534.

CARRIGAN, C. R., HEINLE, R. A., HUDSON, G. B., NITAO, J. J., and ZUCCA, J. J. (1996), *Trace gas emissions on geological faults as indicators of underground nuclear testing*, Nature *382*, 528–531.

D'AMOURS, R., BEAN, M., BOCK, K., HOFFMAN, I., KORPACH, E., MALO, A., STOCKI, T.J., and UNGAR, R.K. (2007), Canada's measurement of the DPRK event. Presentation to informal xenon measurements workshop, 5–9 November 2007, Las Vegas, NV.

DE GEER, L.-E. (1996), *Sniffing out Clandestine tests*, Nature *382* (1996), p. 491.

ENSDF (2009), Evaluated Nuclear Structure Data File (ENSDF) Database version of May 8, 2009, http://www.nndc.bnl.gov/ensdf/.

KALINOWSKI, M.B., BECKER, A., SAEY, P. J. R., TUMA, M., and WOTAWA, G. (2008), *The complexity of CTBT verification. Taking noble gas monitoring as an example*, Complexity *14*, 89–99.

KALINOWSKI, M.B., AXELSSON, A., BEAN, M., BLANCHARD, X., BOWYER, T.W., BRACHET, G., MCINTYRE, J.I., PETERS, J., PISTNER, C., RAITH, M., RINGBOM, A., SAEY, P.R.J., SCHLOSSER, C., STOCKI, T.J., TAFFARY, T., and UNGAR, R.K. (2009), In *Recent Advances in Nuclear Explosion Monitoring* Pure Appl. Geophys. Topical Volume (eds. BECKER, A., SCHURR, B., KALINOWSKI, M.B., KOCH, K., BROWN, D.).

KRAUS, H. (2001), *Die Atmosphäre der Erde: Eine Einführung in die Meteorologie*. 2nd Edition, Springer Verlag, Berlin Heidelberg, ISBN 3-540-41844-X.

KIM, W.Y., and RICHARDS, P. G. (2007), *North Korean nuclear test: Seismic discrimination at low yields*, Eos, Transactions, Am. Geophys. Union *88*, 158–161.

LE BRAS, R., HAMPTON, T., COYNE, J., BOBROV, D., and ZERBO, L. (2007), *CTBTO seismic processing and the announced DPRK nuclear test of October 9, 2006*, Geophys. Res. Abs. *9*, 07286.

PERKINS, R.W. and CASEY, L.A. (1996), *Radioxenons: Their role in monitoring a comprehensive test ban treaty*, Rep. DOE/RL-96-1, Pac. Northwest Natl. Lab., Richland, Washington.

RINGBOM, A., ELMGREN, K., and LINDH, K. (2007), *Analysis of radioxenon in ground level air sampled in the Republic of South Korea on October 11-14, 2006*, FOI report. *FOI-R-2273-SE*.

SAEY, P.R.J., BEAN, M., BECKER, A., COYNE, J., D'AMOURS, R., DE GEER, L.-E., HOGUE, R., STOCKI, T.J., UNGAR, R.K., and WOTAWA, G. (2007), *A long distance measurement of radioxenon in Yellowknife, Canada, in late October 2006*, Geophys. Res. Lett. *34*, L20802, doi:10.1029/2007GL030611.

SCHLITTENHARDT, J., CANTY, M.J., and GRUENBERG, I. (2009), *Satellite Earth observations support CTBT monitoring: A case study of the nuclear test in North Korea of Oct. 9, 2006 and comparison with seismic results*. In *Recent Advances in Nuclear Explosion Monitoring*, Pure Appl. Geophys. Topical Volume (eds. BECKER, A., SCHURR, B., KALINOWSKI, M.B., KOCH, K., BROWN, D.).

SCHULZE, J., AUER, M., and WERZI, R. (2000), *Low level radioactivity measurement in support of the CTBTO*, Appl. Radiat. Isot. *53*, 23–30.

STOHL, A., HITTENBERGER, M., and WOTAWA, G. (1998), *Validation of the Lagrangian Particle Dispersion Model FLEXPART against large-scale tracer experiment data*. Atmos. Environ. *32*, 4245–4264.

STOHL, A., FORSTER, C., FRANK, A., SEIBERT, P., and WOTAWA, G. (2005), *Technical note: The Lagrangian particle dispersion model FLEXPART version 6.2*, Atmos. Chem. Phys. *5*, 2461–2474.

ULIASZ, M. (1994), *Lagrangian particle dispersion modeling in mesoscale applications*. In *Environmental Modeling, Vol. II* (ed. ZANNETTI, P.) Computational Mechanics Publications, Southampton, UK, 1994. 4741, 4749, 4766.

WOTAWA, G., DE GEER, L.-E., DENIER, P., KALINOWSKI, M., TOIVONEN, H., D'AMOURS, R., DESIATO, F., ISSARTEL, J.-P., LANGER, M., SEIBERT, P., FRANK, A., SLOAN, C., and YAMAZAWA H. (2003), *Atmospheric transport modelling in support of CTBT verification–overview and basic concepts*, Atmos. Environ. 37, 18, 2529–2537.

WOTAWA, G., DE GEER, L.-E., BECKER, A., D'AMOURS, R., JEAN, M., SERVRANCKX, R., and UNGAR, K. (2006), *Inter- and intra-continental transport of radioactive cesium released by boreal forest fires*, Geophys. Res. Lett. *33*, L12806, doi:10.1029/2006GL026206.

WOTAWA, G., BECKER, A., KALINOWSKI, M.B., SAEY, P.J.R., TUMA, M., and ZÄHRINGER, M. (2009), *Computation and analysis of the global distribution of the Radioxenon Isotope ^{133}Xe based on emissions from nuclear power plants and isotope production facilities and its relevance for the verification of the Nuclear-Test-Ban Treaty*. In *Recent Advances in Nuclear Explosion Monitoring*, Pure Appl. Geophys. Topical Volume (eds. BECKER, A., SCHURR, B., KALINOWSKI, M.B., KOCH, K., BROWN, D.).

(Received January 26, 2009, revised June 29, 2009, accepted July 25, 2009, Published online December 19, 2009)

Pure Appl. Geophys. 167 (2010), 601–618
© 2010 The Author(s)
This article is published with open access at Springerlink.com
DOI 10.1007/s00024-009-0036-x

█ **Pure and Applied Geophysics**

Satellite Earth Observations Support CTBT Monitoring: A Case Study of the Nuclear Test in North Korea of Oct. 9, 2006 and Comparison with Seismic Results

J. Schlittenhardt,[1] M. Canty,[2] and I. Grünberg[1]

Abstract—The Comprehensive Nuclear-Test-Ban Treaty prescribes the use of seismic stations and arrays as the main measure for verification of Treaty compliance. Since the inception of the Treaty, a vast amount of open source earth observation satellite data has become available. This paper investigates the potential for combining seismic and satellite data for more effective monitoring and response. With data acquired before, during and after the alleged North Korean underground nuclear test on October 9, 2006, wide area change detection techniques using medium resolution optical/infrared satellite sensors are combined with localized high-resolution imagery to attempt to pinpoint the test location within the area identified by the seismic measurements. Problems associated with the timeliness, degree of coverage and ambiguity of the remote sensing data are pointed out, however it is generally concluded that their integration into the CTBT regime would valuably complement the existing seismic observation network.

1. Introduction

Following the October 9, 2006 North Korean nuclear test, seismic signals were recorded around the world at sensitive seismic stations and arrays of the international monitoring system (IMS). This system is being currently built-up by the UN PrepCom (United Nations Preparatory Commission) to monitor compliance with the CTBT (Comprehensive Nuclear-Test-Ban Treaty). Also a number of high quality networks and array stations installed for earthquake monitoring (Ammon and Lay, 2007; Tibuleac et al., 2008) recorded the nuclear test. Different agencies (see Table 1) reported a body-wave magnitude around four for the event, which typically implies

only weak signals at global stations. However, seismic stations at so-called regional distances between 200 and 2,000 km from the epicenter routinely record clear regional phases (e.g., Pg, Sg, Pn, Sn, and Lg) for earthquakes and explosions of that size that can be used for independent epicenter location with seismic velocity models of the Earth's crust.

In order to check the teleseismic (stations at distances greater than 2,000 km) IDC (International Data Centre) and USGS (United States Geological Survey) localization independently, waveform data of stations in the regional distance range were examined for onsets of regional seismic waves suitable for seismic localization. The resulting epicenter location is listed in Table 1: the IDC, USGS and BGR's (Bundesanstalt für Geowissenschaften und Rohstoffe) epicenter locations consistently fall within a radius of 5 km and lie within the area of the seismic confidence ellipse for the IDC location.

Although satellite imagery analysis is not an element of the IMS it can provide, especially in combination with seismology, important information for the verification regime stipulated in the CTBT (Canty and Schlittenhardt, 2001). Past experiences include the investigation of satellite image analysis techniques in combination with seismic detection and localization data for their use in supplementing verification measures provided by the IMS (Fisk, 2002; Gupta, 1995; Thurber et al., 1993). Synergy between seismic and satellite-based data for the characterization of underground nuclear testing in the CTBT monitoring context has been demonstrated through the application of the multispectral MAD (Multivariate Alteration Detection) technique to historical underground nuclear explosions detonated at the Indian

---
[1] Bundesanstalt für Geowissenschaften und Rohstoffe, Stilleweg 2, 30655 Hannover, Germany. E-mail: joerg.schlittenhardt@bgr.de
[2] Forschungszentrum Jülich GmbH, 52425 Jülich, Germany.

Table 1

Epicentral parameters of the North Korean Nuclear Test on October 9, 2006

Agency	Origin time	Latitude	Longitude	Number of stations	Magnitude mb
IDC	01:35:27.6	41.3119 N	129.0189E	22	4.1
USGS	01:35:28.0	41.294 N	129.094E	31	4.2
BGR	01:35:28.9	41.286 N	129.134E	6	4.0[a]

[a] Inferred from GRESS array beam

Test Site and at the Nevada Test Site, USA, (CANTY and SCHLITTENHARDT, 2001; CANTY et al., 2005). In continuation of this work, the ability of DInSAR (Differential Interferometric Synthetic Aperture Radar) techniques to reveal both co-seismic and post-seismic subsidence signals in the cm-range within the damage (spall) zone caused by underground nuclear explosions, as reported by VINCENT et al. (2003), was revisited using ERS-data over the Nevada Test Site (CONG et al., 2007). Subsequent investigations at other test sites, including the nuclear test in North Korea, also showed the limits of the DInSAR technique when mountainous topography and/or temporal and spatial baseline differences become critical factors (SCHLITTENHARDT et al., 2008).

In this paper we investigate the extent to which commercially available satellite data can be used to supplement the seismological monitoring techniques. The critical question that is to be answered under a CTBT verification regime is whether the true epicenter location lies within the seismic confidence ellipse (with an area $<1,000$ km^2), so that an OSI (On-Site Inspection) team can be guided to ground zero of the suspected explosion. In this paper we use available multispectral optical data from the Advanced Spaceborn Thermal Emission and Reflectance Radiometer (ASTER) for a change detection study to find evidence for the explosion. Unfortunately the available data only partly covered the area of the seismic location confidence ellipse. However, the data included the region that has been reputed to be a possible test site. The seismic epicenter locations of the CTBTO, USGS and BGR all lie within distances of a few kilometers from the detected changes. The changes that occurred within a time span of 17 days enclosing the explosion are co-located with an area with mining activity and buildings and small roads in an otherwise uninhabited region. Two high

resolution (0.7 m pixel) images centered on the region with the ASTER wide area change detections were available for a comparative analysis of the area before and after the nuclear test. For two of five areas with detected change signals from the ASTER data, clear evidence for man-made changes was found through the comparative analysis of the high resolution imagery. Some of the changes found have characteristics that can be associated with preparation/excavation activities possibly connected to the underground test. Among these changes are the insertion of a new supporting pillar, changes in the shape of landfill and the spread of deposited material, and leveling of a slope close to the mine entrance.

2. Seismic Monitoring (Results)

2.1. Detection, Identification, Location

Both the International Data Center of the Comprehensive Nuclear-Test-Ban Treaty Organization in Vienna (Austria) and the National Earthquake Information Center (NEIC) of the United States Geological Survey detected and located the nuclear test within hours of its occurrence (KVAERNA et al., 2007; KOPER et al., 2008). Figure 1 shows the geographical location of the test as determined by the IDC together with the location of magnitude > 3.5 shallow seismic events with depths < 50 km. As can be seen from the map, especially north of the Korean Peninsula is characterized by a low seismicity. On the one hand this eases the classification of the event as dubious although on the other hand it complicates seismic discrimination because of the only limited number of known earthquakes in the region. However, in a number of studies (KIM and RICHARDS, 2007; KVAERNA et al., 2007; KOPER et al., 2007; ZHAO et al., 2008) using high-frequency spectral ratios of regional P-and S-wave signals that were archived from previous small earthquakes, the event of October 9, 2006, was identified as an explosion with very high confidence. PATTON and TAYLOR (2008) present an explosion model where tensile failure of surface layers is completely suppressed to explain the poor performance of the teleseismic $m_b - M_s$ discriminant for the North Korean test.

In order to check the IDC and USGS localization, which was mainly based on teleseismic data,

waveform data of stations in the regional distance range were examined for onsets of regional seismic waves. Figure 2 shows the waveforms of the available seismograms together with the detected seismic phases marked with arrows. Due to the geographical circumstances, the use of non-IMS stations of the IRIS-(Incorporated Research Institutions for Seismology) consortium considerably reduced the achievable azimuth gap and aided improvement of the location accuracy (Fig. 3). The resulting location based on the arrival times listed in Table 2 (using the

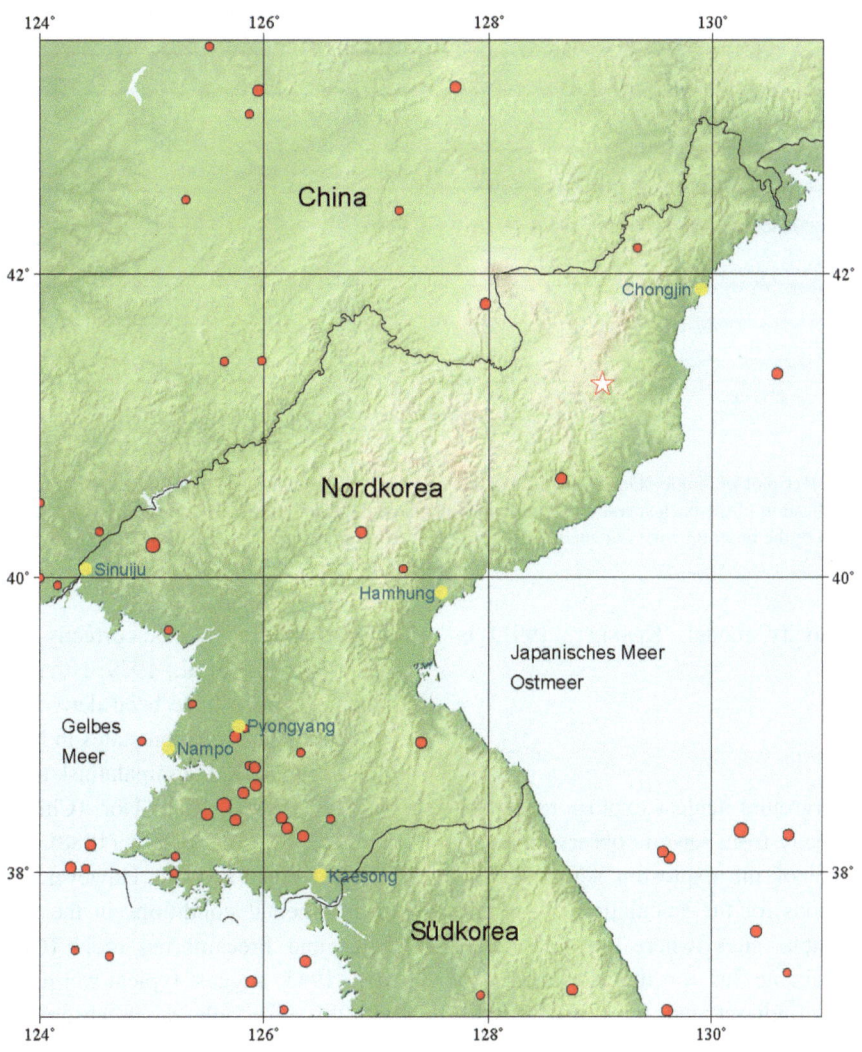

Figure 1

Map of North Korea and adjacent countries together with the epicenters (*red dots*) of seismic events (depth <50 km) with magnitudes >3.5 for the period since 1970 (*Source*: ISC and NEIC, USGS). *Yellow dots* show location of cities. The *star marks* the location of the nuclear test (color figure online)

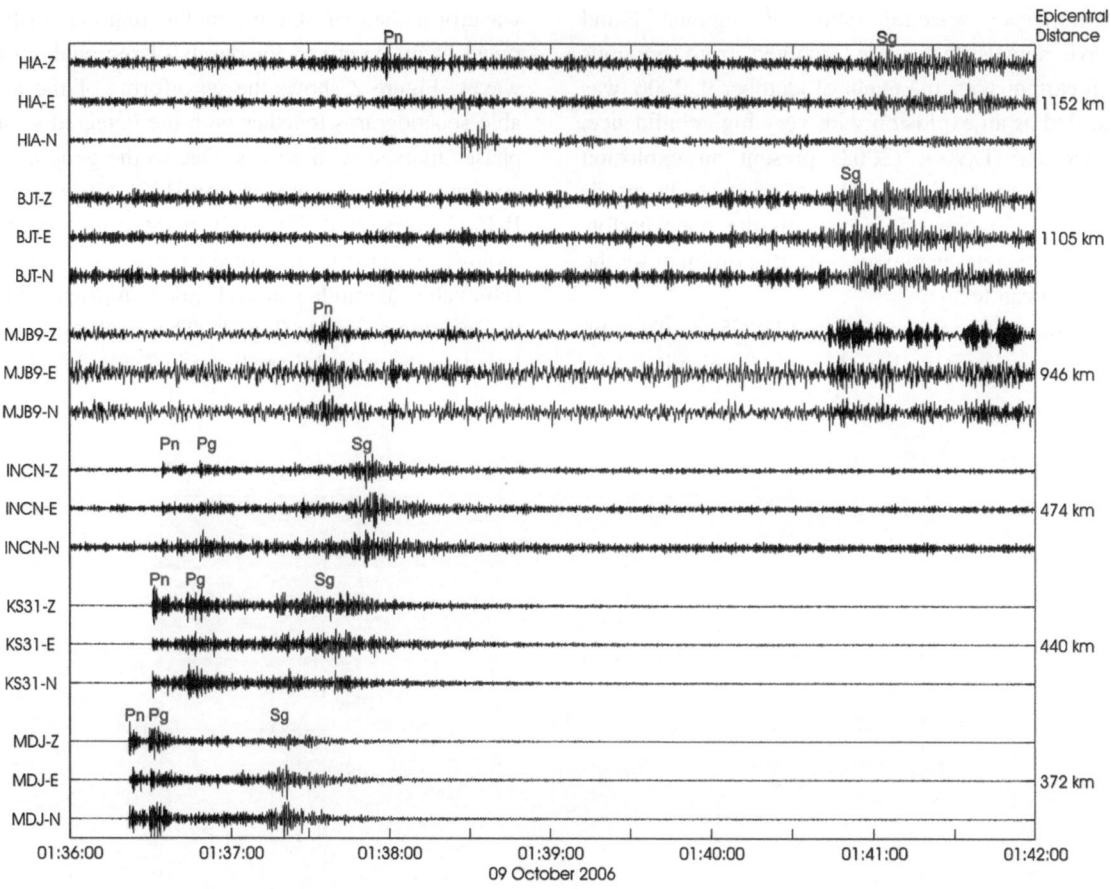

Figure 2

Bandpass-filtered (1–8 Hz) plot of single station seismic recordings (Z, E, N components) of IMS and IRIS stations/arrays at regional distances used for localization of the nuclear test on October 9, 2006. Note the local noise wavetrains on station MJ9B of the MJAR array shortly after 01:40:30. For the onset time measurements, signal-to-noise ratio improvement was achieved through optimal bandpass filtering and array beamforming

IASPEI 1991 velocity model; KENNETT, 1991) is given in Table 1.

2.2. Yield Estimation

Yields of underground nuclear explosions cannot be determined directly from seismic observations of, e.g., the amplitude of the explosion signal on the seismogram. Methods for the calculation of seismic yield at inaccessible sites where no calibration information is available (as for the North Korean test site) are only indirect and must rely on the transportability of known magnitude-yield relations. Such "transportation" becomes possible if the differences in the absorption of seismic body waves in the upper mantle beneath the different test sites

(magnitude bias) can be correctly accounted for (see, e.g., MARSHALL et al., 1979; NUTTLI, 1986). Test-site bias corrections have been derived for the main test sites of nuclear weapon states in Nevada (carried out by USA and UK), Semipalatinsk and Novaya Zemlya (former USSR), Lop Nor (China) and Tuamotu (France), SCHLITTENHARDT (1988), but are not known for the Korean test site. However, the geological and environmental conditions in the region with Cretaceous and Precambrian rocks (General Geological Map, 1945) suggest typical wet hard rock conditions with little or only weak absorption (high seismic quality factor Q) of seismic body waves below the test site (ZHAO et al., 2008).

For the yield estimation at the German NDC (National Data Centre) a body-wave magnitude *mb*

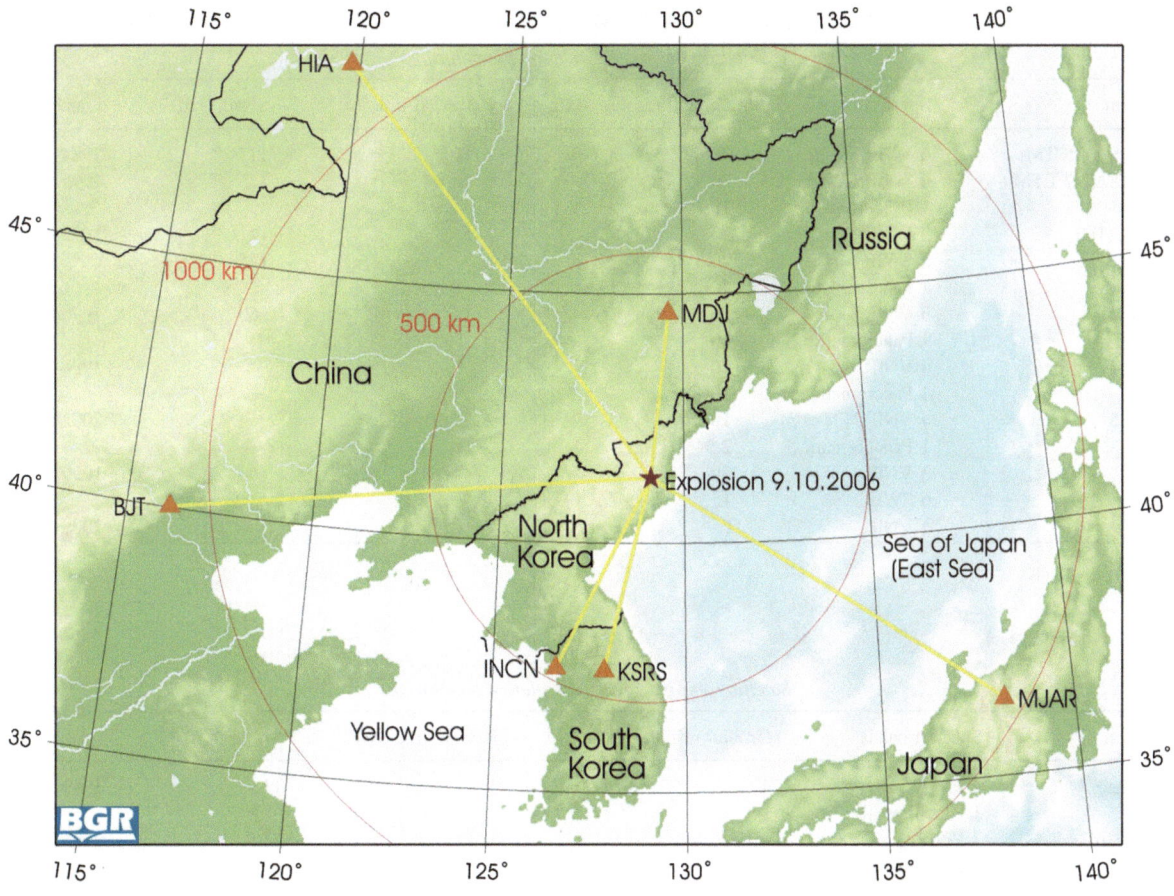

Figure 3

Azimuthally equidistant map of North Korea and adjacent countries showing the locations of the stations used for regional seismic localization (*red triangles* with station codes for IMS (KSRS and MJAR) and IRIS stations (MDJ, HIA, BJT, INCN), respectively) of the explosion epicenter (*star*). *Yellow lines* represent surface trajectories of the ray paths of the various seismic phases used (color figure online)

Table 2

Stations used for regional epicenter determination. Arrival times are on October 9, 2006

Station	Distance (km)	Arrival time	Phase
MDJ	369.1	01:36:21.58	Pn
MDJ	369.1	01:36:29.05	Pg
MDJ	369.1	01:37:13.42	Sg
KSRS (KS31)	442.2	01:36:29.80	Pn
KSRS (KS31)	442.2	01:36:43.10	Pg
KSRS (KS31)	442.2	01:37:31.04	Sg
INCN	474.4	01:36:33.95	Pn
INCN	474.4	01:36:47.69	Pg
INCN	474.4	01:37:45.49	Sg
MJAR (MJB9)	945.2	01:37:31.88	Pn
BJT	1098.7	01:40:42.57	Sg
HIA	1098.7	01:37:57.34	Pn
HIA	1145.3	01:40:59.77	Sg

4.1 with an uncertainty allowance of ± 0.1 magnitude units was adopted. Allowing for the maximum variation in the test site bias occurring for the previously inferred magnitude-yield relations (reflecting a wide range from little to high absorption of seismic waves) gives a yield range of 0.6–1.7 kt, and 0.4–2.2 kt for the additional magnitude uncertainty allowance. This estimate agrees quite well with the yield value derived from regional Lg waves recorded by CENC (China Earthquake Network Center) stations in northeast China and teleseismic estimates made by others (KIM and RICHARDS, 2007; KVAERNA et al., 2007; KOPER et al., 2008; TIBUELAC et al., 2008). The apparent overestimation of the explosion yield based on surface wave empirical Ms/yield relationships can, at least to some degree, be explained by

Table 3

Satellite sensors for wide area monitoring

Platform	Spectral bands	Ground resolution (m)	Revisit time (d)	Swath (km)	Launch date
LANDSAT 5 TM	6 VNIR/SWIR	30	16	175	1984
LANDSAT 7 ETM+	6 VNIR/SWIR	30	16	175	1999
	1 Panchromatic	15			
IRS 1C(1D)	2 VNIR/	23.5	5	141	1995 (97)
	2 SWIR	70.5		148	
	1 Panchromatic	5.8			
SPOT 2	3 VNIR	20	1–4	60	1990
	1 Pancromatic	10			
SPOT 4	4 VNIR	20			1998
	1 Pancromatic	10			
SPOT 5	4 VNIR	10–20	2–3		2002
	1 Panchromatic	2.5–5			
ASTER	3 VNIR	30	4–16	60	1999
	6 SWIR	15			
RapidEye 1-5	5 VNIR	5	1–5	77	2008

Table 4

Satellite sensors for high resolution monitoring

Platform	Spectral bands	Ground resolution (m)	Revisit time (d)	Swath (km)	Launch date
Ikonos	4 VNIR	4	3	11	1999
	1 Panchromatic	1			
QuickBird	4 VNIR	2.4	1–3	16.5	2001
	1 Panchromatic	0.6			
WorldView 1	1 Panchromatic	0.55	2–6	17.6	2007
EROS A	1 Panchromatic	1.8	2–10	12	2000
EROS B	4 VNIR	3.5	2–10	13	2006
	1 Panchromatic	0.7		7	

emplacement medium effects on surface wave magnitude and moment (BONNER et al., 2008).

3. Available Optical Satellite Data Overview

Here we only mention the panchromatic, visual and near infra-red (VNIR) and short-wave infra-red (SWIR) capabilities of the respective sensors; thermal radiation detection not being relevant in the present context. The information is summarized in Tables 3 and 4 for two classes of sensor: Those suitable for wide-area monitoring and change detection, and those having very high ground resolution for detailed image interpretation.

3.1. Wide-Area Monitoring

In the context of CTBT verification, we can define "wide area" as being roughly equivalent to the geographical extent of a nuclear test site. For example, the NTS extends over an area of about 3,500 km². Alternatively, for purposes of verification of seismic events, we can choose the extent of a typical error ellipse as reference, about 1,000 km². These areas are covered for example roughly by ASTER full scenes (60×60 km²) or LANDSAT full scenes (170×183 km²). Any of these or comparable instruments can be used for wide-area monitoring. At this scale (pixel sizes ranging from 5 m (RapidEye 2) to 30 m (Landsat 5, 7), see Table 4), pixel-oriented image processing methodologies—as opposed to feature-based approaches—are more appropriate. In particular for unsupervised change detection, that is, for change detection in the absence of ground reference data, various change detection techniques are available. These include thresholding of simple image differences and ratios, change vector analysis, post-classification comparison and iterated principal components analysis. For reviews of change detection methods in remote sensing see SINGH (1989). For more recent reviews of change detection in a more general context see RADKE et al. (2005) or COPPIN et al. (2004).

The Iteratively Re-weighted Multivariate Alteration Detection (IR-MAD) algorithm (NIELSEN et al., 1998; NIELSEN, 2007) was chosen in the present investigation. This is due to its robustness to variations in acquisition conditions such as instrument gain, atmospheric scattering and absorption, solar illumination or seasonal changes in vegetation.

3.1.1 LANDSAT

Two LANDSAT satellites are currently operational:

- LANDSAT 5 with 6 VNIR/SWIR spectral bands with a ground resolution of 30 m.
- LANDSAT 7 also with 6 VNIR/SWIR spectral bands with a ground resolution of 30 m and a panchromatic band with 15 m resolution.

The swath width in both cases is 175 km and revisit time is 16 days. The LANDSAT 7 satellite is operational despite a scan line corrector failure in 2003.

3.1.2 IRS

Several IRS (Indian Remote Sensing) satellites are in orbit and provide panchromatic and multispectral images with ground resolutions between about 5 and 30 m, respectively. The IRS C,D pan sensor can be pointed off the orbit path which allows 2–4 day revisits to specific sites.

3.1.3 SPOT

Three SPOT (Satellite Pour l'Observation de la Terre) satellites are currently operational:

- SPOT 2: The panchromatic band has a resolution of 10 m, and the 3 multispectral bands have resolutions of 20 m. It has an image swath of 3,600 km^2 and a revisit interval of 1–4 days depending on the latitude.
- SPOT 4: This platform has an additional 20 m resolution band at mid-infrared wavelengths (1.58–1.75 μm). Two HRVIR imaging instruments are programmable for independent image coverage, increasing the number of imaging opportunities.
- SPOT 5: The satellite has two high resolution geometrical (HRG) instruments offering a higher resolution of 2.5–5 m in panchromatic mode and 10 m in multispectral mode. The pointing capabilities of the panchromatic camera allow acquisition of near simultaneous stereo pairs.

3.1.4 ASTER

ASTER (Advanced Spaceborne Thermal Emission and Reflection Radiometer) is one of five remote sensory devices on board the Terra satellite launched into Earth orbit by NASA in 1999. It provides stereo and VNIR images with ground resolution of 15 m as well as 6 SWIR spectral bands with ground resolutions of 30 m. The pointing capability of the VNIR camera allows revisit times of about 4–16 days.

3.2. High Resolution Monitoring

3.2.1 Ikonos

Ikonos is a commercial Earth observation satellite, and was the first to collect publicly available high-resolution imagery at 1- and 4-meter resolution. It offers 4 multispectral bands and one panchromatic band with revisit times of 3–5 days off-nadir and 144 days for true-nadir. Single scenes are 11×11 km^2.

3.2.2 QuickBird

QuickBird is a high-resolution commercial Earth observation satellite, owned by DigitalGlobe and launched in 2001 as the first satellite in a constellation of three. The satellite collects panchromatic imagery at 60–70 cm resolution and provides 4 multispectral bands at 2.4- and 2.8-m resolutions. Revisit frequency is 1–4 days.

3.2.3 WorldView

WorldView-1 is also a commercial Earth observation satellite owned by DigitalGlobe and launched September 18, 2007. With an average revisit time of 1.7 days, WorldView-1 is capable of collecting up to 750,000 km^2 per day at 0.5 m resolution.

3.2.4 EROS

EROS (Earth Resources Observation Satellite) is a series of Israeli commercial Earth observation satellites.

- EROS A was launched on December 5, 2000 and provides panchromatic images with an optical resolution of 1.8 m.
- EROS B was launched on April 25, 2006. The satellite offers an panchromatic resolution of 70 cm.

4. IR-MAD

To summarize the IR-MAD algorithm briefly, consider two N-band images registered to one another with sub-pixel accuracy. The observations in the first image may be represented by a random vector $F = (F_1...F_N)^T$, those in the second image by $G = (G_1...G_N)^T$. The components F_i and G_i correspond to the original spectral bands and are conventionally ordered by wavelength. The MAD algorithm determines transformation matrices A and B such that the components of the transformed random vectors $U = AF$, $V = BG$ are ordered by similarity, where similarity is measured by positive linear correlation. The transformation matrices are obtained by applying standard Canonical Correlation Analysis (CCA) (HOTELLING, 1936). The components of the transformed images $U = [U_1...U_N]^T$ and $V = [V_1...V_N]^T$ are called the canonical variates (CVs). They are mutually orthogonal and have unit variance. The pair (U_1, V_1) is maximally correlated, the pair (U_2, V_2) is maximally correlated subject to being orthogonal to (uncorrelated with) both U_1 and V_1, and so on.

Performing paired differences (in reverse order) of the canonical variates generates a sequence of transformed images

$$M_i = U_{N-i+1} - V_{N-i+1}, \ i = 1...N, \qquad (1)$$

referred to as the MAD variates. The MAD variates have nice statistical properties which make them very useful for visualizing and analyzing change information. Thus for instance they are uncorrelated, and the sum of the squared standardized MAD variates is approximately chi-square distributed with N degrees of freedom.

An improvement of the sensitivity of the MAD transformation is obtained by establishing an increasingly better background of no change against which to detect change. This can be done in an iteration scheme in which observations are weighted by the probability of no change, as determined on the preceding iteration, when estimating the sample means and covariance matrices for the next iteration (NIELSEN, 2007). No-change probabilities are conveniently estimated from the chi-square values mentioned above.

The change detected in this fashion is invariant (or insensitive) to separate linear (affine) transformations in the originally measured spectral signals at the two points in time, such as changes in gain and offset in the measuring device used to acquire the data, data normalization or calibration schemes that are linear (affine) in the gray values of the original spectral signals, or any orthogonal or other affine transformations used in preprocessing. IR-MAD typically isolates different change signatures in different change components.

5. Imagery and Preprocessing

Two ASTER images covering parts of the presumed test area, acquired on September 30 and October 16, 2006, and thus bracketing the event, were available on the Land Processes Distributed Active Archive Center (LPDAAC) for wide area change detection. Although both full scenes covered an area of 60 × 60 km^2, the overlap was unfortunately quite small. Moreover, the September image was partly obscured by clouds. Figure 4 shows part of the image acquired on October 16. The ASTER images were downloaded from the LPDAAC EOS Data Gateway website as Level 1B registered radiance at the sensor. DEMs were then generated from the 3 N and 3B

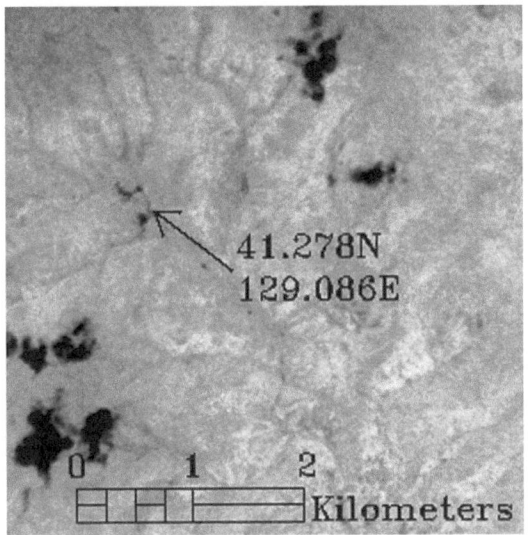

Figure 4

IR-MAD variate 2. *Bright* and *dark pixels* signify change; *middle gray pixels* indicate no-change

infrared stereo bands and the images were ortho-rectified with the commercial ENVI add-on ASTER-DTM (SULSOFT, 2003). The ortho-rectified images were registered to one another using ENVI's automatic image-image registration algorithm (linear polynomial, nearest neighbor resampling). The RMS error was about 0.6 pixels.

5.1. Results

Figure 4 shows the second IR-MAD variate obtained for the ASTER data. This IR-MAD component showed the changes most clearly. Bright and dark pixels signify change; middle gray pixels indicate no-change. The large, irregular change signals are due to cloud cover in the September 30 image. The changes indicated at coordinates 41.278 N, 129.086 E may be associated with preparation/excavation activities connected to the underground test. This will be discussed in detail in the section below.

6. Comparative Analysis of High Resolution Optical Satellite Images

Two high resolution images (Quickbird and EROS) taken approximately 3 weeks before (17

September, 2006) and 1 week after (16 October, 2006) the test on October 9, 2006 were available for a detailed analysis of the region covered by the wide area change detection study. The scenes were centered on the region identified in Fig. 4 as possibly associated with activities connected to the underground test and are each 5-km wide. In the following we present a comparative analysis of the two before and after images, emphasizing the areas for which change signals were detected with the ASTER data. Figure 5 shows a composite plot in which the areas with significant changes are marked with five colored circles/ellipses together with two sections of the high resolution images taken before and after the nuclear test. In the following Figs. 6a–6e a detailed comparison of visually detected changes in the high resolution images is discussed for each of the five encircled areas of Fig. 5. Figure 6a shows such a comparison for the yellow encircled area labeled "Mine entrance" in Fig. 5. The upper left part of Fig. 6a shows an ASTER MAD overview and detail plot (left) as well as a section of the Quickbird image (right) for orientation. Below are two detail sections showing the "Mine entrance" area on 17.09.2006 (left) and 16.10.2006 (right). The two columns panel on the right of Fig. 6a shows a detailed comparison of four sub-areas where changes between the times of recording of the two images become clearly visible. Among these changes are the insertion of a new supporting pillar, changes in the shape of landfill and the spread of dumped material, and leveling of a slope close to the mine entrance. Figure 6b shows the detailed comparison for the "Mine camp" area. For this sub-area no visible changes could be found. Changes that led to detections in the ASTER data might be caused by bulk mass transport along the existing roadway (marked by arrows). Alternatively they may be false positives due to nonlinear reflectance effects from the sloped rooftops. Figure 6c shows the detail comparison for a sub-area with mining activity. Regarding the change detection results the small open pit seems to be active. However, due to shadows in the EROS image no visual changes can be identified. Figure 6d shows the detail comparison for the southernmost sub-area. Again no visible changes could be found. Changes are possibly caused by bulk mass transport along the existing roadway marked by an arrow. Similar change

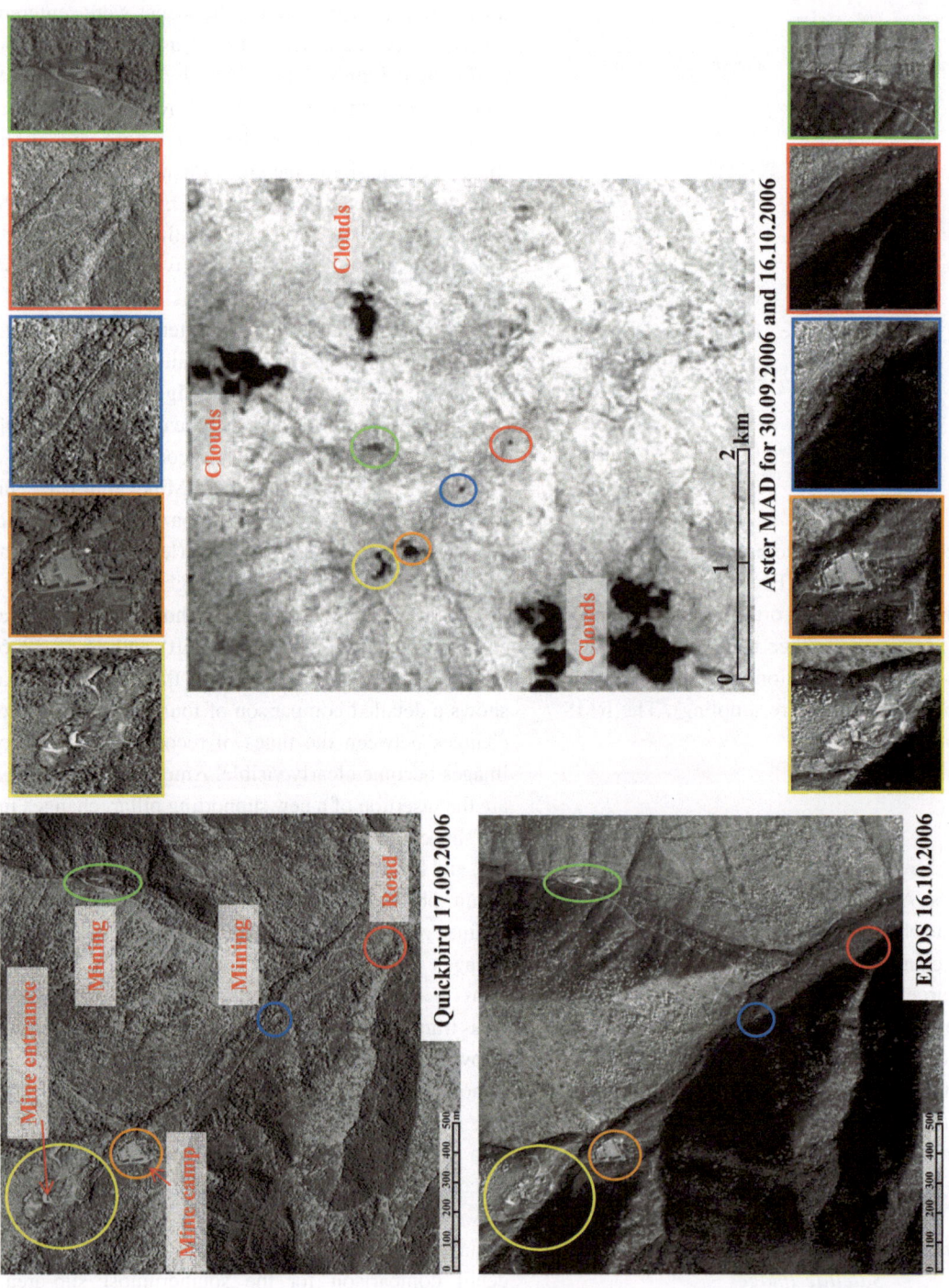

Figure 5

Composite plot of ASTER data change detections (*middle, right*) and sections of Quickbird and EROS high resolution images (*left, top* and *bottom*). Areas with significant changes are marked with corresponding *colored circles/ellipses* in the images and the *colored frames* (*right, top* and *bottom*) give enlarged views of the corresponding Quickbird and EROS sections. Local time of data acquisition (UTC + 9 h) is 11:33:51 and 14:38:35 for the Quickbird and EROS image, respectively. Large change signals resulting from clouds are annotated correspondingly

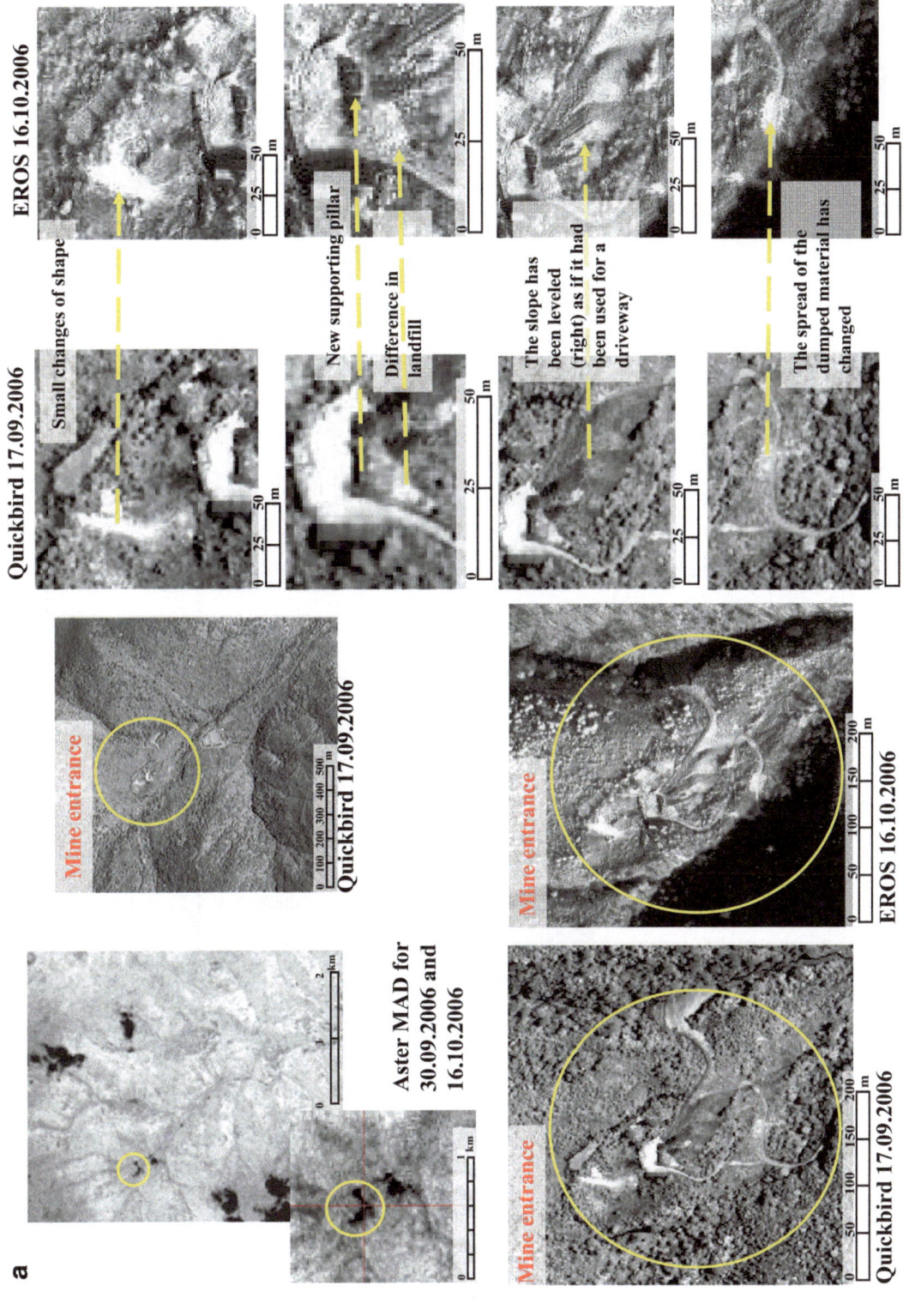

Figure 6a

Detailed comparison of visually detected changes for the *yellow* encircled area labeled "Mine entrance" in Fig. 5. The *upper left* part shows an ASTER MAD overview and detail plot (*left*) as well as a section of the Quickbird image (*right*) for orientation. Below on the left are two detailed sections showing the "Mine entrance" area on 17.09.2006 (*left*) and 16.10.2006 (*right*). For the two columns panel on the *right* see text

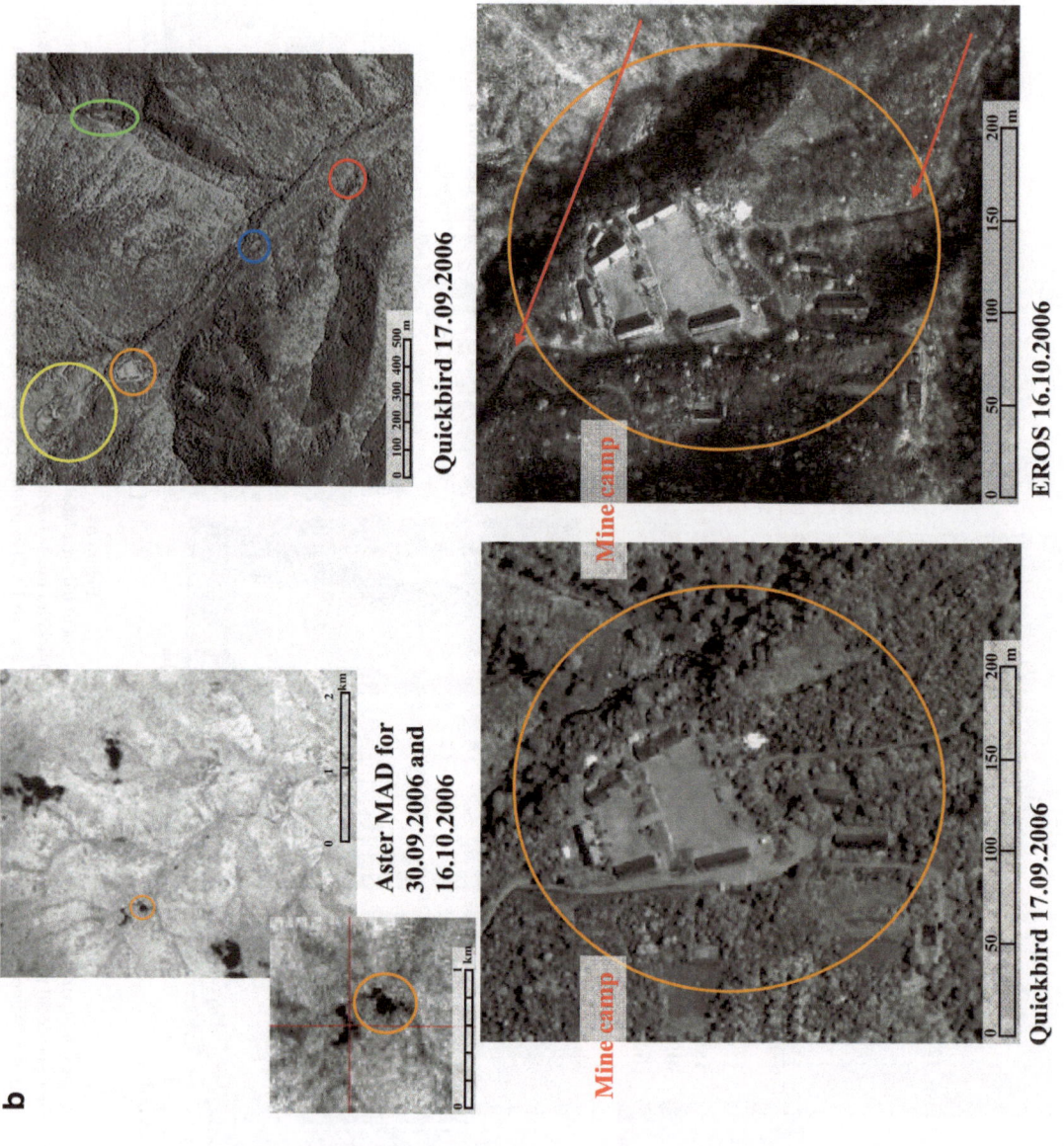

Figure 6b

Same as 6a but for the orange encircled area labeled "Mine camp" in Fig. 5. For this sub-area no visible changes could be found. Changes that led to detections in the ASTER data might be caused by bulk mass transport along the existing roadway (marked by *arrows*)

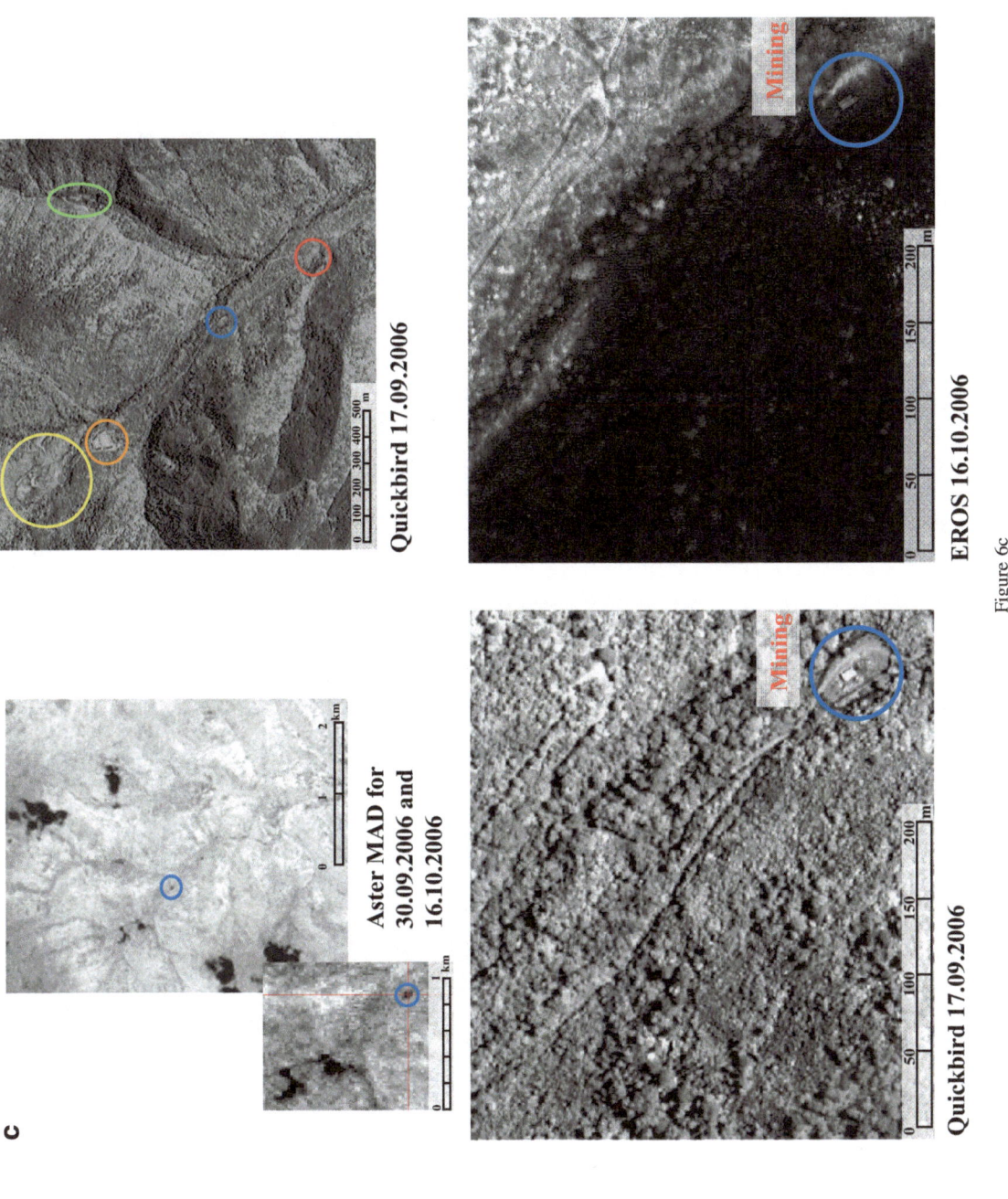

c

Quickbird 17.09.2006

Aster MAD for
30.09.2006 and
16.10.2006

EROS 16.10.2006

Quickbird 17.09.2006

Figure 6c

Same as 6a but for the *blue* encircled area labeled "Mining" in Fig. 5. Regarding the change detection results the small pear-shaped open pit seems to be active. However, due to shadows in the EROS image no visual changes can be found

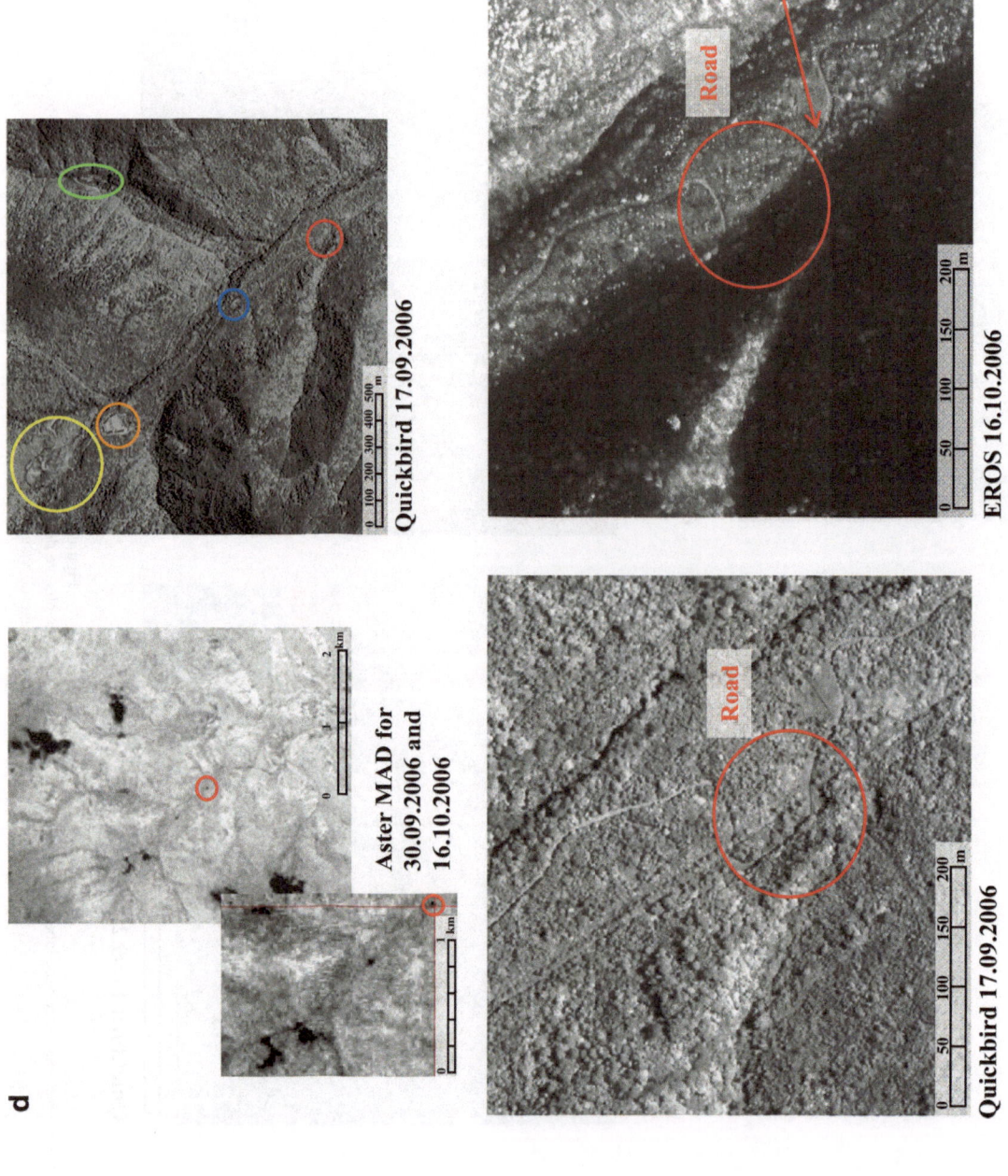

Figure 6d

Same as 6a but for the red encircled area labeled "Road" in Fig. 5. No visible changes could be found for this sub-area. Changes from the ASTER data are possibly caused by bulk mass transport along the existing roadway marked by an *arrow*

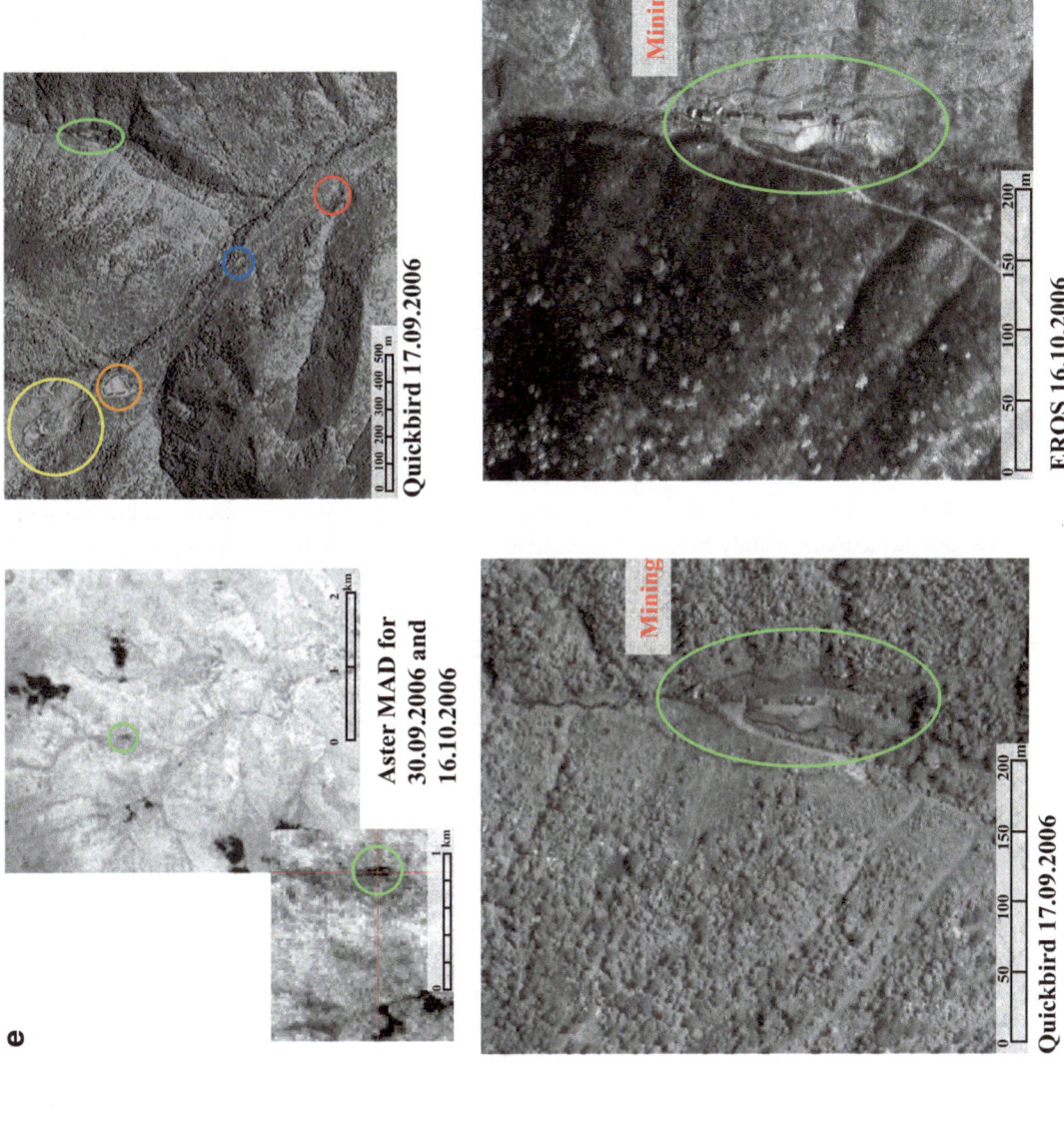

Figure 6e

Same as 6a but for the *green* encircled area labeled "Mining" in Fig. 5. The existing open pit was visibly advanced. Further excavations are clearly visible in the northern part of the ellipse

signals due to preparatory activities for underground tests at the Nevada Test Site have been detected with Landsat imagery (CANTY et al., 2005) Figure 6e displays again the detailed comparison for a sub-area with mining activity. The existing open pit was visibly advanced. Further excavations are clearly visible in the northern part of the ellipse.

To summarize, for two of five areas with changes detected with the ASTER data, clear evidence for man-made changes was found through the comparative analysis of high resolution images recorded before and after the nuclear test. (All five areas involve locations with man-made structures such as buildings, open pits, roadways). As opposed to these man-made changes, no significant changes of large-scale morphological features or surface changes such as displaced rocks or boulders, which might be expected, e.g., for tensile failure of surface layers (spallation), could be found for the two high resolution images.

7. Summary and Conclusions

The IDC and USGS localization of the North Korean underground nuclear test, which was mainly based on teleseismic data, was checked independently using regional waveform data of a combination of stations of the IMS and the IRIS consortium. The addition of non-IMS stations around the epicentral region considerably reduced the achievable azimuth gap and aided in improvement of the location accuracy using regional data. The resulting epicenter locations of the IDC, USGS and BGR consistently fall within a radius of 5 km and lie within the area of the seismic confidence ellipse for the IDC location. A yield range of 0.6–1.7 kt TNT equivalent is inferred from previously derived magnitude-yield relations (SCHLITTENHARDT, 1988), allowing for a maximum variation in the test site bias.

Wide-area change detection techniques using medium resolution optical/infrared satellite sensors were combined with localized high-resolution imagery to attempt to pinpoint the test location within the area identified by the seismic location measurements, i.e., the seismic confidence ellipse. Multispectral optical data from the Advanced Spaceborn Thermal Emission and Reflectance Radiometer (ASTER) were used for a wide area change detection study to find

evidence for the explosion. Unfortunately the available data only partly covered the area identified by the seismic measurements, however, included the region that has been reputed to be a possible test site. The changes inferred from the ASTER data occurred within a time span of 17 days, enclosing the explosion, and were found to be co-located with an area with mining activity and buildings and small roads in an otherwise uninhabited region. The seismic epicenter locations of the CTBTO, USGS and BGR all lie within a maximum distance of 6.2 km from the detected changes. Two high resolution (0.7 m pixel) images centered on the region with the ASTER wide-area change detections were available for a comparative analysis of the area before and after the nuclear test. All of the five identified areas with ASTER wide-area change detections show man-made structures in the high resolution imagery. Some of the changes found have characteristics that can be associated with preparation/excavation activities possibly connected to the underground test, like e.g., the insertion of a new supporting pillar, changes in the shape of landfill and the spread of deposited material, and leveling of a slope close to the mine entrance. Apart from these man-made change signals no significant changes of large-scale morphological features or surface changes such as displaced rocks or boulders, expected, e.g., for explosion induced tensile failure of surface layers (spallation), could be found for the two images, which were taken approximately 4 weeks apart.

Altogether three categories of changes: (a) Changes caused by construction work possibly in preparation for the test, (b) geomorphologic changes as direct consequences of the test and (c) ongoing changes due to mining activities, have been discussed on the basis of Figs. 6a–6e. However, due to the problems with the incomplete spatial coverage and rather poor temporal resolution, the interpretation of the observed changes is tied with a certain degree of ambiguity. Rather than having the character of forensic useful interpretations, this study clearly revealed the potential of satellite data in the context of an interdisciplinary case study, even if only data with the mentioned restrictions are available.

Due to its small size the North Korean nuclear test may be typical for problem events under the CTBT monitoring regime. This made it an ideal test case to explore how commercially available satellite data can be used to supplement seismological monitoring techniques stipulated in the treaty. However, it should be mentioned that the data available for this study do not by far have the necessary temporal resolution needed in a CTBT verification scenario. Such data could become available in the near future since there are promising system developments under way (e.g., TerraSAR-X and RapidEye for radar and optical data, respectively) that greatly reduce the revisit time for a given point on Earth with an improved spatial resolution. Clearly, such information is important for nuclear monitoring and, given the required timeliness, could be used in a CTBT verification scenario together with the seismically detected explosion signals to initiate and guide an on-site-inspection by the CTBTO authority.

Acknowledgments

This work has been carried out in part within the framework of the Global Monitoring for Security and Stability (GMOSS) Network of Excellence initiated by the European Commission. The facilities of the IRIS Data Management System, and specifically the IRIS Data Management Center, were used for access to waveform data required in this study. ASTER images Courtesy NASA/JPL-Caltech. EROS image © 2002 ImageSat International N.V., Licensed by ImageSat International N.V. Includes QuickBird and/or Wold-View-1 Products © DigitalGlobe™, distributed by Eurimage. Certain figures of this paper were generated using the software by "Wessel, P., and W. H. F. Smith, New version of the Generic Mapping Tools released, EOS Trans. Amer. Geophys. U., vol. 76 (33), pp. 329, 1995".

References

AMMON, C. J., and LAY, T. (2007), *Nuclear test illuminates USArray data quality*, Eos Trans. AGU *88*, 37.

BONNER, J., HERRMANN, R. B., HARKRIDER, D., and PASYANOS, M. (2008), *The surface wave magnitude for the 9 October 2006 North Korean nuclear explosion*, Bull. Seismol. Soc. Am. *98*, 2498–2506.

CANTY, M. J., NIELSEN, A. A., and SCHLITTENHARDT, J. (2005), *Sensitive change detection for remote monitoring of nuclear Treaties.* Proc. 31st Int. Symp. on *Remote Sensing of Environment, Global Monitoring for Sustainability and Security*, St. Petersburg, Russia, 20–24 June 2005.

CANTY, M. J., and SCHLITTENHARDT, J. (2001), *Satellite data used to locate site of 1998 Indian nuclear test*, Eos Trans. AGU *82*(3), 25–29.

COPPIN, P., JONCKHEERE, I., NACKAERTS, K., and MUYS, B. (2004), *Digital change detection methods in ecosystem monitoring: A review*, Int. J. Remote Sens. *25*(9), 1565–1596.

CONG, X., SCHLITTENHARDT, J., GUTJAHR, K., SOERGEL, U., CANTY, M., and NIELSEN, A. (2007), *Using differential SAR interferometry for the measurement of surface displacement caused by underground nuclear explosions and comparison with optical change detection results. In Global Monitoring for Security and Stability (GMOSS)—Integrated Scientific and Technological Research Supporting Security Aspects of the European Union* (eds. G. Zeug and M. Pesaresi), European Commission—Joint Research Centre, pp. 282–293.

FISK, M. D. (2002), *Accurate locations of nuclear explosions at the Lop Nor test site using alignment of seismograms and IKONOS satellite imagery*, Bull. Seismol. Soc. Am. *92*, 2911–2925.

GENERAL GEOLOGICAL MAP of KOREA (1945), 1:1000000, multicolor, Bibl.-Magazin BGR, Hannover.

GUPTA, V. (1995), *Locating nuclear explosions at the Chinese test site near Lop Nor*, Sci. Global Security *5*, 205–244.

HOTELLING, H. (1936), *Relations between two sets of variates.* Biometrika *28*, 321–377.

KENNETT, B. L. N. (1991), *IASPEI 1991 seismological tables*, Research School of Earth Sciences, Australian National University, Canberra, Australia.

KIM, W. Y., and RICHARDS, P. G. (2007), *North Korean nuclear test: seismic discrimination at low yields*, Eos Trans AGU *88*, 158–161.

KOPER, K. D., HERRMANN, R. B., and BENZ, H. M. (2008), *Overview of open seismic data from the North Korean event of 9 October 2006*, Seismol. Res. Lett *79*, 178–185.

KVAERNA, T., RINGDAL, F., BAADSHAUG, U. (2007), *North Korea's nuclear test: The capability for seismic monitoring of the North Korean test site*, Seismol. Res. Lett. *78*, 487–497.

MARSHALL, P. D., SPRINGER, D. L., and RODEAN, H. C. (1979), *Magnitude corrections for attenuation in the upper mantle.* Geophys. J. R. Astr. Soc. *57*, 609–638.

NIELSEN, A. A. (2007), *The regularized iteratively reweighted MAD method for change detection in multi- and hyperspectral data.* IEEE Trans. Image Process, *16*(2), 463–478. http://www.imm.dtu.dk/pubdb/p.php?4695.

NIELSEN, A. A., CONRADSEN, K., and SIMPSON, J. J. (1998), *Multivariate alteration detection (MAD) and MAF post-processing in multispectral, bitemporal image data: New approaches to change detection studies*, Remote Sens. Environ. *64*, 1–19. http://www.imm.dtu.dk/pubdb/p.php?1220.

NUTTLI, O. W. (1986), *Yield estimates of Nevada test site explosions obtained from seismic Lg waves.* J. Geophys. Res. *91*, 2137–2151.

PATTON, H. J., and TAYLOR, S. R. (2008), *Effects of induced tensile failure on m_b – M_s discrimination: Contrasts between historic nuclear explosions and the North Korean test of 9 October 2006*, Geophys. Res. Lett. *35*, L14301, doi:10.1029/2008GL034211.

RADKE, R. J., ANDRA, S., AL-KOFAHI, O., and ROYSAM, B. (2005), *Image change detection algorithms: A systematic survey*, IEEE Trans. Image Process. *14*(4), 294–307.

SCHLITTENHARDT, J. (1988), *Seismic yield estimation using teleseismic P- and PKP-waves recorded at the GRF-(Gräfenberg) array*, Geophys. J. *95*, 163–179.

SCHLITTENHARDT, J., CONG, X., CANTY, M., GUTJAHR, K., and SOERGEL, U. (2008), *Satellite Earth observations support CTBT monitoring. In Remote Sensing for International Stability and Security, Integrating GMOSS Achievements in GMES* (eds. G. Zeug, T. Kemper, A. Steel and M. Pesaresi), JRC Ispra, 19–20 February 2008, pp. 83–84.

SINGH, A. (1989), *Digital change detection techniques using remotely-sensed data*, Internat. J. Remote Sens. *10*(6), 989–1002.

SULSOFT (2003), *AsterDTM 2.0 installation and user's guide*, Technical report, SulSoft Ltd, Porto Alegre, Brazil.

TIBULEAC, I. M., VON SEGGERN, D. H., ANDERSON, J. G., SMITH, K. W., ABURTO, A., and RENNIE, T. (2008), *Location and magnitude estimation of the 9 October 2006 Korean nuclear explosion using the southern Great Basin digital seismic network as a large-aperture array*, Bull. Seismol. Soc. Am. *98*, 756–767.

THURBER, C., QUIN, H., and RICHARDS, P. (1993), *Accurate locations of nuclear explosions at Balapan, Kazakhstan, 1987 to 1989*, Geophys. Res. Lett. *20*, 399–402.

VINCENT, P., LARSEN, S., GALLOWAY, D., LACZNIAK, R.J., WALTER, W.R., FOXALL, W., and ZUCCA, J. J. (2003), *New signatures of underground nuclear tests revealed by satellite radar Interferometry*, Geophys. Res Lett. *30*(22), 2141.

ZHAO, L.-F., XIE, X.-B., WANG, W.-M., and YAO, Z.-X. (2008), *The regional seismic characteristics of the October 9, 2006 North Korean nuclear test*, Bull. Seismol. Soc. Am. *98*, 2571–2589.

(Received October 31, 2008, revised May 20, 2009, accepted June 10, 2009, Published online January 22, 2010)